DEVELOPMENTAL NEUROCHEMISTRY

Edited by Richard C. Wiggins, David W. McCandless, and S. J. Enna

University of Texas Press, Austin

Printed in the United States of America

First Edition, 1985

Library of Congress Cataloging in Publication Data
Main entry under title:
Developmental neurochemistry.
Includes bibliographies.
1. Neurochemistry. 2. Developmental neurology.
3. Brain chemistry. I. Wiggins, Richard C. (Richard
Calvin), 1945– . II. McCandless, David W., 1941–
III. Enna, S. J.
QP356.3.D484 1985 599.01'88 84-25613
ISBN 0-292-71548-X

Contents

DEVELOPMENTAL NEUROCHEMISTRY

Preface

Research in the neurosciences has been expanding in an exponential fashion during the past two decades, and progress has been especially rapid in the field of neurochemistry. Development is an especially complex application of neurochemistry, offering challenges of enormous magnitude and diversity to clinicians and basic scientists alike. Given the breadth of the neurochemical and developmental sciences, the great clinical relevance, and the rapid flux of scientific thought and effort in these areas, we intend to present a concise overview of essential and interesting topics.

The first three chapters describe various morphological correlates underlying neurochemical development of the brain. The next four chapters review the energy requirements and transitions in intermediary metabolism that characterize fetal and postnatal brain development, including the coupling of energy metabolism to the development of function. Following these, two chapters describe the biochemical development of neurotransmission, including differentiation of receptor sites and transmitter metabolism.

The processes of development, differentiation, and growth involve a series of rapid transitions in cellular and biochemical events in addition to the usual metabolic flux of adult homeostasis. The opening chapter provides a perspective on so-called critical events in development. The developing fetal and postnatal nervous system is dramatically different from the nervous system of the adult with regard to the regulation of dynamic events and the vulnerability of these processes to genetic and environmental perturbation. Accordingly, a concluding chapter discusses the clinical relevance of basic developmental neurochemistry.

We encouraged chapter authors to stimulate interest by addressing the limits as well as the depth of knowledge. We placed rather strict page limits on each chapter to keep the book affordable and to strengthen thematic messages. Aiming at a general audience in the developmental sciences and neurosciences, we intended to summarize current (chapters were submitted during the early part of 1983) knowledge about maturation in the nervous system. Although each chapter was authored independently, we structured the complete volume to provide balance of detail and information through the selection of topics.

Critical Periods in Development: Editorial and Historical Perspectives

Richard C. Wiggins

Introduction

The concept that there are critical periods in development is quite old (Dareste, 1891) and is inseparable from the origins of teratology as an experimental science (Stockard, 1921). The original hypothesis stated that developmental processes proceed at different rates and that during periods when a given rate process is predominant, it is subject to permanent arrest by environmental conditions outside a normal range. Restoring a normal environment would not necessarily restore normal development of the organism. Instead, the opportunity for certain critical events to occur would have passed, causing a distortion in the balance of development. Although Stockard's studies involved abnormalities of morphological development, he believed that the underlying causes were biochemical.

Since Stockard's work, the features of the critical interval hypothesis have been widely applied within the neurosciences, especially in behavior (Scott, 1962; Bronson, 1965; Rodier, 1976). Flexner, Flexner, and Straus (1941) observed that growth and differentiation of the fetal pig sensory-motor cortex proceeds by stages rather than by continuous uniform processes. They described the period of neuroblast proliferation and differentiation as a critical period when cells show morphological change and the activities of such enzymes as cytochrome oxidase and succinic dehydrogenase increase rapidly. Similarly, Kavaler and Kimel (1952) demonstrated that acetylcholinesterase activity rises sharply at the midpoint of gestation, a finding from which they speculated that the biochemical event was precursor to a cellular differentiation. Davison and Dobbing (1966) have described myelination as a critical, or vulnerable, period. These early studies of neurochemical development demonstrated the first appearance of enzyme activities or of specific compounds in the developing brain and demonstrated biochemically that at specific periods in development, certain metabolic processes do occur at predominant rates.

History

Physicians of the early nineteenth century generally believed that birth defects were the result of some pathological lesion intervening to produce deformity. However, Dareste (1891) demonstrated with chick embryos that abnormal development can actually result from an alteration of the normal plan of embryonic development and that an early appearing abnormality of development produced a more grave deformity. From the work of Charles Darwin, Dareste believed that embryonic development recapitulated evolutionary stages in the ascent of animal groups. Thus, vestigial characteristics and atavistic characters, such as supernumerary organs, were readily explained by the shared ancestry of animal species. Dareste also recognized that the clitoris and penis were, respectively, in females and males, homologous structures present in varying degrees of development. The clitoris was a rudimentary or arrested penis; the prostate, a rudimentary uterus. Thus, the general theory of teratology offered by Dareste is that the formation of birth defects generally follows a plan in which development is abnormally arrested at a rudimentary stage.

Stockard (1921) further elaborated the theme that abnormal development, if not of hereditary origin, was caused by developmental arrests. However, Stockard surpassed Dareste by showing through experimentation with fish embryos that there are "indifferent" and "critical" moments of development. He observed that the embryonic organs arose with initial moments of extremely high activity and that "such moments of supremacy for the various organs occurred at different times during development" (p. 139). Such particularly sensitive periods are in Stockard's words "critical moments" (p. 139). Thus, it follows that the earlier the arrest the more numerous will be the types of defects found, and the later the arrest the more limited will be the variety of deformity. The complete interruption of development at a critical stage will prevent those organs or parts having supremacy at the moment from resuming their development at their relatively accelerated rate. In other words, organs developing at the most rapid rate, or having the highest metabolic rates at the time of an arrest, will be affected more than the organs or parts developing at a less rapid rate. When development is only slowed but not stopped, all organs lower their rate more or less relative to one another, and on resuming a normal rate of development, the more rapidly developing parts may be able to maintain their supremacy. Stopping or slowing development has no effect on organs at indifferent moments in their development.

Stockard also recognized that the embryonic formation of an organ was restricted to a limited moment of development. He believed that this time limitation on when an organ could form was a function of growth competition between organs. Interestingly, Stockard explained that the heightened activity of a group of cells reflected increased metabolic activity, possibly linked to the formation of certain specific and labile compounds. Oxygen or temperature insufficiencies, for example, could at such a moment block or suppress the level of activity normally featured by the driven cellular group. In such a case, their moment for normal growth or differentiation would be lost.

The problem that Stockard posed for the future was to identify the critical moments for the different organs. One would predict (and Stockard partly supports) a hypothesis that a type of defect in one organ can be induced by many types of treatments and that different defects in many organs can be induced by the same treatment. One of Stockard's important conclusions regarding the experimental induction of abnormalities is that the result (of a perturbation) depends on the developmental time during which a treatment acts and not on the properties of the substance used.

Applications

Through various processes, including cell multiplication, cell migration, and the differentiation of cells into unique types, a single fertilized egg gives rise to the multicellular gastrula, germinal layers, and eventually the organs and patterns of growth recognizable in each species. Because the expression of genes is temporarily ordered, development proceeds as a series of ordered events. Development is characterized more as a series of stages rather than as a continuum. When the brain is viewed as a whole organ, several stages of particular rapid growth characterize it (Gottlieb, Keydar, and Epstein, 1977), and at the cellular level, the cytoplasmic architecture of individual neurons matures in stages (Lavelle, 1964). During such critical periods, development is interactive with the environment.

Interestingly, the nervous system is one of the first organ systems to develop from the gastrula and thus plays a major role in organizing the induction of other organs and tissues, yet it is probably the last to reach maturity. Because of its intrinsic complexity and the required long period of development, the nervous system is uniquely vulnerable to nongenetic influences during development and growth.

The next chapter provides a comparatively detailed accounting of cellular development of the nervous system. Biochemical differentiation (treated in various other chapters) is more closely linked with the divergence of unique cell types and consequently, is not as readily subject to useful generalization. However, it is useful to roughly identify stages of brain development as a preface to discussing critical or vulnerable events. Useful stages are (1) proliferation and migration of cells (see the chapter by Bayer), (2) development of specific recognition between cells during outgrowth (Edelman, 1983), (3) arborization of dendrites and development of receptors for neurotransmitters (see the chapters by Telang and Enna, and by Jones) and (4) myelination (see the chapter by Gould). These events are neither entirely sequential nor synchronous throughout the brain. However, it is clear that interactions *between* cells and cell contacts are fundamental to all these processes.

A major part of brain development occurs postnatally during a period called the *brain growth spurt*. Thus, during much of its development the brain does not have the protection afforded the fetus by the homeostatic mechanisms of the dam. The postnatal brain growth spurt has been called *the vulnerable period* in development because of these circumstances (Dobbing, 1968; Davison and Dobbing, 1966). The critical period in organ development has usually been described (incorrectly) as the time of most rapid cell division, and as one might suspect, deviations from normality become increasingly subtle the later an insult or perturbation occurs in the organism's approach to final maturity. However, biochemical differentiation is a prerequisite to morphological differentiation; consequently, the critical event leading to an abnormality may occur at a time prior to the first appearance or complete development of the structure.

References

Bronson, G. (1965). The hierarchial organization of the central nervous system: Implications for learning processes and critical periods in early development. *Behav Sci* 10, 7–25.

Dareste, M. C. (1891). Recherches sur la production artificielle des monstruosités. Paris: C. Reinwald.

Davison, A. N., and Dobbing, J. (1966). Myelination and vulnerable period in brain development. *Br Med Bull* 22, 40–44.

Dobbing, J. (1968). Vulnerable periods in developing brain. In *Applied neurochemistry*, ed. A. N. Davison and J. Dobbing, pp. 287–316. Philadelphia: F. A. Davis.

Edelman, G. M. (1983). Cell adhesion molecules. *Science* 219, 450–457.

Flexner, J. B., Flexner, L. B., and Straus, W. L. (1941). The oxygen consumption, cytochrome and cytochrome oxidase activity and histological structure of the developing cerebral cortex of the fetal pig. *Journal of Cellular and Comparative Physiology* 18, 355–368.

Gottlieb, A., Keydar, I., and Epstein, H. T. (1977). Rodent brain growth stages: An analytical review. *Biol Neonate* 32, 166–176.

Kavaler, F., and Kimel, V. M. (1952). Biochemical and physiological differentiation during morphogenesis: XV. Acetylcholinesterase activity of the motor cortex of the fetal guinea pig. *J Comp Neurol* 96, 113–119.

Lavelle, A. (1964). Critical periods of neuronal maturation. *Prog Brain Res* 9, 93–96.

Rodier,P. (1976). Critical periods for behavioral anomalies in mice. *Environ Health Perspect* 18, 79–83.

Scott, J. P. (1962). Critical periods in behavioral development. *Science* 138, 949–958.

Stockard, C. R. (1921). Developmental rate and structural expression: An experimental study of twins, "double monsters" and single deformities, and the interaction among embryonic organs during their origin and development. *Am J Anat* 28, 115–277.

The Development of the Central Nervous System

Shirley A. Bayer

Development and Characteristics of the Neuroepithelium

Formation

The nervous system begins its long journey to maturity with the formation of a specialized layer of epithelial cells, the neuroepithelium. Largely because of the pioneering work of the experimental embryologist Hans Spemann during the early part of this century (summarized in Spemann, 1938) it is known that in vertebrate embryos, the initial stimulus to form the neuroepithelium occurs during gastrulation. As cells from the dorsal lip of the blastopore (in amphibian embryos) or from the primitive streak (in avian and mammalian embryos) grow anteriorly to form the dorsal mesoderm, they induce the overlying ectoderm to become the neural ectoderm, whose cells constitute the neuroepithelium. Toivonen et al. (1975) found that the dorsal mesoderm retains its inductive capacity when a millipore filter is sandwiched between it and the overlying ectoderm. Since subsequent electron microscopic examination did not reveal any cytoplasmic processes in the filter pores, these workers concluded that morphogenetic signals for the nervous system are carried by biochemical compounds and that no cytoplasmic contact between the inducer and the responding tissue is required.[1]

Initially, the neural ectoderm is a flat sheet of cuboidal cells. Then, two shape changes take place. First, the cells become columnar, each cell being a cylinder of nearly constant diameter throughout its length. In this stage, the columnar ectodermal tissue is called the neural plate (see *A* of fig. 1). Next, the apical end of each cell becomes constricted. Because the cells are firmly attached to each other (by tight and gap junctions) at their apical ends, a constriction of all the cells just at one end will cause them to curve inward to form the neural groove (*B* and *C*, fig. 1; Burnside, 1973; Karfunkel, 1974; Bancroft and Bellairs, 1975). According to Burnside (1973), end-to-end alignment of microtubules probably accounts for the columnar shape change; colchicine, which interferes with microtubule alignment, will block the columnar shape

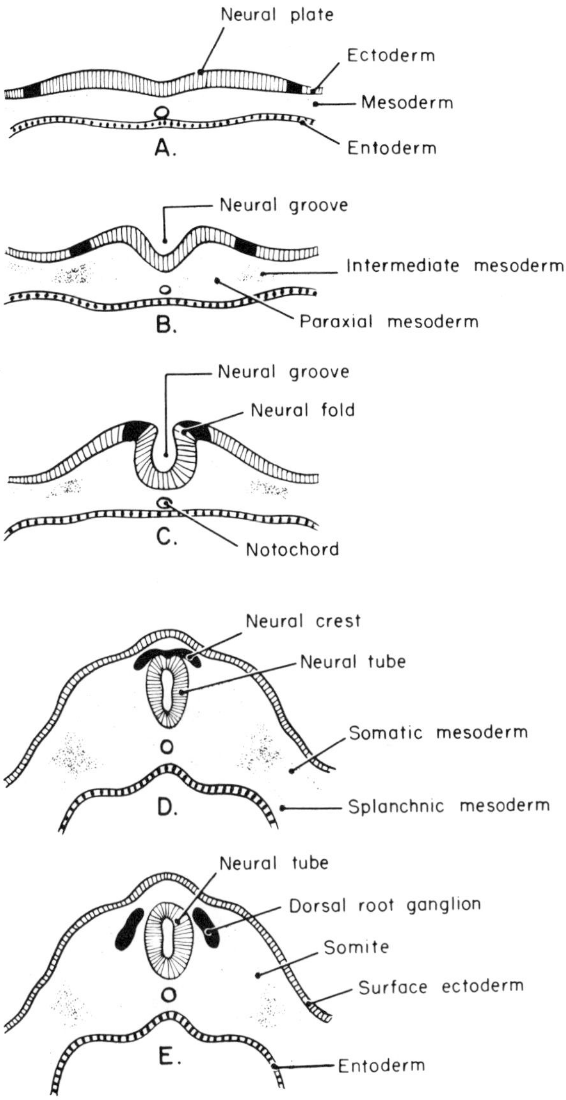

Figure 1. Diagrams of coronal sections (cut is perpendicular to the midline) of mammalian embryos at various stages of early embryonic life to show the initial development of the nervous system. The neural tube differentiates into structures of the central nervous system (see fig. 2), while the neural crest produces all peripheral neurons (sensory neurons in spinal and cranial ganglia and autonomic motor neurons) and glial cells (Schwann cells). (From Truex and Carpenter, 1969; copyright © 1969 by the Williams and Wilkins Co., Baltimore.)

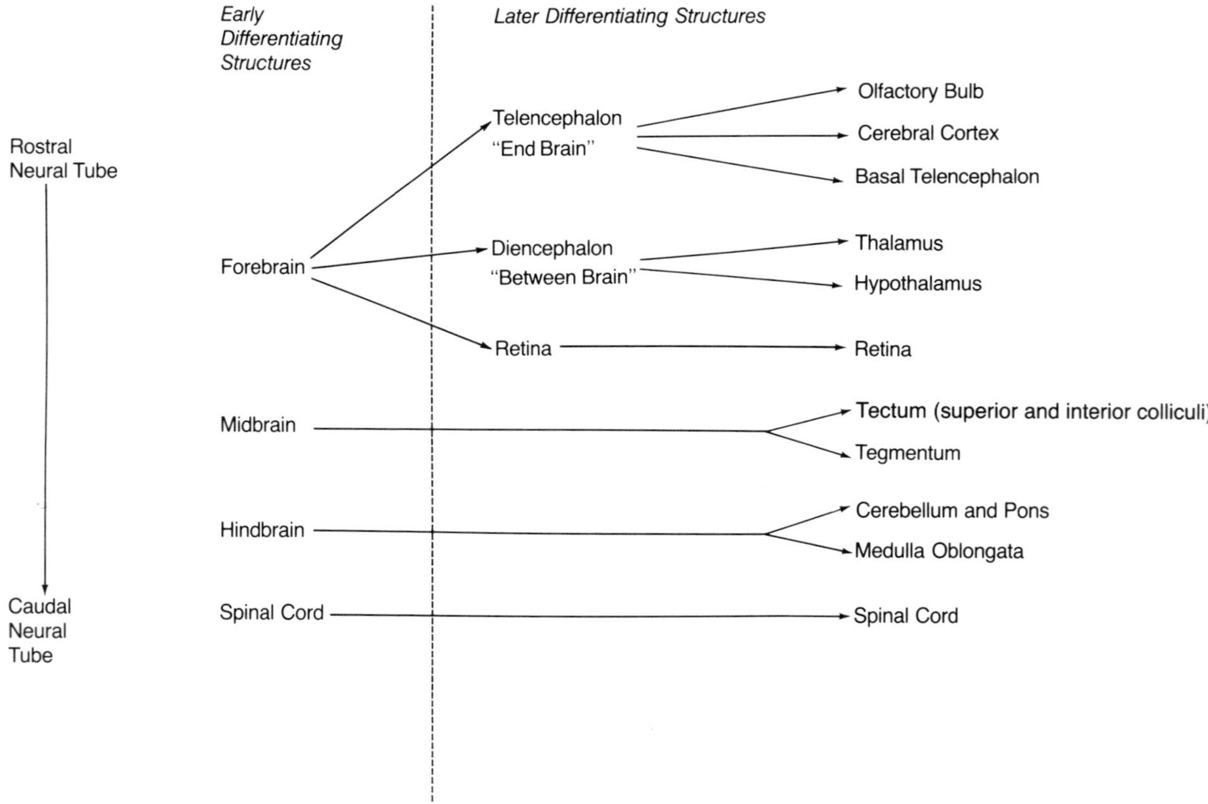

Figure 2. Parts of the nervous system that differentiate from the neural tube.

change. Burnside (1973) also found that microfilaments at the apical end slide past each other to form thick bundles, a formation that decreases the cell diameter much like the contraction of a sphincter muscle. Certain drugs such as cytochalasin B and vinblastine will inhibit apical constriction (Schroeder, 1973).

As the leading edges of the neural groove, called the *neural folds*, approach each other in the midline (*C*, fig. 1), they fuse to form the neural tube (*D* and *E*, fig. 1). During this process, extracellular material (such as carbohydrate residues) accumulate on the apical surfaces of the cells, especially in the zone of future fusion (Moran and Rice, 1975; Lee et al., 1976). If chick embryos are exposed to concanavalin A, a drug that interferes with cell-surface binding of carbohydrate residues, the folds of the neural groove meet but do not fuse (Lee et al., 1976). These results lend support to the idea that the accumulation of extracellular material is important during neural tube formation.

The neural tube is composed of a layer of cells called the *neuroepithelium*. These cells are the source of all structures included in the central nervous system (CNS) (brain and spinal cord). Figure 2 is a diagram showing early and late differentiating structures.

Mosaic Nature of the Neuroepithelium

During formation of the neural tube, the neuroepithelium is mitotically active, and young neurons begin to accumulate along its perimeter. This process continues at a rapid pace throughout embryonic life as the neural tube is formed into the structures of the brain and spinal cord (fig. 2). A given region of neuroepithelium will eventually develop into a highly specialized part of the CNS, each part having distinct structural features. The induction of the presumptive neural ectoderm to form the CNS takes place in a two-step process. In a classical experiment on amphibian embryos, Toivonen and Saxén (1968) removed a portion of the neural plate, which would eventually become the forebrain, along with a portion of trunk mesoderm during early embryonic life. The cells of both explants were separated and then cultured together in various ratios. The results (fig. 3) showed that forebrain neuroepithelium developed into hindbrain and spinal cord structures as the ratio of neural to mesodermal cells became 50:50 or less. During the first stage of induction, neuroepithelial cells are determined to produce neurons, but not those in a particular part of the CNS. Further regional specificity is induced by the proportion of mesodermal cells acting on various zones of neuroepithelium throughout the neural tube. Figure 2 lists the final products of neural tube differentiation throughout its rostral to caudal extent. After the secondary induction, the neuroepithelium is determined to develop into a specific portion of the nervous system. For example, if a "patch" of neural plate cells is stained with a vital dye, the structures these cells begin to produce will retain the stain for several

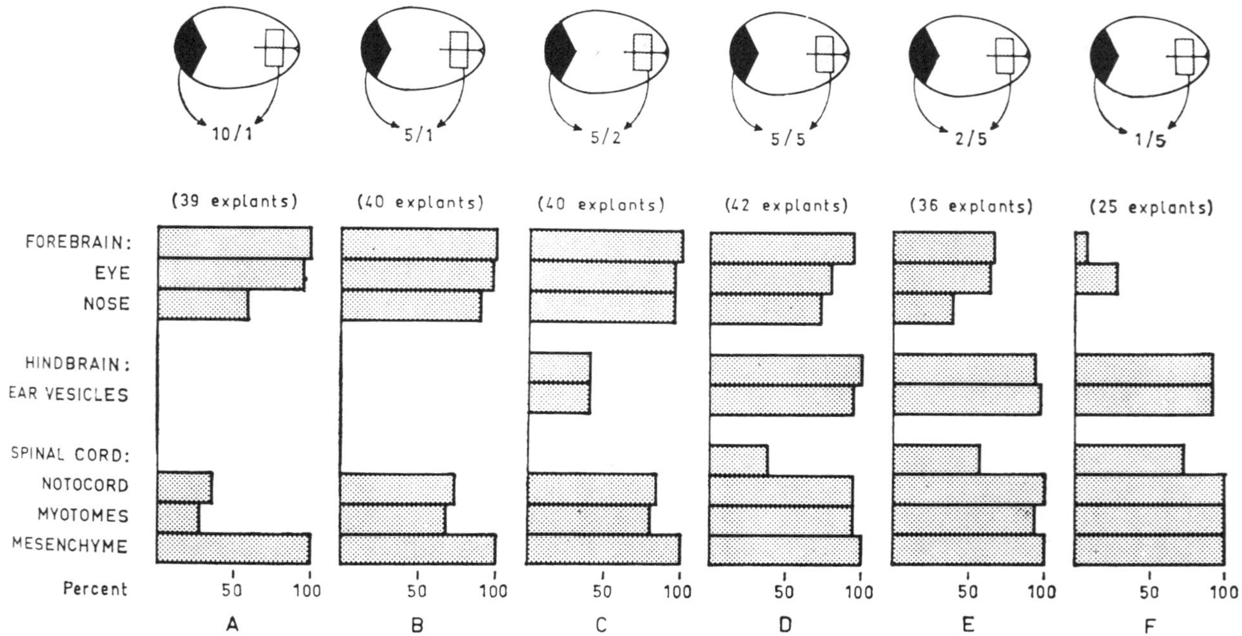

Figure 3. The egg-shaped segments represent amphibian embryos from which both forebrain neural tissue (black) and axial mesoderm (stippled) were removed, dissociated, and together cultured in vitro according to the ratios shown. The graphs show the percentages of structural types that differentiated from the reaggregated neural and mesodermal cells after 14 days. Note that as the ratio of neural to mesodermal cells declines, the proportion of hindbrain and spinal cord structures increases. (From Toivonen and Saxén, 1968; copyright © 1968 by the American Association for the Advancement of Science.)

days. This method has shown that the same region of the neural plate always gives rise to a specific region in the CNS (Jacobson, 1959). Thus, the "fate" of a neuroepithelial cell is sealed very early in development. This predetermination will be expressed outside the nervous system. For example, if a specific portion of the hindbrain neuroepithelium is removed after an early stage in certain fish and amphibian embryos, a giant Mauthner cell[2] will not develop; but a Mauthner cell will be produced by this neuroepithelium if it is transplanted into the body cavity of the embryo (Stefanelli, 1951). In spite of the future diversity of its progeny, the neuroepithelium throughout the neural tube has a very homogeneous structure. Biochemical differences may exist in various regions, but their discovery awaits future research.

Cellular Dynamics of the Neuroepithelium

If one looks at a histological section of the neuroepithelium (fig. 4*A*), it appears to be composed of many cell layers because cell nuclei are scattered throughout a broad band. Prior to 1935, the generally held view was that the neuroepithelium was a multilayered tissue. Using only a light microscope, Sauer (1935) confirmed that mitotic figures are always located near the central cavity, or lumen, of the neural tube (*arrows*, fig. 4*A*). But Sauer also found that nuclei in the basal rim of the neuroepithelium were consistently larger than those located nearer the lumen. He hypothesized that the basally positioned nuclei synthesize a duplicate set of chromosomes and that then they translocate to the apical portion of the cell (near the lumen) to divide. During mitosis the cytoplasm loses its attachment to the basal surface and rounds up at the apical surface. Following division, the cytoplasm of each daughter cell elongates to reach the basal surface, and the nucleus subsequently translocates basally to synthesize another set of chromosomes. Thus, the neuroepithelium is a single layer of cells.

Sauer's hypothesis has been confirmed with ^{3}H-thymidine autoradiography (Sidman, Miale, and Feder, 1959). With this technique, one exposes an embryo to ^{3}H-thymidine, which is incorporated into the DNA molecules during the S phase (chromosomal duplication) of the cell cycle. The neuroepithelium in figure 4*B* is from a rat embryo with one exposure to ^{3}H-thymidine two hours before death. Notice that only nuclei in the basal rim are labeled. These are the nuclei of the cells in S phase. In contrast, most nuclei located nearer the lumen and in the mitotic phase are unlabeled because their chromosomes have already been assembled. If the survival time is increased to three or four hours, mitotic figures start to become labeled, while other nuclei in the basal rim are unlabeled. The positional changes in labeled and unlabeled nuclei and mitotic figures within the neuroepithelium have allowed some investigators to estimate the total cell-cycle time. Langman, Guerrant, and Freeman (1966) estimated that the cell cycle takes approximately

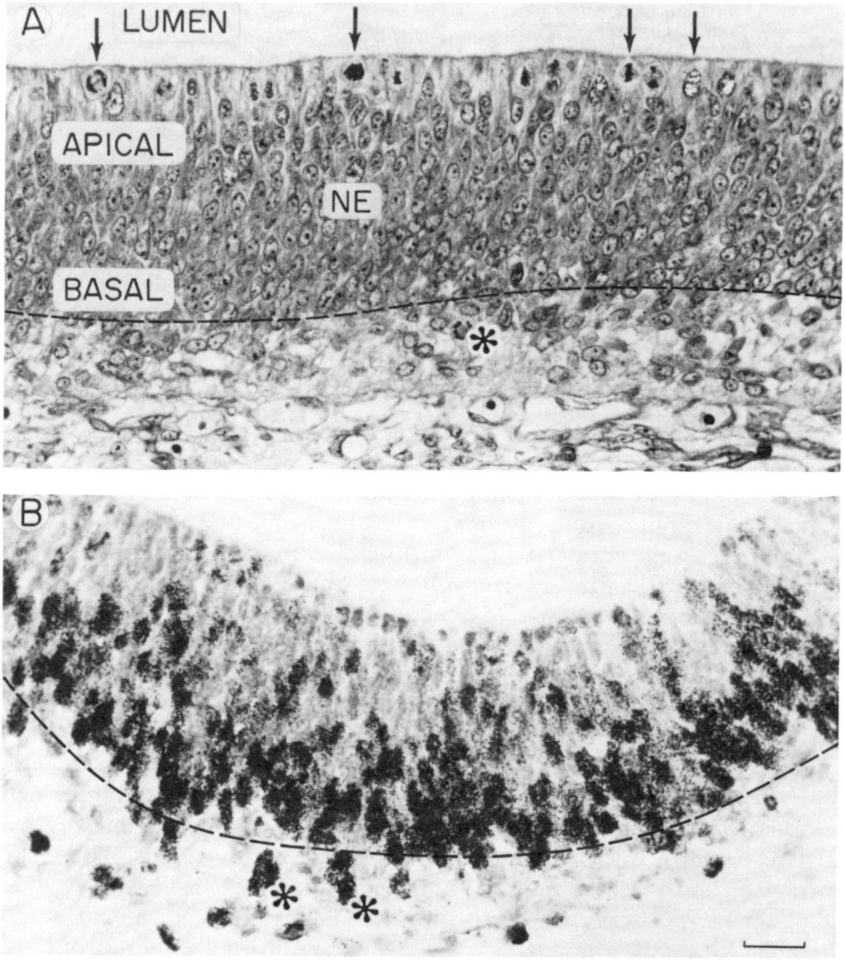

Figure 4. Both *A* and *B* are sections of the neuroepithelium in an E15 rat embryo. The apical surfaces of the cells are facing up; the basal surfaces (*dashed line*) are down (bar = 25 μm). *A*, methacrylate section (3 μm) showing that mitotic figures (*arrows*) are preferentially located near the lumen of the neural tube. Asterisk indicates a mitotic figure outside the neuroepithelium. *B*, paraffin section (6 μm) of the neuroepithelium six hours after a single exposure to ^{3}H-thymidine. Nuclei that have incorporated labeling are covered with reduced silver grains (black). Generally, nuclei nearer the basal surface (*dashed line*) are labeled, but those near the apical surface are either very lightly labeled or unlabeled. This indicates that the majority of the cells in S phase have basally positioned nuclei, just as Sauer (1935) had predicted. Asterisks indicate groups of cells proliferating outside the neuroepithelium.

eight hours in chick embryos 18–22 hours of age. With increasing embryonic age, the cell cycle lengthens (Wilson, 1973; Kauffman, 1968; Hoshino, Matsuzawa, and Murakami, 1973). Denham (1967) found that retinal neuroepithelial cells in the rat on postnatal day 2 have a total generation time of twenty-eight hours.

Neurogenesis

After the introduction of the ^{3}H-thymidine autoradiographic method during the early 1960s, it soon became a tool for dating the time of origin of cells in the nervous system. Unlike the cell-cycle studies mentioned above in which the survival time is usually a few hours or days (short-survival autoradiography), survival times for ^{3}H-thymidine autoradiographic cell-dating studies are usually weeks or months (long-survival autoradiography). Once a neuron has been produced by a labeled neuroepithelial cell, it no longer goes through cell division. After a variable delay, depending on cell type, the young neuron migrates to its final destination within the nervous system and begins to grow characteristic axonal and dendritic processes. Throughout the entire period of maturation and adult life, the neuronal nucleus retains the label that was incorporated by its mitotic precursor cell. The longer survival time simply allows these neurons to be studied in their mature state. It is important to realize that at the time of a ^{3}H-thymidine injection, young neurons already generated by the neuroepithelium will not incorporate the label into their cell nuclei. If enough ^{3}H-thymidine is made available to the neuronal precursors after the initial injection (cumulative labeling), one can ensure that most neurons originating after a specified age will be labeled (Bayer and Altman, 1974). The general finding with the cumulative labeling technique is that neurons are produced over very specific time periods.

In order to illustrate these principles, detailed examples will be discussed concerning the sequence of neurogenesis in the septum and cerebral cortex, specialized areas of the telencephalon. A photomicrograph of a section through the telencephalon on embryonic day (E) 15 in the rat (fig. 5) shows the future septum lies in the ventromedial wall, while the cerebral cortex composes the roof. At this stage of development, the lateral ventricle represents the lumen of the neural tube. By E19, the neuroepithelium giving rise to the septum is considerably thinner, while that producing the cerebral cortex is still prominent. Many young neurons have already accumulated in the lateral septal nucleus and in the cortical plate, the future cerebral cortex. The other two photomicrographs in fig. 5 are autoradiographs of the postnatal day (P) 60 rat brain in the regions of the lateral septal nucleus and the cerebral cortex. Cells with black dots overlying the nucleus have been labeled by two consecutive ^{3}H-thymidine injections given during the embryonic period.

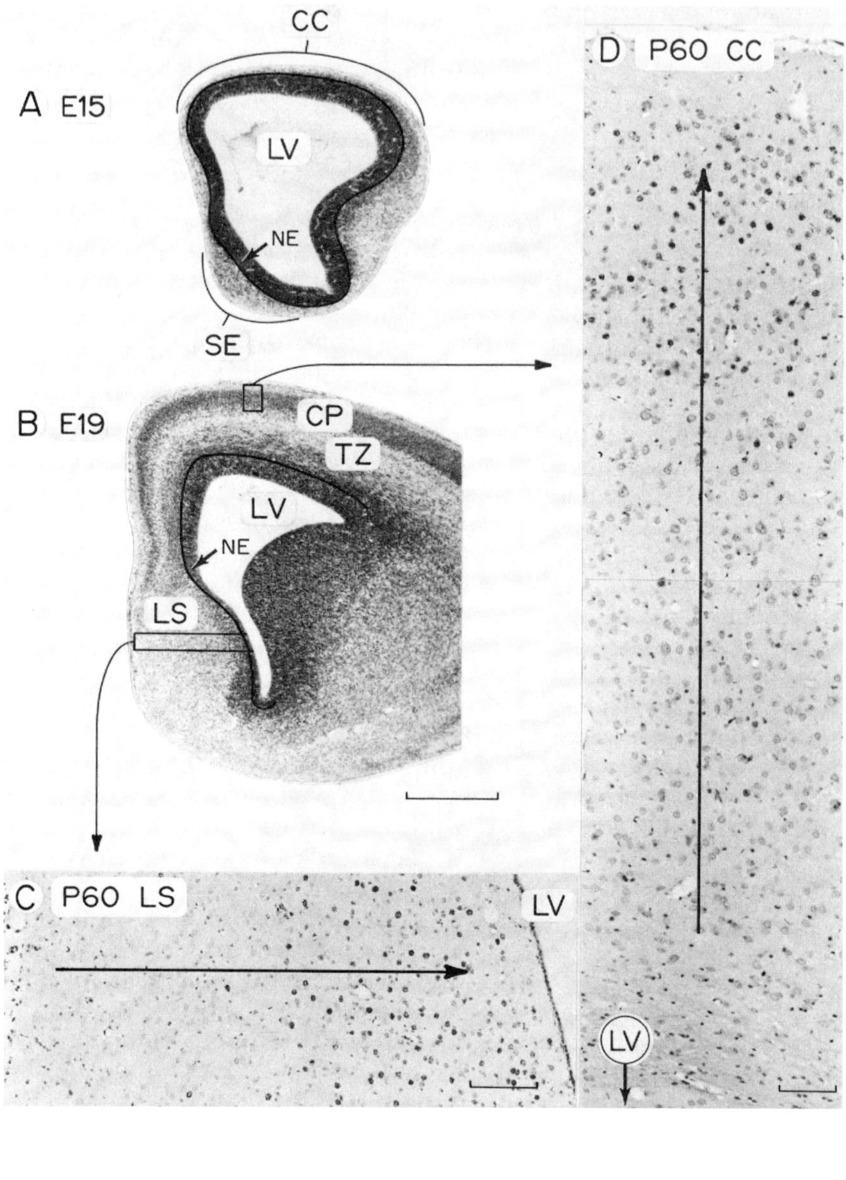
CC
A E15
LV
NE
SE
B E19
CP
TZ
LV
NE
LS
C P60 LS
LV
D P60 CC
LV

These are younger neurons than those which are not labeled. The nonlabeled cells had already originated before the injections began. The large arrows indicate that there is a gradient of older to younger neurons in both structures, showing that final position is related to time of origin. Notice that the lateral septal nucleus and the cerebral cortex have gradients going in opposite directions. The youngest cells in the lateral septal nucleus are located nearest the lateral ventricle, which was once lined by the neuroepithelium (Bayer, 1979). On the other hand, the youngest cells in the cortex are located farthest away from the lateral ventricle in the superficial layers (Angevine and Sidman, 1961).

Figure 6 diagrams what happens to produce these two gradients. In the "outside-in" gradient, the oldest neurons produced by the neuroepithelium migrate to the layer just outside; as younger neurons are produced, they also migrate out and "push" the older neurons farther away. This is the type of gradient found in the lateral septal nucleus. In the "inside-out" gradient, older neurons remain closest to the neuroepithelium, while younger ones migrate past them and pile up progressively farther away. This is the type of gradient found in the cerebral cortex.

The regularity in the positioning of neurons with respect to time of origin has prompted studies to see if the order of neurogenesis in a given system of brain structures correlates with their patterns of anatomical interconnections. The findings continually show a positive relationship. Figure 7 illustrates just one example. The hippocampus, a specialized

Figure 5. *A* and *B* are sections of the telencephalon in the embryonic rat brain (6 μm, paraffin, hematoxylin-eosin stain, bar = 0.5 mm). On E15 (*A*) the future septum (*SE*) lies in the ventromedial telencephalon; the future cerebral cortex (*CC*) occupies the roof. Notice that the neuroepithelium (*NE*) is quite thick as it lines the lateral ventricle (*LV*). By E19 (*B*) the septum has expanded, and a lateral septal nucleus (*LS*) has formed. The neuroepithelium is considerably thinner here. The cortical plate (*CP*) contains young neurons that have migrated through the transition zone (*TZ*) to form the future cerebral cortex; the neuroepithelium here is still quite thick, indicating that cortical neurons are still being produced. *C* and *D* are autoradiographs from rats on postnatal day 60 (P60) that had two consecutive ^{3}H-thymidine injections during embryonic life (6 μm, paraffin, hematoxylin-eosin stain, bars = 0.1 mm). Labeled nuclei are neurons that have originated after the injections began and are therefore younger than the unlabeled cells. Arrows indicate the direction of chronological gradients from older to younger neurons. Notice that the gradient in the lateral septal nucleus points toward the lateral ventricle, while that in the cerebral cortex points away from it. (The lateral ventricle is not in the photomicrograph of the P60 cerebral cortex. It lies out of the field of view in the lower left corner.)

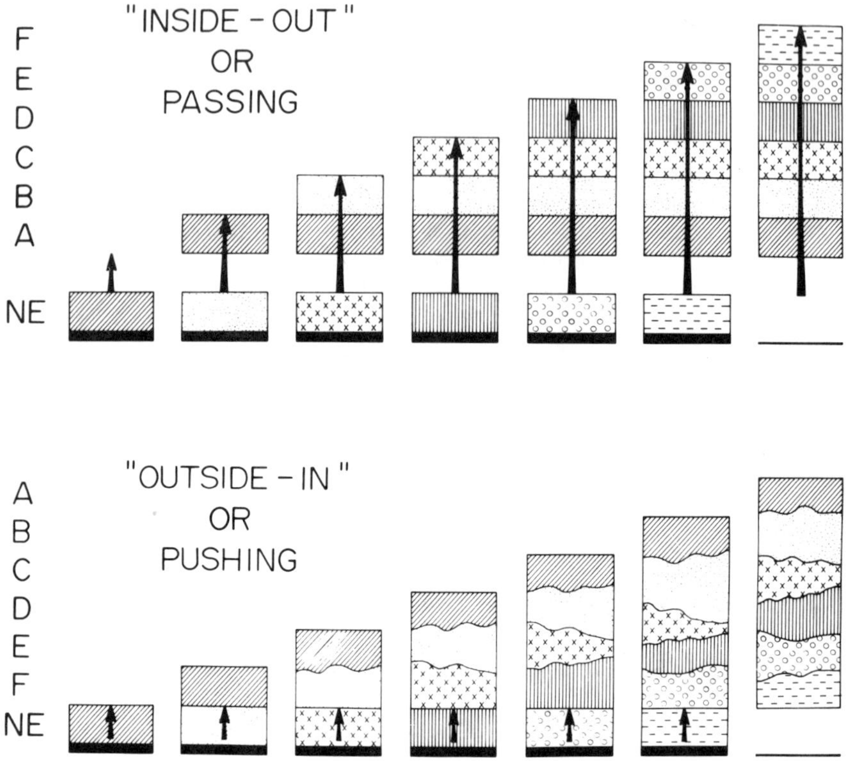

Figure 6. Diagram summarizing two different types of gradients found with ^{3}H-thymidine autoradiography. The patterned segments *A–F* represent groups of neurons sequentially produced (*A* is oldest and first; *F* is youngest and last) by the neuroepithelium (*NE, heavy black line*). In the "inside-out" gradient, younger neurons migrate past older neurons to settle progressively farther away from the neuroepithelium. In the "outside-in" gradient, younger neurons pile up beneath older ones, the youngest group (*F*) remaining closest to the neuroepithelium.

part of the telencephalon, is represented by the curved structure on the left. Its neurons send fibers (axons) to two main target areas: (1) the lateral septal nucleus in the telencephalon and (2) the mammillary body in the hypothalamus (part of the diencephalon). (See fig. 2.) The projection is organized so that dorsal and ventral parts of the hippocampus terminate in different areas of the target structures; such an organized projection is called *topographic* by neuroanatomists. The pattern of the topographic projection shown in figure 7 can be related to the patterns of neurogenesis in all of the interconnecting structures. Throughout the entire length of the hippocampus, neurons are generated nearly at the same time (Bayer, 1980), and it is reasonable to assume that their fibers start to grow toward the target areas simultaneously. Notice that the dorsal hippocampus always projects to older neurons, while the ventral hippocampus projects to younger ones. Since the dorsal hippocampus is closer to the targets, its fibers have a shorter distance to grow and probably arrive before those from the ventral hippocampus farther away. Older neurons are more likely to be mature enough to form connections with the earliest arriving fibers on a "first-come, first-served" basis. This and other similar examples throughout the CNS have led to the formulation of a hypothesis that the temporal order of neurogenesis may be the initial prerequisite for the later establishment of appropriate anatomical connections.

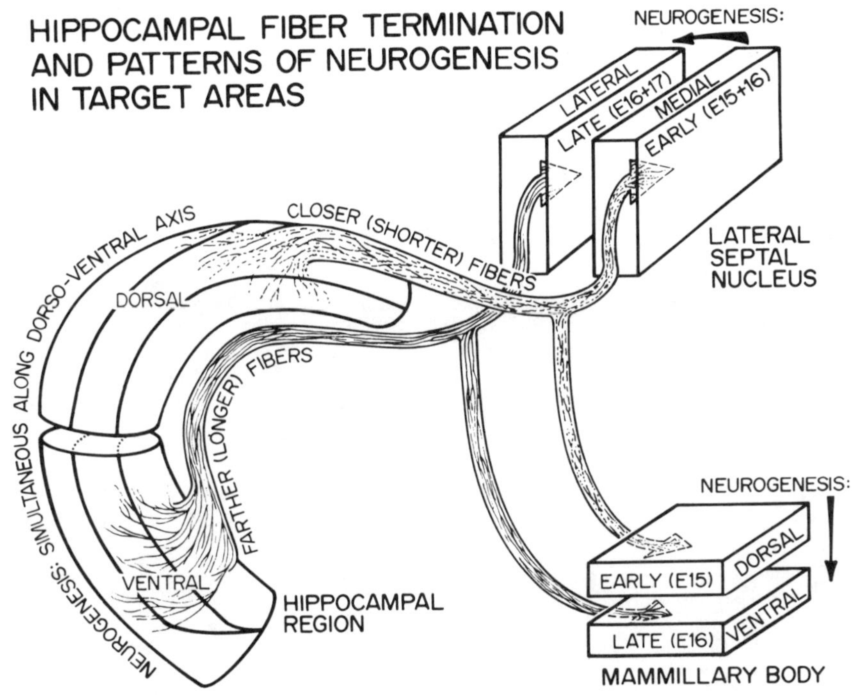

Figure 7. Diagram showing that the topographic projection of the hippocampus to the lateral septal nucleus and the mammillary body in the hypothalamus can be related to patterns of neurogenesis. Gradients are indicated by the solid black arrows. The dorsal hippocampus is closer to the targets, and its fibers probably arrive first (they have a shorter distance to grow) and preferentially contact cells generated earlier. The later arriving fibers from the ventral hippocampus contact later generated cells in each target. (From Bayer, 1980.)

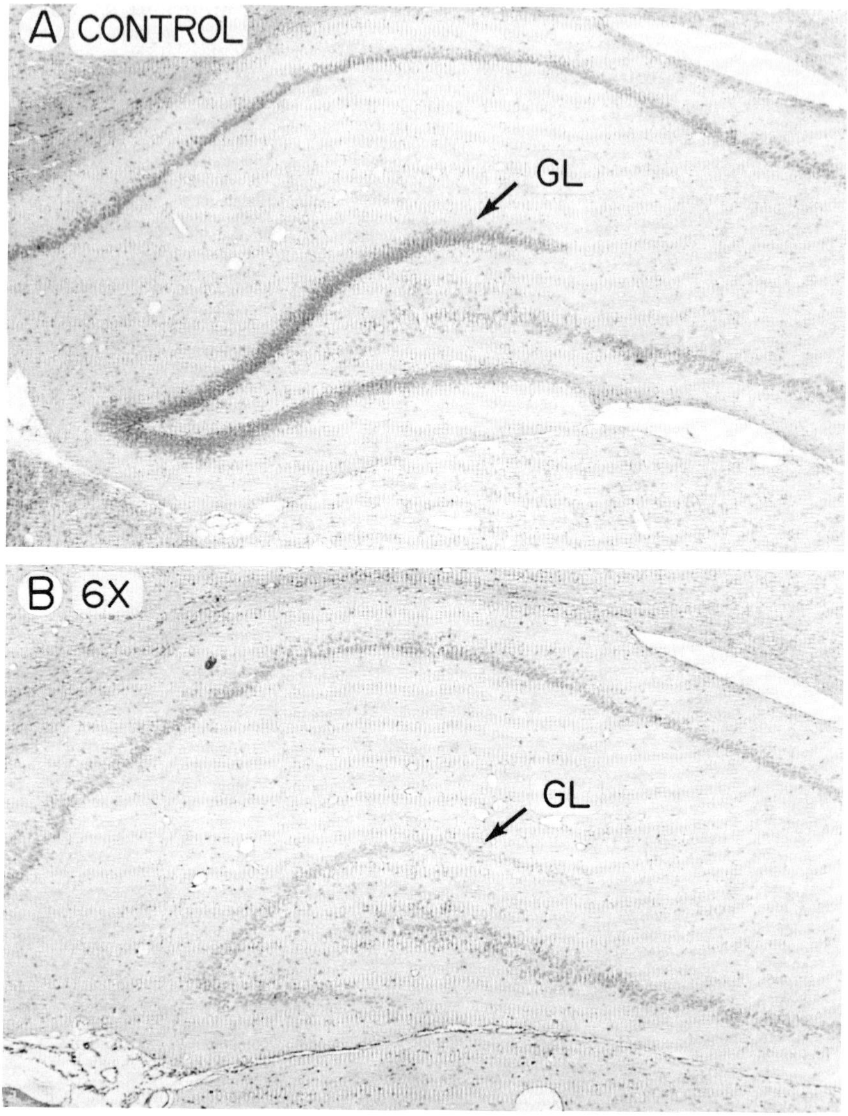

Figure 8. The hippocampus in a normal rat (*A*) and in an X-irradiated rat (*B*, six exposures between P2 and P11) on P60. Notice that the granular layer (*GL*) is greatly reduced in *B* because of the X-ray exposures during the time when granule cells are being generated (×40.5). (From Bayer and Altman, 1975.)

Another important feature of neurogenesis is that the rate at which the neuroepithelium will produce a specific kind of neuron is quite rigid. This point will be illustrated by discussing an experiment on the hippocampus in the rat brain. Figure 8*A* shows a section through the hippocampus of a P60 normal rat brain. The V-shaped layer (*GL* in fig. 8) contains densely packed neurons, the granule cells. In altricial mammals such as the rat, which are very immature at birth, the majority of these neurons originate during postnatal life (Bayer and Altman, 1974). If that part of the head containing the developing hippocampus is exposed to low-level X radiation (150–200 rad), most of the precursors of the granule cells are killed, but the surrounding young neurons that have started to differentiate are spared. Bayer and Altman (1975) exposed baby rats to various schedules of X radiation to see whether or not the few surviving precursor cells would increase their proliferative rates to compensate for the losses in their population. Figure 8*B* shows a section through the hippocampus in a P60 rat that received six exposures to X radiation between P2 and P11. Notice that the granular layer is drastically reduced. Later survival times show this effect to be permanent. Although there is evidence that hippocampal granule cells are generated in appreciable numbers during adulthood in normal rats (Bayer, Yackel, and Puri, 1982), their permanent reduction after X irradiation indicates that the few surviving granule cell precursors do not play "catch up" by producing neurons at a faster rate until the normal number is finally reached. Since the precursor cells are remarkably unresponsive to reductions in their populations, they may be genetically programmed to go through only so many cell divisions to produce a predetermined number of neurons.

Gliogenesis

The production of glial cells follows quite a different pattern from the production of neurons in three main ways. First, many glial cells are generated by proliferative cells lying outside the neuroepithelium near sites (or at the site) where they will be located in the adult. The preceding section showed that neurons usually migrate away from proliferative zones to their final locations. In figure 4*B*, the labeled cells lying outside the basal margin of the neuroepithelium (two groups are marked with asterisks) are probably glial precursors. These proliferative cells were ultimately derived from the neuroepithelium (for a review, see Altman, 1969b). Second, the time during which glial cells are normally produced is quite prolonged, extending throughout adult life (Hommes and Leblond, 1967; Dalton, Hommes, and Leblond, 1968; Paterson et al., 1973). There is additional evidence that some types of partially differentiated glial cells can go through cell division (Paterson et al., 1973), a phenomenon never shown by differentiating neurons. Third, damage to a glial population is rarely permanent. There is nearly complete recovery after low-level X irradiation (Bayer and Altman, 1975; Bayer, 1977), and glial cells rapidly proliferate after brain injury (Sjöstrand, 1965).

Cell Migration and Aggregation

Once neurons are generated, they migrate away from the neuroepithelium toward their final locations. On the way, they meet and mingle with other young neurons, finally becoming arranged into typical laminar structures (cortex) or clumps (nuclei). The autoradiographs in figure 5 give examples of a cortical and a nuclear structure. The proper spatial arrangement of neurons must be successfully completed if normal functional activity is to result. Thus, the developmental program of the nervous system must contain mechanisms for young neurons to accurately find their destinations.

Cell Recognition

The cells that make up the CNS are very heterogeneous. Despite this diversity, the nervous system reliably develops into its characteristic complex form. Experimental evidence is accumulating that the cell surfaces of specific types of young neurons become "tagged" with molecules that are capable of recognizing other cells of the same type. These compounds are called *cognins* (Moscona, 1980). Probably one of the structurally simplest parts of the nervous system is the eye's retina which originates from an outpocketing of the neural tube in the forebrain (see fig. 2). Within this microscopically thin sheet lie several different cell types. Moscona (1976, 1980) and his coworkers have studied the mechanisms by which retinal cells can recognize each other. It is quite possible to remove the retina from the chick embryo and maintain it in tissue culture. When trypsin is briefly added to an isolated retina, the enzyme breaks down proteins on the cell surface, and live cells are dissociated to freely float in the culture medium. After trypsin is removed from the culture, the cells begin to resynthesize their surface proteins and eventually reaggregate into a tissue that takes on the original organization of the embryonic retina. After a suitable length of time, the reaggregated retina becomes biochemically and morphologically differentiated, including the formation of synapses. Moscona's hypothesis is that the resynthesized macromolecules are cell ligands responsible for cell recognition. A glycoprotein (molecular weight 50,000 ± 6,000) has been isolated that will enhance the reaggregation of freshly dissociated cultures of retinal cells. This macromolecule, the retina cognin, is specific for retinal cells. An antibody has been raised against the retina cognin, and when it is added to dissociated retinal cell cultures, it inhibits its reaggregation. If such antibody-inhibited cultures are exposed to a

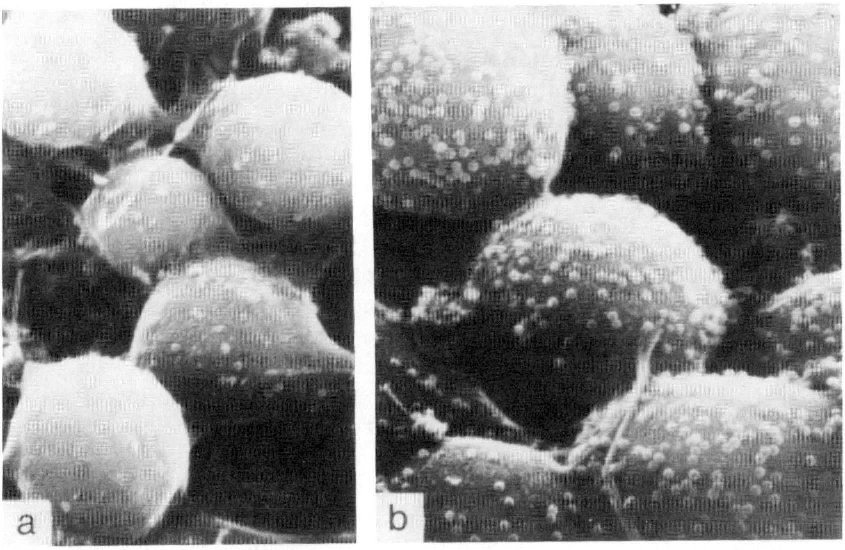

Figure 9. Scanning electron micrographs of dissociated retinal cells (from E10 chicks) labeled with immunolatex microbeads to detect cognin sites on the cell surface. *A*, freshly trypsinized cells, showing few cognin sites (×8,350). *B*, labeling after incubation of the cells for six hours to allow cognin regeneration (×25,050). (From Moscona, 1980.)

suspension of polystyrene latex microbeads precoated with antirabbit goat gamma globulin, the beads will preferentially attach to the sites of the retina cognin antigen-antibody complex. Figure 9 shows scanning electron micrographs of the dissociated retinal cells in these cultures. Notice that when the antibody is exposed to freshly trypsinized cultures (fig. 9*A*), there are fewer cognin molecules than on those cells that have been cultured for six hours after trypsinization (fig. 9*B*) to allow time for some resynthesized cognin molecules to reappear (Moscona, 1980). There is additional information that the retina cognin may be genetically encoded (Moscona et al., 1981). In the place of thymidine, 5′-bromodeoxyuridine (BrdU) will be incorporated into the DNA molecule and garble the genetic message. If dissociated retinal cells are cultured in the presence of BrdU, the cells will not reaggregate, even if BrdU is later removed. Thus, the effect is permanent. When the DNA synthesis inhibitor cytosine arabinoside is added to the culture medium with BrdU, the effect of BrdU is neutralized, and retinal cells will reaggregate after both compounds are removed from the culture medium. Finally, and most interesting from a developmental point of view, the ability of dissociated cells to reaggregate declines with age (Moscona,

1976), indicating that the cognins may be present on the cell surface only during a "critical period."

Working with mouse embryos, DeLong (1970) found that cells dissociated from either the cerebral cortex or the hippocampus would reaggregate into histotypical structures. The reeler mutant mouse has a cerebral cortex with improperly aligned cells. Cultures of dissociated cells from reeler embryos will reaggregate, but the cells are not lined up properly within the aggregate, and no definite cortical layers are formed (Sidman, 1970). Thus, there is evidence that cell-surface macromolecules may be under genetic control in mammals.

In regard to the above discussion, it should be pointed out that retina cognin shows tissue-specific activity. For example, when two different tissues are dissociated in the same culture (e.g., retina and cartilage cells), they sort themselves out and reaggregate into clumps of either retina or cartilage. But when cells from two regions of the brain are dissociated in the same culture, the cells from the same structure show a preferential tendency to reaggregate; however, there is some cross-reaction with cells of the other brain structure (Moscona, 1976). This indicates that the cell ligands on the surface of young neurons may not be entirely cell-type specific. Still, the fact that there is a preferential tendency for aggregation of cells of the same brain region means that the nervous system may use these cell ligands, along with other mechanisms, to achieve its final form.

Cell Migration

The discovery that neurons in the cerebral cortex settle in an "inside-out" pattern (Angevine and Sidman, 1961) has sparked tremendous interest in analyzing the mechanisms by which young neurons migrate from the neuroepithelium up into the cortical plate. Various theories have been put forth, but no one is yet sure how the process is accomplished. Berry, Rogers, and Eayrs (1964) and Morest (1970) hypothesized that cortical neurons do not migrate. In their Golgi studies, the young neurons on the way to the cortical plate appeared to have very long processes that extended from the neuroepithelium all the way up to the pial surface of the brain. (The pia mater is one of the meninges covering the central nervous system.) During the "migration" phase, the neuronal nucleus simply translocates upward within the cell process. Rakic (1972) also noted these radial processes, but his electron micrographs indicated that they were bipolar glial cells. Figure 10 shows Golgi impregnations of these radial glial cells in an E18 rat brain. Cell processes of the migrating neurons are much shorter. Rakic proposed that the young neurons migrate straight upward to the cortical plate using the radial glial fibers as guides. This hypothesis has been very popular, but it still is unable to explain all the known observations. Figure 11 shows drawings of Golgi-impregnated radial glial cells (on the left of the dotted line)

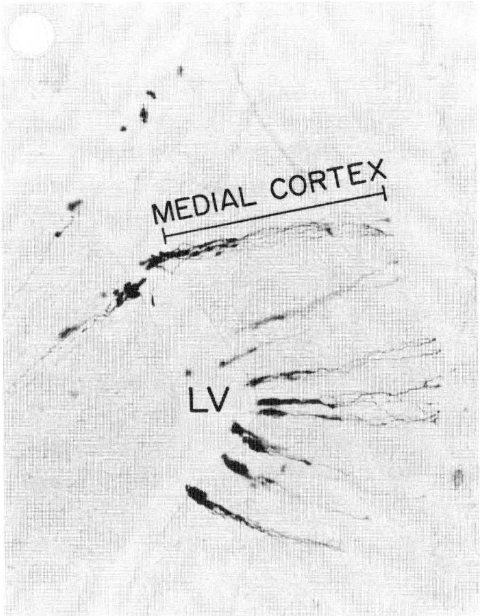

Figure 10. Radial glia in the medial part of the cerebral cortex in an E18 rat brain (100-μm Golgi-stained section).

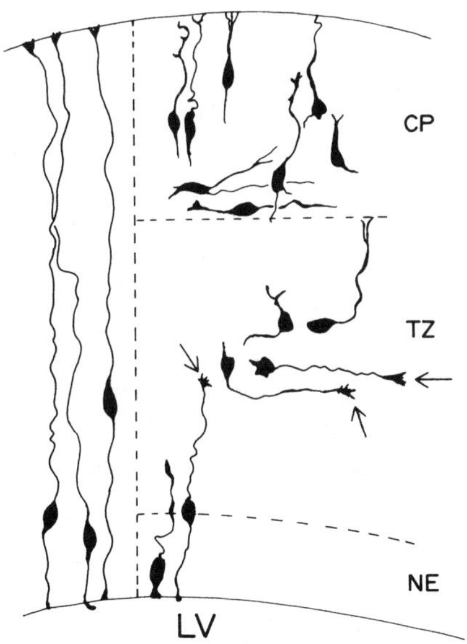

Figure 11. Drawing of impregnated radial glia (*left of vertical dashed line*) and young neurons (*right of vertical dashed line*) either migrating toward or in the cortical plate (*CP*). Orientation in figure 11 is flipped 90 degrees in relation to figure 10. Notice that the transitional zone (*TZ*) contains many migrating neurons that have a horizontal orientation. Arrows indicate presumptive axonal growth cones (*NE*, neuroepithelium; *LV*, lateral ventricle).

and migrating young cortical neurons (on the right). Notice that the young neurons do not maintain a radial orientation throughout all phases of their migration. When they reach the transitional zone, many cells develop horizontal processes. Perhaps these are the axons growing toward other structures in the brain. Once the cells reach the cortical plate, they again have a vertical orientation, except for a few horizontally oriented cells at the very base of the plate. The horizontal cells are probably cortical neurons that have migrated several days before and have already taken on their final differentiated form in the bottom (and oldest) layer of the cortex. Taken together, all of these observations suggest that instead of directly migrating upward along the radial glial cells, the future cortical neurons begin their differentiation in the transition zone. Only after sending out an axon does the cell body start migrating radially once more to reach the pia and begin to grow its dendrites. This two-stage pattern has been noted by other investigators, especially by Stensaas (1967). Another recent line of investigation using immunocytochemical staining of glial fibrillary acidic protein (Woodhams et al., 1981) also casts doubt on the validity of the radial glial hypothesis. The results are summarized in figure 12. Notice that the cortical plate matures in a lateral-to-medial direction; on the other hand, the radial glial fibers mature in the opposite direction. Also, radial glial fibers do not mature until after neurons have begun to migrate into the medial cortical plate. This means that neurons migrating to the lateral parts of the cortical plate get there without the guidance of the radial glial fibers.

Axonogenesis

One of the most important but least understood aspects of the development of the nervous system is how axons find their targets. Sensory information relayed through fibers from the skin, muscles, and the special sense organs (such as the eye and ear) have to reach appropriate targets in the CNS; motor fibers from the CNS to muscles and other effector mechanisms (autonomic) must grow toward their specific targets in the body. Finally, within the CNS, relay structures processing sensory and motor information have to be connected to one another, often in a precise topographic order.

The growth of fibers from primary sensory neurons in the trunk into the CNS is relatively simple. Since sensory neurons are located in the spinal ganglia that lie next to the neural tube, their fibers have only a short distance to grow to reach the CNS. Similarly, the motor fibers destined to reach either the trunk or limb muscles have to travel only a short distance in the embryo to contact either the somites (precursors of trunk muscles) or the limb buds (precursors of limb muscles). Weiss (1941) has done many important experiments elucidating the mecha-

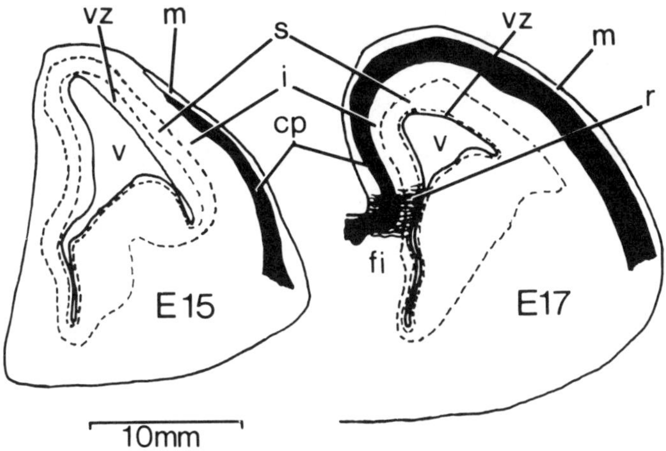

Figure 12. Drawings of one side of the coronally sectioned brain to show the development of the cortical plate (*cp*). Dorsal is at the top; ventral at bottom; medial on left; lateral on right. On E15, the cortical plate appears first in the lateral part of the cortex; by E17, it has grown medially. Immunoreactive radial glia (*r*) are first stained on E17 in the medial cortex; they do not stain until P7 in the lateral cortex. This means that many neurons arrive in the cortical plate before the radial glia are present (*v*, lateral ventricle; *vz*, ventricular zone [neuroepithelium]; *m*, mantle layer; *s*, subependymal zone; *i*, intermediate zone; *fi*, fimbria). (From Woodhams et al., 1981.)

nisms by which these initially short paths are established. He hypothesized that there are three stages in this process (fig. 13). First, a pioneering fiber reaches the target structure when it is relatively close. Next, the fiber establishes contact with it (the application stage in fig. 13). Thereafter, the structure grows farther away as the embryo enlarges, and the fiber grows with it (the towing stage in fig. 13). Weiss further hypothesized that the pioneering fiber reaches the target by *contact guidance*; that is, it grows upon existing surfaces in the embryo that lead to the target. Later on, as more fibers reach the same target, they grow along the pioneering fiber, a process called *fasciculation*.

The path from the retina to visual relay stations in the diencephalon and midbrain (optic tectum) involves a considerable distance and requires appreciable re-sorting of the fibers. This fiber tract has been one of the most intensively studied pathways, probably because of the discovery that adult fish and amphibians will regenerate this pathway when the optic nerve is cut. Therefore, much of this research is based on studies of regrowing axons rather than on embryonic development. It is known that the retinal fibers project to the optic tectum in a point-to-point topography. In an ingenious experiment, Sperry (1965) removed one half of the retina and damaged the optic fibers from the remaining half to initiate regeneration. The results are shown in figure 14. The intact half of the retina regrew its axons back to its original tectal targets, avoiding the portion of the tectum that was normally innervated by the ablated half of the retina. Sperry postulated that retinal fibers can locate their targets by chemospecificity. Other mechanisms have been postulated in addition to chemospecificity in the growth of the optic pathway. For example, there is evidence that the timetable of neurogenesis in the retina sets the stage for a sequential outgrowth of axons into the optic

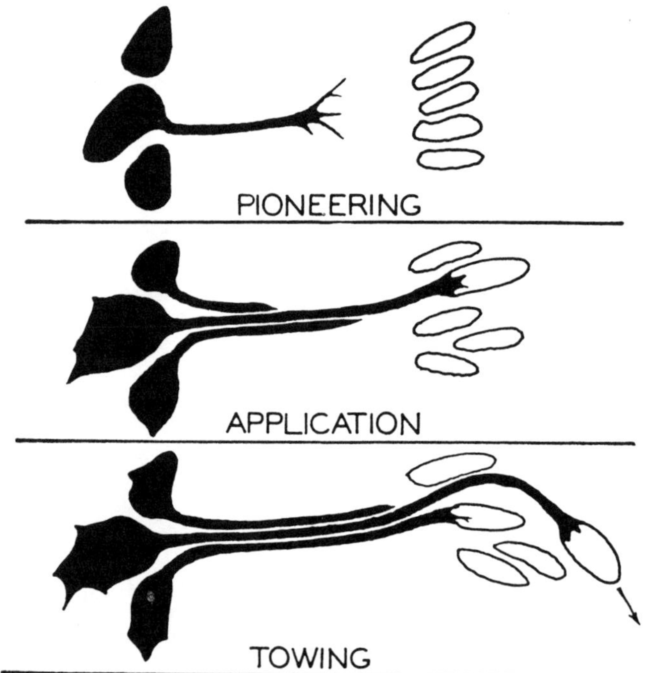

Figure 13. Groups of young neurons in three successive stages of axonal growth. In the pioneering stage, an axon finds a nearby target using contact guidance of existing surfaces in the embryo. In the application stage, the axon makes contact with the target while other axons grow toward nearby targets along the pioneering fiber (fasciculation). In the towing stage, the target structures and young neurons get farther apart as the embryo grows. The axons elongate and are "towed" by the target to their final locations. (From Weiss, 1941.)

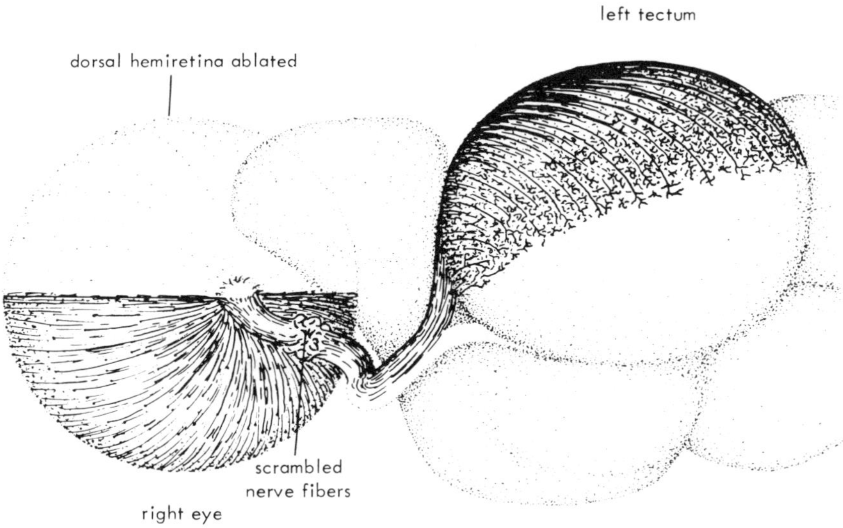

Figure 14. A drawing of the right eye and a portion of the amphibian brain to demonstrate the results of Sperry's experiment. Following ablation of the dorsal half of the retina and optic nerve section, regenerating fibers from the intact ventral half of the retina cross the midline and grow back to their original terminals in the dorsal half of the left optic tectum. (From Sperry, 1965.)

nerve; presumably resulting in a sequential arrival of axons in the tectum. In a way similar to the connections between the hippocampus and the lateral septal nucleus (see fig. 9 and the "Neurogenesis" section above), the axons would establish contact on a "first-come, first-served" basis. This temporal order may account for some of the specificity in the optic projection (Horder and Martin, 1978). In another line of research, Silver and Sapiro (1981) have noted that open glial channels develop along the surface of the retina and along the future course of the optic nerve before the first retinal axons start growing to the brain. These channels are filled with growing fibers according to a specific order, which may also account for their finally establishing orderly connections.

Not only chemospecificity, but also contact guidance, fasciculation, and glial channels imply some type of chemical recognition betwen fibers and (1) their pathways to the target and (2) the target itself. Mechanisms involving the search for cell recognition surface macromolecules on growing axons are prime candidates for the future studies of axonogenesis. Indeed, there is morphological evidence, first noted by Cajal (1911) and later confirmed in the first tissue culture studies of Harrison (reviewed in Harrison, 1969) that growing axon tips have growth cones (*arrows*, fig. 11) that send out minute filopodia that advance and retreat as if sampling their environment.

Growth of the Dendrite and Synaptogenesis

Dendritic Maturation

The shape of a dendrite in adult neurons takes on a tremendous variety of different forms. In some cases, there are dendrites radiating in all directions from the cell body. In other neurons, dendrites are polarized either to leave from only one side of the cell body or to exist in two tufts on opposite sides. The dendritic surface can either be smooth or have small protrusions called spines. If dendrites contain spiny surfaces, they can be either densely or sparsely covered. The variation is so vast that often specific types of neurons can be identified solely by the shape and surface characteristics of the dendrite.

Since the chief function of a dendrite is to receive synaptic input from axon terminals of other neurons, no doubt the development of the dendrite is closely correlated with the orientation of various afferent fibers. In some cases, the fibers come from distant brain regions in long bundles. In other cases, the fibers are randomly arranged and come from neurons only a few microns away. On the other hand, short fibers may also be stacked in regular bundles while long fibers are randomly branched. Thus, each brain region has a specific collection of afferent axons that could interact with growing dendrites during neuronal differentiation.

Because of its relative simplicity and regularity, the internal structure of the cerebellar cortex is probably one of the best understood regions of the brain. The orientation of most afferent fibers is well known, along with the structure of each cell type within the cerebellar cortex. The time of origin of all these neurons is also known (Altman, 1969a). Altman (1976, 1982) has worked out some of the developmental mechanisms that shape the dendrite of the Purkinje cell, the largest neuron in the cerebellar cortex. A summary of this work will be presented to give an example of how a dendrite may be shaped by its environment to take on a particular form. It should be emphasized that these mechanisms apply only to Purkinje cells; neurons in different microenvironments will necessarily be subject to other influences.

A photomicrograph of the cerebellar cortex in the rat brain is shown in figure 15. There are three layers: molecular layer, Purkinje cell layer, and granule cell layer (the granular layer is shaded in fig. 16). The entire cortical sheet is thrown into a series of folds, or lobules, as it is packed into a small space in the brain. Figure 17 is a drawing of a typical Purkinje cell with its fan-shaped dendrite from a Golgi-stained section.[3] Both figures 16 and 17 are from brain sections parallel, or sagittal, to the mid-

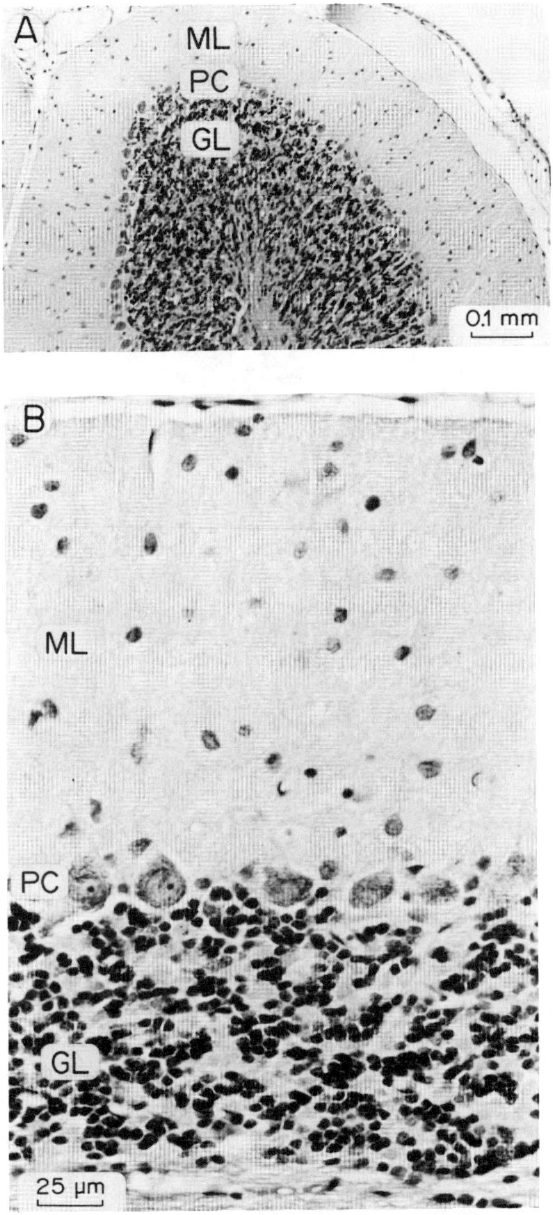

Figure 15. The cerebellar cortex in the adult rat brain. *A*, low magnification view of a portion of the cortex showing that the three layers are sharply folded (*ML*, molecular layer; *PC*, Purkinje cell layer; *GL*, granular layer). *B*, high magnification view.

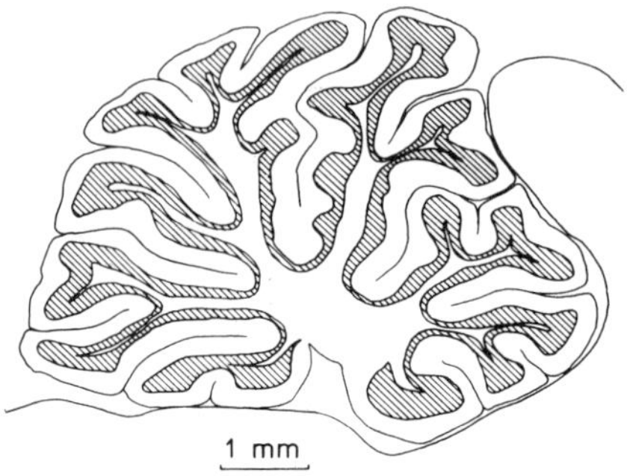

Figure 16. Drawing of a sagittal section of the adult rat cerebellum in the midline; the granular layer is shaded. The cerebellar cortex has many sharp folds called *lobules*.

Figure 17. Drawing of a typical Purkinje cell from a sagittal Golgi-stained section. The dendrites fan out in the molecular layer (*dashed line*, surface).

line. In a Golgi-stained section perpendicular, or coronal, to the midline, the Purkinje dendrites look like telephone poles extending from the cell body up to the surface of the molecular layer (see fig. 18). Thus, the Purkinje cell dendrite is flattened (much like a fan blade) in the sagittal plane.

The cerebellar cortex also has three populations of smaller neurons called *interneurons*. The most numerous are the granule cells. The cell body is located in the granular layer, and small club-shaped dendrites extend from it. The axon projects into the molecular layer and bifurcates into two long unbranched processes, giving the entire axon a shape like the capital letter *T*. The fibers are oriented so that the cross bar of the *T* is perpendicular to the midline. In coronal Golgi-stained sections of the cerebellar cortex, the axons of the granule cells look like thin parallel lines running horizontally in the molecular layer (see fig. 18). For this reason, the granule cell axons are called *parallel fibers*. Notice that the parallel fibers are always perpendicular to the flattened Purkinje cell dendrite (see fig. 18). Basket cell bodies are located in the lower half of the molecular layer. The dendrite arborizes randomly throughout the molecular layer. The axon wraps around the short stem dendrite of the Purkinje cell, and several branches extend down to form a "basket" around the Purkinje cell body. (See the drawing in fig. 18 for day 8 and that in fig. 19 for day 30 of normal development.) A single basket cell axon will have a long process with several sets of shorter branches dropping down to form baskets around several Purkinje cells in a row. The stellate cell bodies are located in the upper half of the molecular layer. They have dendrites that are similar in structure to those of basket cells; the shape of the axon is variable, but it is known to preferentially contact the secondary branches of the Purkinje cell dendrite (Eccles, Ito, and Szentágothai, 1967). (See the drawing in fig. 18 for day 12 of normal development.)

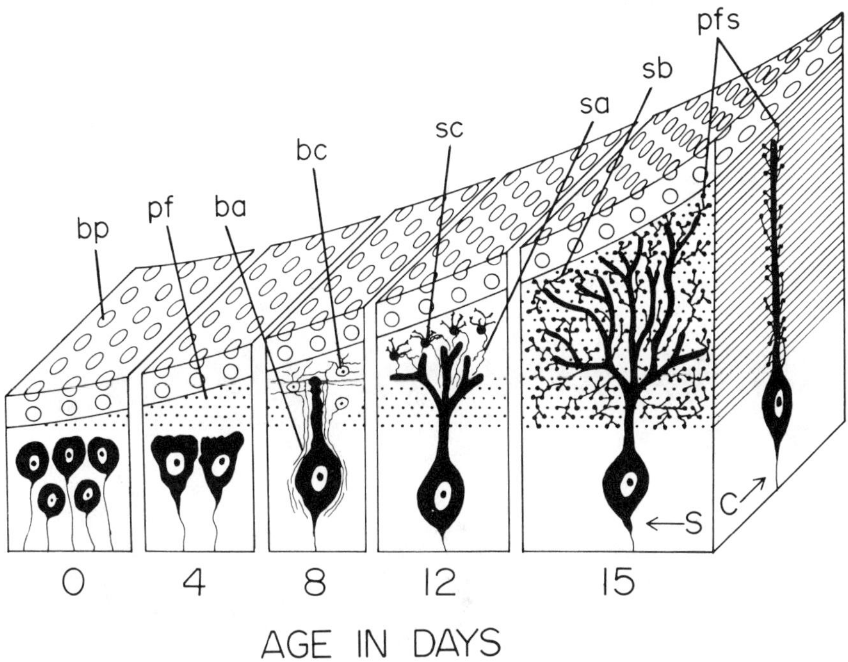

Figure 18. Scheme of the major steps in the postnatal morphogenesis of rat Purkinje cells and of the proposal for the growth of spiny branchlets in a single plane and at a right angle to the long axis of the stacked parallel fibers. An "exclusion principle" allows only a single contact between a given Purkinje cell and a parallel fiber. This situation confines outgrowth in the coronal plane (*C*) but does not affect growth in the sagittal plane (*S*) because contacts can be made with a large number of different parallel fibers (*BA*, basket cell axon; *BC*, basket cell; *BP*, bipolar cell in the external germinal layer; *PF*, parallel fibers; *PFS*, parallel fiber synapses [contacts]; *SA*, stellate cell axon; *SB*, spiny branchlets; *SC*, stellate cells). (From Altman, 1976.)

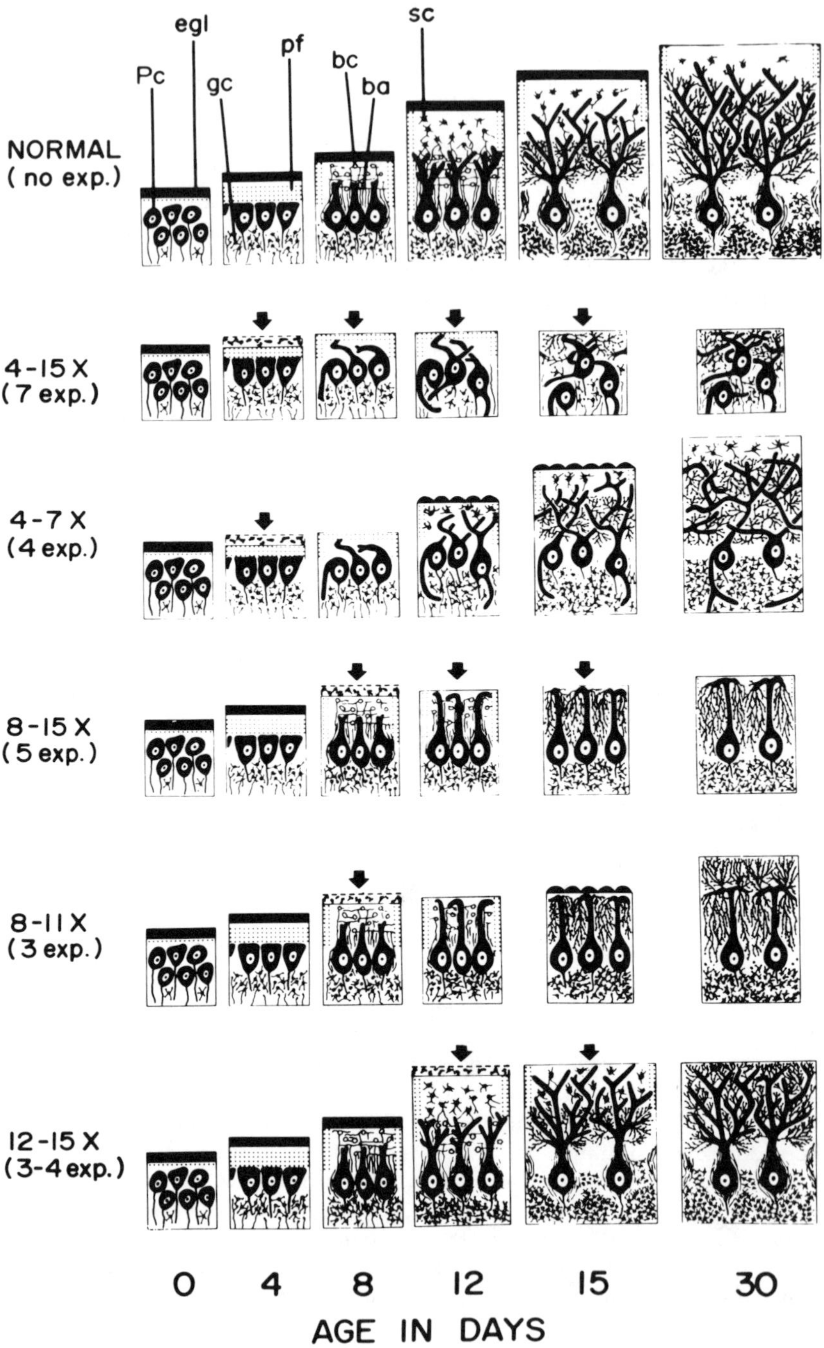
egl
sc
pf
bc
Pc
gc
ba
NORMAL
(no exp.)
4-15 X
(7 exp.)
4-7 X
(4 exp.)
8-15 X
(5 exp.)
8-11 X
(3 exp.)
12-15 X
(3-4 exp.)
0
4
8
12
15
30
AGE IN DAYS

Over 90 percent of the interneurons in the cerebellar cortex originate after birth in the rat in the following sequential order: basket cells during the first week, stellate cells during the second week, and over 50 percent of the granule cells during the last part of the second and into the third weeks. By P 21, all of these cell types have originated (Altman, 1969a). These cerebellar interneurons are produced by mitotic cells lying on the surface of the cerebellar cortex in the external germinal layer (diagrammed in fig. 18 as the most superficial layer of cells). Earlier in embryonic development, these cells migrated from the neuroepithelium to set up a secondary germinal matrix (Altman and Bayer, 1978). Once all the interneurons have originated, the external germinal layer disappears. This is why the adult cerebellar cortex in figure 15 does not have an external germinal layer.

If the part of a rat's head containing the cerebellum is exposed to low-level X radiation at various times during the first 21 days of the rat's life, a cerebellar cortex devoid of all or of a specific population of interneurons is produced (Altman and Anderson, 1972). The X rays kill mitotic cells in the cerebellum just as they would in the developing hippocampus mentioned above. By examining the shape of the Purkinje cell dendrite in these X-irradiated cerebella, Altman (1976, 1982) found that the axons of interneurons are responsible for the production of various structural features in the Purkinje cell dendrite.

The results are summarized in figure 19. In the normal animal, the dendrite begins to grow from a prominent apical cone of cytoplasm. Initially there is only a single heavy stalk or stem dendrite. After growing a short distance, the stem dendrite forms smooth secondary branches. Finally, thin spiny branchlets grow out from the smooth secondary branches. If the cortex is irradiated so that either most interneurons are eliminated (4–15X, fig. 19) or just basket cells are killed (4–7X, fig. 19), the Purkinje cell dendrite has no definite orientation. Thick dendrites grow in all directions, some up, others down; many Purkinje cells show more than one large dendrite exiting from the cell body. If the irradia-

Figure 19. Schematic summary of the stages and time course of Purkinje cell dendritic development in normal rats (top row) and those receiving X irradiation during early postnatal life. The number of exposures is given in parentheses under each experimental group (*PC*, Purkinje cells; *EGL*, external germinal layer; *GC*, granule cells; *PF*, parallel fibers; *BC*, basket cells; *SC*, stellate cells; *X*, postnatal days during which animals were exposed to X rays.) Arrows indicate X irradiation and killing of interneuron precursors in the external germinal layer. Corrugated heavy black lines (4–15X, days 12 and 15; 8–11X, day 15) symbolize the external germinal layer that regenerates after X-ray exposures are completed in these experiments. (From Altman, 1982.)

tion is between P8 and P15 (8–15X), all basket cells and some granule cells are present, but stellate cells are missing. The Purkinje cell dendrites in these animals are oriented, but, curiously, they have very long stem dendrites that grow straight upward toward the surface of the molecular layer. Thin spiny branchlets run back down into the molecular layer as they exit from the tip of the stem dendrite, giving the appearance of a weeping willow. Since basket cells are the most prominent interneuronal population present, their axons can exert the greatest influence on the developing dendrite. Apparently these axons ensure that only one dendrite leaves from the Purkinje cell body, and that this dendrite will grow upward. Possibly this influence is brought about by the tendency for a basket cell axon to wrap around the Purkinje cell stem dendrite before its branches extend downward over the cell body. If X-ray exposures are given between P12 and P15 (12–15X), both stellate and basket cells are present while some granule cells are eliminated. In these animals, the dendrite is nearly normal in shape, except it has less depth, which is most likely due to a considerable loss of granule cells. Only when stellate cell axons are present will the smooth secondary branches be produced. Since stellate cell axons form synapses on secondary branches in the mature Purkinje cell, their presence during development probably influences the degree of secondary branching. If X irradiation is begun on the day of birth and continued until P15, granule cells are almost entirely eliminated. In these animals, the Purkinje cell dendrites resemble those in the 4–15X experiment (second line, fig. 19) except that no spiny branchlets are present. In the adult, parallel fibers form synapses only with the spiny branchlets of the Purkinje cell dendrite. Thus, parallel fibers probably induce the formation of the spiny branchlets.

If rats are exposed to X rays only three times between P8 and P11, the external germinal layer will regenerate by day 15 and continue to produce granule cells. The parallel fibers of the granule cells formed before X-ray exposures began will be normally oriented, perpendicular to the midline. However, the parallel fibers of granule cells formed by the regenerated germinal layer do not have the normal orientation. In many cases, the later parallel fibers may run parallel to the midline. In the normal molecular layer, the parallel fibers are stacked according to age: axons of the earliest granule cells are near the bottom; those from the latest are near the top. In the 8–11X experiment, the lower part of the Purkinje cell dendrite associates with normally oriented parallel fibers, while the upper part associates with those abnormally oriented. Altman (1976, 1982) found that the Purkinje cell dendrites in these animals are also oriented in two planes (fig. 20). The lower part of the dendrite has a normal orientation parallel to the midline. Depending on the orientation of the later parallel fibers, the upper part of the dendrite will rotate. In the cell shown in figure 20, the upper part is rotated 90 de-

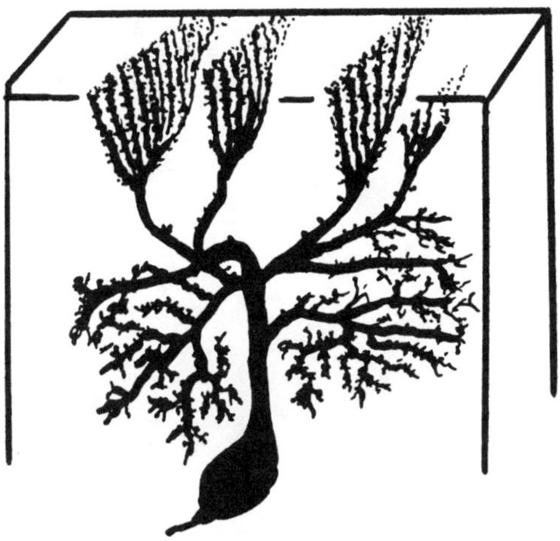

Figure 20. Drawing of a Golgi-impregnated Purkinje cell with the dendrites oriented in two planes. The lower branches are aligned normally in the sagittal plane and associate with normally arranged parallel fibers produced before X irradiation began. The upper branches are oriented at a right angle to the abnormally arranged parallel fibers produced by the regenerated external germinal layer. (From Altman, 1982.)

grees. Throughout all of these planar shifts, the flat part of the Purkinje cell dendrite is always perpendicular to the long axis of the parallel fibers. This observation led Altman (1976, 1982) to formulate a theory that the parallel fibers induce the Purkinje cell dendrite to fill a two-dimensional rather than a three-dimensional space. Figure 18 shows that if a Purkinje cell dendrite grows crosswise through the bundles of parallel fibers, the probability of making contacts with the greatest number of different axons is increased. On the other hand, growing the dendrite lengthwise between the parallel fibers would mean that many contacts are made with the same group of parallel fibers. Altman's theory is that an "exclusion principle" is active such that once a Purkinje cell dendrite has formed a few synapses with a specific parallel fiber, further contacts with the same parallel fiber have a low probability of occurrence. Figure 18 shows that such a pattern would inhibit growth in the coronal plane and enhance it in the sagittal plane.

Synaptogenesis

Neurons communicate with each other through specialized junctions called *synapses* (in most cases with chemical signals called neurotransmitters). The formation of a synapse represents the culmination of a long series of developmental events. First, both the sending and receiv-

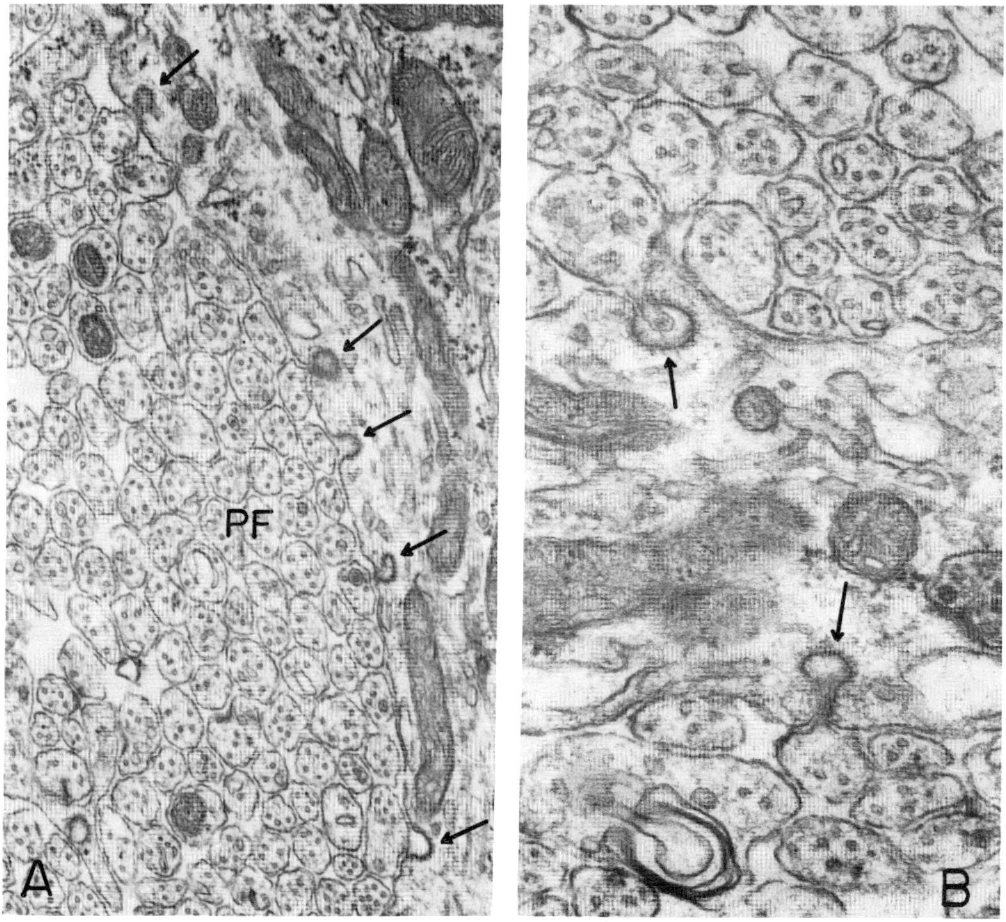

Figure 21. *A*, high concentration of open coated vesicles (*arrows*) in the primary dendrite of a Purkinje cell opposite parallel fibers (*PF*) (×32,000). *B*, portion of the parallel fiber may be sucked into the open coated vesicles (×48,000). Electron micrographs of the cerebellar cortex in a P12 rat. (From Altman, 1982.)

ing neurons originate during precise times of development. Second, the arrival of the incoming axons at the target zone is most likely timed to correlate with the beginning of dendritic outgrowth from the cell body. Finally, the two interacting neurons probably "test" each other (perhaps chemically) to see if they are compatible. When all of these prerequisites are met, a synapse is likely to form. Since dendritic maturation is closely correlated to synaptogenesis, the preceding section provides a basis for the following discussion of synaptogenesis on the Purkinje cell dendrite. Once again it should be mentioned that synaptic formation patterns on other neurons are probably different, reflecting the sequence of a specific set of developmental events not necessarily similar to the maturation of Purkinje cells.

First it is necessary to provide more detail about the structure of synapses. Synaptogenesis is the formation of specialized membrane contacts between the axon terminal of a "sender" or presynaptic neuron and the cell body and dendrites of a "receiver" or postsynaptic neuron. Altman (1971) has studied the ultrastructural changes during the formation of synapses in the cerebellar cortex. Unusual organelles are routinely found in immature Purkinje cells. They are membrane-bound round vesicles covered with a dense filamentous coat (coated vesicles). In some instances, the coated vesicles appear to "pinch off" from the Golgi apparatus. As the Purkinje cells mature and start to form dendrites, the coated vesicles move to the cell membrane, fuse with it, and unfold, opening themselves to the extracellular space (fig. 21*A*) resembling pinocytotic vesicles. Often an adjacent parallel fiber appears to be sucked into the coated vesicle cavity (fig. 21*B*). Perhaps chemicals diffuse between the two cells as they "sample" each other. This may be the ultrastructural evidence for Altman's exclusion principle described above. Later on, dense thickenings of both parallel fiber and dendrite are apposed to each other across a cleft of extracellular space. Altman (1971) hypothesizes that a dense postsynaptic thickening in the Purkinje cell was once the zone where a coated vesicle fused with the plasma membrane. These attachment sites may be transformed into mature synapses (fig. 22). This process is repeated countless times as synapses are formed on the Purkinje cell dendrite.

Purkinje cells originate before birth, migrate to the future cerebellar cortex, and line up beneath the external germinal layer. Also before birth, fibers from another brain structure outside the cerebellum grow into the zone where the young Purkinje cells are accumulating (Altman and Bayer, 1978). Figure 23 summarizes the sequence of events as synapses appear on the Purkinje cell body and dendrite. On P3, synapses are absent. By P7, the prenatally arriving fibers have formed synapses on the Purkinje cell body. Later, these first synapses are translocated upward and climb into the molecular layer because of their permanent attachment to the growing Purkinje cell dendrite; these axons are aptly called

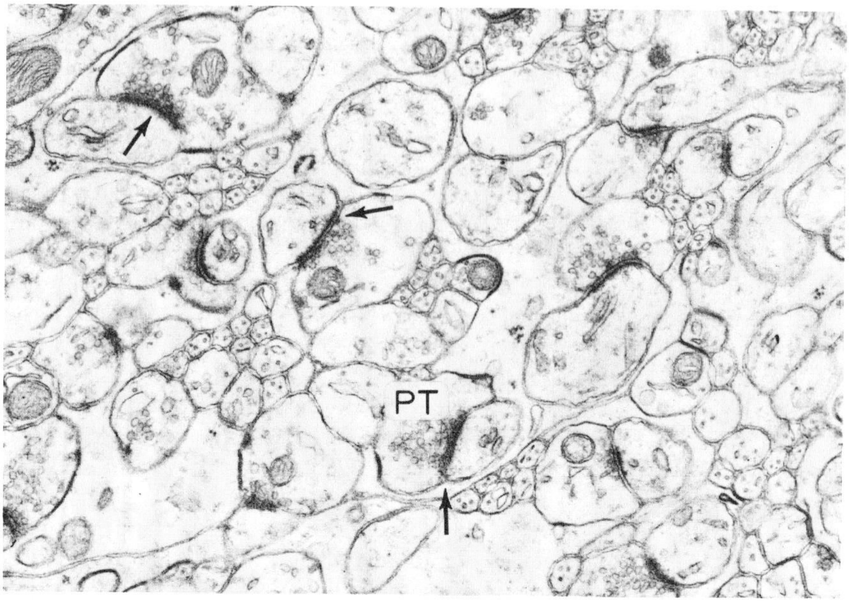

Figure 22. An electron micrograph of mature synaptic contacts (*arrows*) between parallel fiber terminals (*PT*) and Purkinje cell dendrites in the cerebellar cortex of a P21 rat (×36,000). (Photo courtesy of Dr. Joseph Altman.)

climbing fibers. Around P12, mature synapses between the Purkinje cell body and basket cell axons appear. By P15, stellate cell axons have formed some synapses on the secondary branches, and the oldest parallel fibers in the lower molecular layer (axons of the earliest originating granule cells) synapse with spiny branchlets of the lower Purkinje cell dendrite. At P21, younger parallel fibers have formed synapses in the middle molecular layer on the newly developing spiny branchlets. By P30, most of the synapses have formed between the most superficial parallel fibers (axons of the latest originating granule cells) and the most superficial (and youngest) spiny branchlets. After specific regions of the Purkinje cell body and dendrite have synaptically matured, glial processes (represented as heavy black deposits in fig. 23) cover the free space, first on the cell body and then gradually from the bottom to the top of the molecular layer. Figure 24 shows the histochemical staining of succinate dehydrogenase during this time. As synaptogenesis occurs, there is an increase in oxidative metabolism. The lower, more mature zones of the Purkinje cell dendrite show the earliest increases in enzyme activity, possibly correlating with the onset of synaptic transmission.

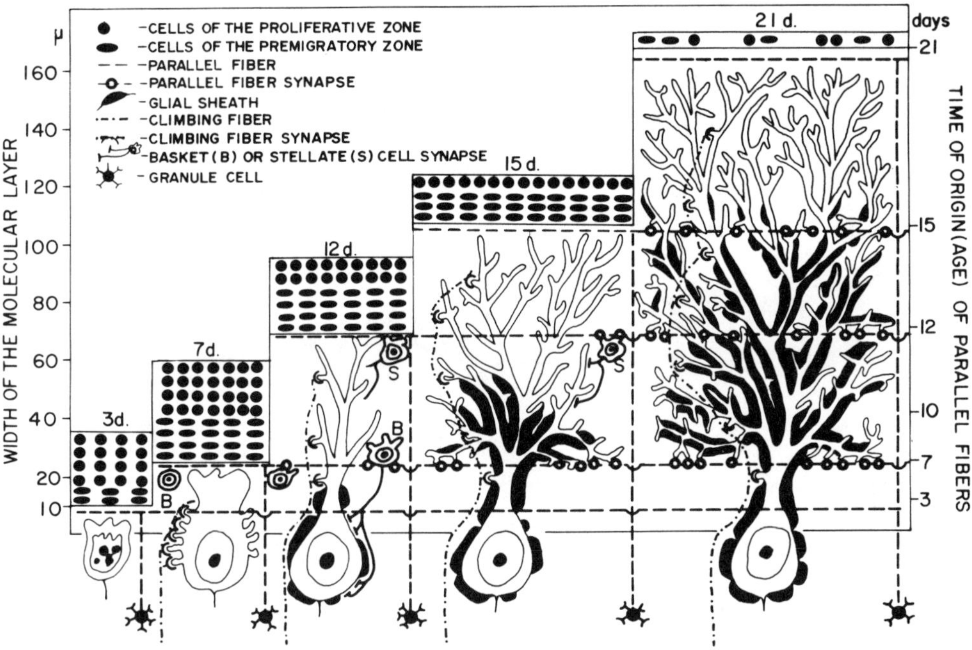

Figure 23. Diagrammatic illustration of the major events during synaptogenesis in the molecular layer of the cerebellar cortex. Notice that synapses first appear in the Purkinje cell layer and then march upward toward the superficial part of the molecular layer. (From Altman, 1972.)

Notice that the zone of heightened activity continues to advance upward, correlating with the pattern of synaptic appearance.

Notes

1. This is not true of all tissues. For example, contact between the spinal cord and the presumptive kidney primordium is required to induce the formation of kidney tubules (Lehtonen et al., 1975).
2. Mauthner's cells are two giant neurons whose cell bodies are on either side of the medulla in many fish and amphibian brains. They have large axons that cross the midline to run down the full length of the spinal cord (Stefanelli, 1951).
3. With the Golgi stain, the entire cell, including all its processes, is completely filled with a black precipitate. Since only about 2–3 percent of the cells are impregnated, they stand out against a clear background of unstained cells.

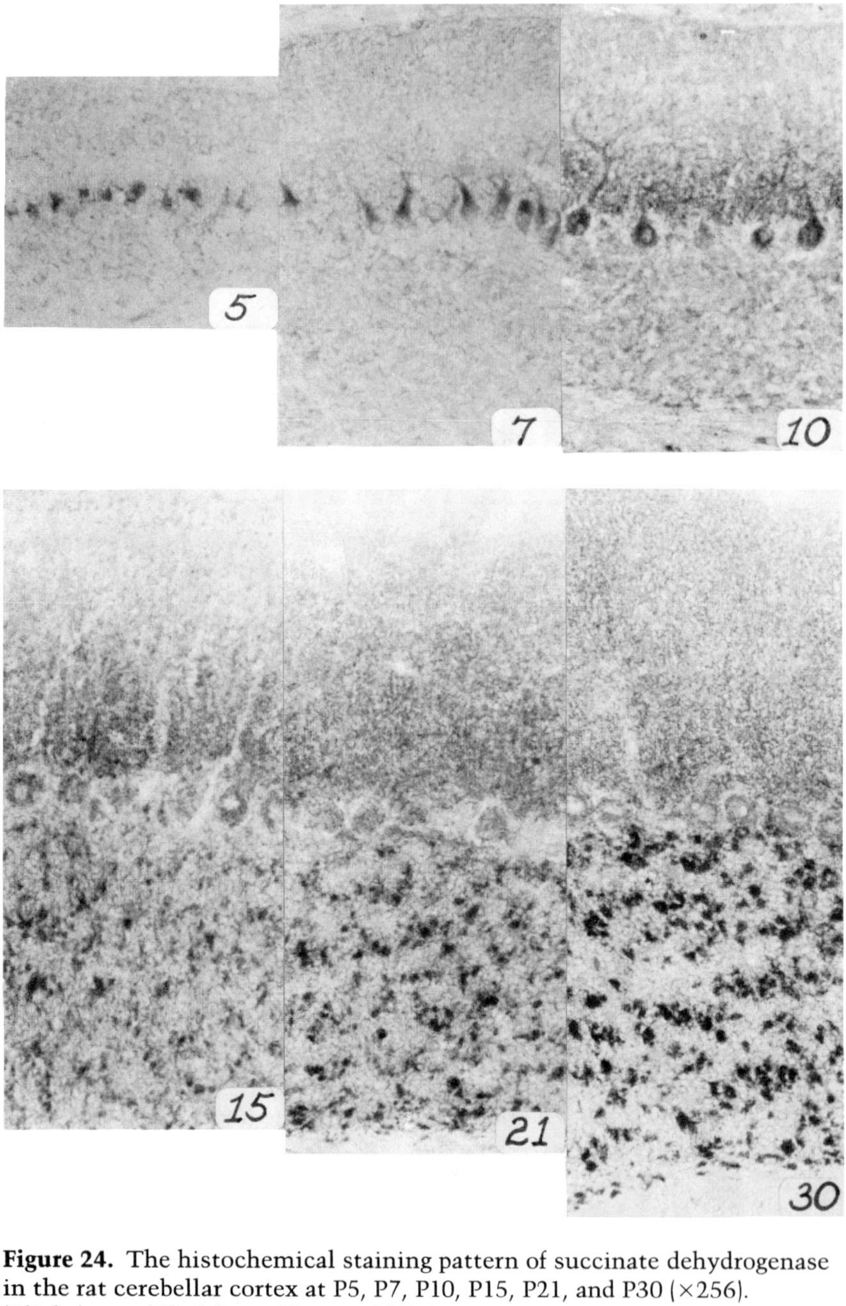

Figure 24. The histochemical staining pattern of succinate dehydrogenase in the rat cerebellar cortex at P5, P7, P10, P15, P21, and P30 (×256). (Slightly modified from Altman, 1972.)

References

Altman, J. (1969a). Autoradiographic and histological studies of postnatal neurogenesis: III. Dating the time of production and onset of differentiation of cerebellar microneurons in rats. *J Comp Neurol* 136, 269–294.

Altman, J. (1969b). DNA metabolism and cell proliferation. In *Handbook of neurochemistry*. Vol. 2, ed. A. Lajtha, pp. 137–182. New York: Plenum.

Altman, J. (1971). Coated vesicles and synaptogenesis: A developmental study in the cerebellar cortex of the rat. *Brain Res* 30, 311–322.

Altman, J. (1972). Postnatal development of the cerebellar cortex in the rat: II. Phases in the maturation of Purkinje cells and of the molecular layer. *J Comp Neurol* 145, 399–464.

Altman, J. (1976). Experimental reorganization of the cerebellar cortex: VII. Effects of late X-irradiation schedules that interfere with cell acquisition after stellate cells are formed. *J Comp Neurol* 165, 65–76.

Altman, J. (1982). Morphological development of the rat cerebellum and some of its mechanisms. In *The cerebellum: New vistas*, ed. S. L. Palay. *Exp Brain Res* (Suppl. 6), 8–49.

Altman, J., and Anderson, W. J. (1972). Experimental reorganization of the cerebellar cortex: I. Morphological effects of elimination of all microneurons with prolonged X-irradiation started at birth. *J Comp Neurol* 146, 355–406.

Altman, J., and Bayer, S. A. (1978). Prenatal development of the cerebellar system in the rat: I. Cytogenesis and histogenesis of the deep nuclei and cortex of the cerebellum. *J Comp Neurol* 179, 23–48.

Angevine, J. B., Jr., and Sidman, R. L. (1961). Autoradiographic study of cell migration during histogenesis of cerebral cortex in the mouse. *Nature* 192, 766–768.

Bancroft, M., and Bellairs, R. (1975). Differentiation of the neural plate and neural tube in the young chick embryo. *Anat Embryol (Berl)* 147, 309–335.

Bayer, S. A. (1977). Glial recovery patterns in rat corpus callosum after X-irradiation during infancy. *Exp Neurol* 54, 209–216.

Bayer, S. A. (1979). The development of the septal region in the rat: I. Neurogenesis examined with ^{3}H-thymidine autoradiography. *J Comp Neurol* 183, 89–106.

Bayer, S. A. (1980). The development of the hippocampal region in the rat: I. Neurogenesis examined with ^{3}H-thymidine autoradiography. *J Comp Neurol* 190, 87–114.

Bayer, S. A., and Altman, J. (1974). Hippocampal development in the rat: Cytogenesis and morphogenesis examined with autoradiography and low level X-irradiation. *J Comp Neurol* 158, 55–80.

Bayer, S. A., and Altman, J. (1975). Radiation-induced interference with postnatal hippocampal cytogenesis in rats and its long-term effects in the acquisition of neurons and glia. *J Comp Neurol* 163, 1–20.

Bayer, S. A., Yackel, J. W., and Puri, P. S. (1982). Neurons in the rat dentate gyrus granular layer substantially increase during juvenile and adult life. *Science* 216, 890–892.

Berry, M., Rogers, A. W., and Eayrs, J. T. (1964). Pattern of cell migration during cortical histogenesis. *Nature* 203, 591–593.

Burnside, B. (1973). Microtubules and microfilaments in amphibian neurulation. *American Zoology* 13, 989–1006.

Cajal, S. R. (1911). Histologie du système nerveux de l'homme et des vertébrés. Vol. 1. Madrid: Instituto Ramon Y. Cajal.

Dalton, M. M., Hommes, O. R., and Leblond, C. P. (1968). Correlation of glial proliferation with age in the mouse brain. *J Comp Neurol* 134, 397–400.

DeLong, G. R. (1970). Histogenesis of fetal mouse isocortex and hippocampus in reaggregating cell cultures. *Dev Biol* 22, 563–583.

Denham, S. (1967). A cell proliferation study of the neural retina in the two day rat. *J Embryol Exp Morphol* 18, 53–66.

Eccles, J. C., Ito, M., and Szentágothai, J. (1967). *The cerebellum as a neuronal machine*. New York: Springer.

Harrison, R. G. (1969). *Organization and development of the embryo*, ed. S. Wilens. New Haven, Conn.: Yale University Press.

Hommes, O. R., and Leblond, C. P. (1967). Mitotic division of neuroglia in the normal adult rat. *J Comp Neurol* 129, 269–278.

Horder, T. J., and Martin, K. A. C. (1978). Morphogenetics as an alternative to chemospecificity in the formation of nerve connections. In *Symposia of the Society for Experimental Biology: Cell-cell recognition*, pp. 275–358.

Hoshino, K., Matsuzawa, T., and Murakami, V. (1973). Characteristics of the cell cycle of matrix cells in the mouse embryo during histogenesis of telencephalon. *Exp Cell Res* 77, 89–94.

Jacobson, C. O. (1959). The localization of the presumptive cerebral regions in the neural plate of the axolotl larva. *J Embryol Exp Morph* 7, 1–21.

Karfunkel, P. (1974). The mechanisms of neural tube formation. *Int Rev Cytol* 38, 245–271.

Kauffman, S. L. (1968). Lengthening of the generation cycle during embryonic differentiation of the mouse neural tube. *Exp Cell Res* 49, 420–424.

Langman, J., Guerrant, R., and Freeman, B. (1966). Behavior of neuroepithelial cells. *J Comp Neurol* 127, 399–412.

Lee, H., Sheffield, J. B., Nagele, R. G., Jr., and Kalmus, G. W. (1976). The role of extracellular material in chick neurulation: I. Effects of concanavalin A. *J Exp Zool* 198, 261–266.

Lehtonen, E., Wartiovaara, J., Nordling, S., and Saxén, L. (1975). Demonstration of cytoplasmic processes in millipore filters premitting kidney tubule induction. *J Embryol Exp Morph* 33, 187–203.

Moran, D., and Rice, R. W. (1975). An ultrastructural examination of the role of the cell membrane surface coat material during neurulation. *J Cell Biol* 64, 172–181.

Morest, D. K. (1970). A study of neurogenesis in the forebrain of opossum pouch young. *Zeitschrift für Anatomie und Entwicklungsgeschichte* 130, 265–305.

Moscona, A. A. (1976). Cell recognition in embryonic morphogenesis and the problem of neuronal specificities. In *Neuronal recognition,* ed. S. H. Barondes, pp. 205–226. New York: Plenum Press.

Moscona, A. A. (1980). Embryonic cell recognition: Cellular and molecular aspects. In *Membranes, receptors, and the immune response,* pp. 171–188. New York: Alan R. Liss.

Moscona, M., Degenstein, L., Byun, K. Y., and Moscona, A. A. (1981). Development of differential affinities and positional information in embryonic retina cells: Inhibition by BrdU. *Cell Differ* 10, 317–327.

Paterson, J. A., Privat, A., Ling, E. A., and Leblond, C. P. (1973). Investigation of glial cells in semithin sections: III. Transformation of subependymal cells into glial cells, as shown by radioautography after ^{3}H-thymidine injection into the lateral ventricle of the brain of young rats. *J Comp Neurol* 149, 83–102.

Rakic, P. (1972). Mode of cell migration to the superficial layers of fetal monkey neocortex. *J Comp Neurol* 145, 61–83.

Sauer, F. C. (1935). Mitosis in the neural tube. *J Comp Neurol* 62, 377–405.

Schroeder, T. E. (1973). Cell constriction: Contractile role of microfilaments in division and development. *American Zoology* 13, 949–960.

Sidman, R. L. (1970). Cell proliferation, migration, and interaction in the developing mammalian central nervous system. In *The neurosciences: Second study program,* ed. F. O. Schmitt, pp. 100–106. New York: Rockefeller University Press.

Sidman, R. L., Miale, I. L., and Feder, N. (1959). Cell proliferation and migration in the primitive ependymal zone: An autoradiographic study of histogenesis in the nervous system. *Exp Neurol* 1, 322–333.

Silver, J., and Sapiro, J. (1981). Axonal guidance during development of the optic nerve: The role of pigmented epithelia and other extrinsic factors. *J Comp Neurol* 202, 521–538.

Sjöstrand, J. (1965). Proliferative changes in glial cells during nerve regeneration. *Zeitschrift für Zellforschung* 68, 481–493.

Spemann, H. (1938). *Embryonic development and induction.* New Haven, Conn.: Yale University Press.

Sperry, R. W. (1965). Embryogenesis of behavioral nerve nets. In *Organogenesis,* ed. R. L. DeHaan and H. Ursprung, pp. 161–186. New York: Holt, Rinehart and Winston.

Stefanelli, A. (1951). The Mautherian apparatus in the Ichthyopsida; its nature and function and correlated problems of histogenesis. *Q Rev Biol* 26, 17–34.

Stensaas, L. J. (1967). The development of hippocampal and dorsolateral pallial regions of the cerebral hemisphere in fetal rabbits: II. Twenty millimeter stage, neuroblast morphology. *J Comp Neurol* 129, 71–84.

Toivonen, S., and Saxén, L. (1968). Morphogenetic interaction of presumptive neural and mesodermal cells mixed in different ratios. *Science* 159, 539–540.

Toivonen, S., Tarin, D., Saxén, L., Tarin, P. J., and Wartiovaara, J. (1975). Transfilter studies on neural induction in the newt. *Differentiation* 4, 1–7.

Truex, R. C., and Carpenter, M. B. (1969). *Human neuroanatomy*. Baltimore, Md.: Williams & Wilkins.

Weiss, P. A. (1941). Nerve patterns: The mechanics of nerve growth. In *Dynamics of development: Experiments and inferences*. New York: Academic Press.

Wilson, D. B. (1973). Chronological changes in the cell cycle of chick neuroepithelial cells. *J Embryol Exp Morphol* 29, 745–751.

Woodhams, P. L., Basco, E., Hajos, F., Csillág, A., and Balázs, R. (1981). Radial glia in the developing mouse cerebral cortex and hippocampus. *Anat Embryol (Berl)* 163, 331–343.

Myelin Development

Robert M. Gould

Introduction

Myelin, a major constituent of human brain, spinal cord, and peripheral nerve, is a specialized form of the plasma membranes of those glial cells that associate with nerve processes. The glial cells that make myelin in the central nervous system (CNS) are called oligodendroglia; those that make peripheral nervous system (PNS) myelin are called Schwann cells. The occurrence of myelination is dependent on the development of the neurons and their axonal processes and the development and differentiation of glial cells that become associated with these processes.

The sequence of events that leads to myelination and is essential for its occurrence are (*a*) *neurogenesis*, the proliferation of neurons and their migration from the embryonic neural tube and neural crest; (*b*) *axonogenesis*, the development of nerve processes; (*c*) *dendritic maturation* and *synaptogenesis*, the formation of neuronal connections; (*d*) *gliogenesis*, the proliferation of glial cells and their migration along the developing nerve processes; and (*e*) *myelinogenesis*, the establishment of those interactions between nerve processes and glia that lead up to and include myelin sheath formation. In addition, there are destruction and loss of some neuronal and glial elements that become redundant during development. The preceding chapter by Shirley A. Bayer covers the first four topics, which will only be briefly touched upon here. The focus will be on myelinogenesis.

The presence of myelin sheaths around large-caliber CNS and PNS axons (and the accompanying nodal axon differentiation [see "Myelination and Its Effect on Axon Development" below]) allows them to conduct impulses far more rapidly and efficiently than nonmyelinated axons of comparable caliber. In myelinated axons, the energy-requiring electrochemical events are restricted to small specialized portions (approximately 1/100 to 1/1000) of the axonal surface, the nodes of Ranvier. The rapid passage of impulses from one node to the next occurs by a process called *saltatory conduction*. The following reviews are recommended for further reading on the role of myelin in nerve conduction:

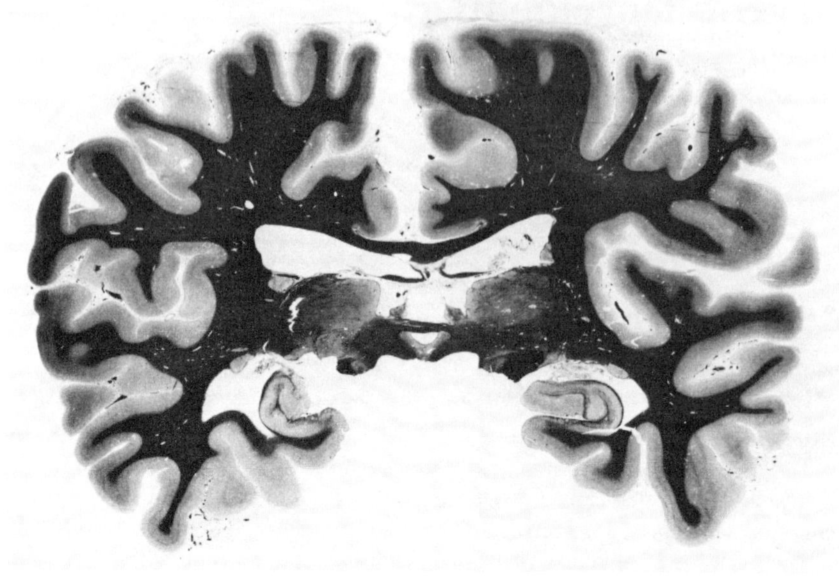

Figure 1. Coronal section of human brain taken through the lateral ventricles. The white matter (stained in black) accounts for greater than 50 percent of the cross-sectional area of this photograph.

Hodgkin, 1971; Rogart and Ritchie, 1977; Waxman, 1978; the chapter by Stampfli and other chapters in Waxman and Ritchie, 1981; Tasaki, 1982.

The magnitude of myelination (as an event of brain development) can be appreciated by viewing a section of human brain or by determining biochemically the levels of myelin in brain. The abundance of white matter (as opposed to myelin-poor gray matter) seen by lipophilic myelin staining (fig. 1) is evident. The total amount of myelin present in human brain can be determined from isolating a subcellular fraction rich in myelin. A normal human brain of 1,400 g (315 g dry weight) contains 110 g of myelin, the equivalent of 35 percent of the brain's dry weight (Norton, 1981b).

The formation of myelin in the CNS does not occur simultaneously; different regions of the nervous system become myelinated on a time scale dependent on the regional development of axons and the glia that associate with them. As a general rule, myelin appears in the PNS before the CNS, and it develops in phylogenetically older regions before newer regions (Gilles, 1976; Richardson, 1982). Myelination can be considered a yardstick of brain development. In grazing animals, whose PNS and CNS carry out quite complex activities at birth, myelination commences earlier than in nesting animals, whose systems are quite immature at birth. In man, myelin has been observed as early as 18 weeks in the sural nerve (PNS) of the human fetus (Ochoa, 1971). Generally, the most active period of myelination in man occurs between the fifth month of fetal development and the second year (see Richardson, 1982, for review). Not only the onset, but also the rate of myelination varies from region to region and possibly from one internode to the next. The slowdown of myelination (to a maintenance function, for myelin constituents are continually renewed) is a gradual process because individual myelin sheaths continue to enlarge with the growth of the individual. For example, peripheral myelin sheaths must still enlarge while the growth of peripheral axons in arms and legs persists (more than 15 years).

The development of PNS myelin is the responsibility of Schwann cells. Since each Schwann cell's task is to cover a single unit of axon (approximately a millimeter in length) with myelin, forming the internode, many Schwann cells must work in concert to provide the entire length of each axon with myelin. The nodes of Ranvier (approximately a micrometer in length) are located at the junctions between the internodes of individual myelin-forming Schwann cells. For the long motoneurons of humans that extend axonal processes from cell bodies in the lower spinal cord to muscles in the foot, each meter-long axon has approximately 1,000 internodes of myelin formed by 1,000 Schwann cells. For impulses to traverse one of these axons, action potentials must be generated at 1,000 nodes of Ranvier.

Unlike Schwann cells, which have the limited responsibility of myelinating single internodes, oligodendroglial cells each myelinate many segments of axons. To do this, every oligodendroglial cell sends out many processes that individually contact sections of one to several axons and produce internodes of myelin sheaths around each one. It has been calculated that the average oligodendroglial cell (i.e., in the optic nerve of the rat, Peters and Vaughn, 1970) forms 30 to 50 internodes of myelin. Presumably, in other white matter tracts, the oligodendroglia have a comparable responsibility (see figs. 2 and 3).

Myelination—The Product of Neuron-Glia Interaction

A complex series of events precedes myelination. First, the neurons (whose axons are to be myelinated) must be formed, reach an appropriate location, and send out the axonal processes to suitable targets. Then the glial cells that will make the myelin sheaths must be formed and distribute themselves to those regions where the axons to be myelinated are located (i.e., the future white matter tracts). Then they must multiply. Finally, the glia must stop proliferating and differentiate so they will start forming myelin.

The process of myelination is the formation of many individual myelin

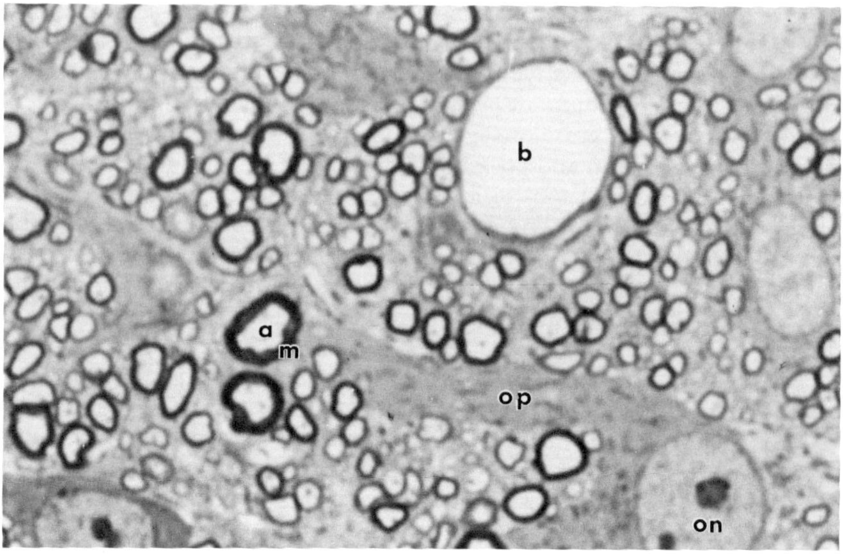

Figure 2. Light micrograph of developing rat spinal cord (anterior funiculus) showing the general relationships between growing axons (*a*) and oligodendroglial cells (*on*, nucleus; *op*, process). A blood vessel (*b*) is cleared of cells by perfusion with fixative. It is likely that the few oligodendroglia seen here are synthesizing many of the myelin sheaths (*m*) present in this figure.

sheaths throughout central and peripheral nervous systems. The numbers and locations of the myelin sheaths formed depend on many localized interactions between glial cells and the axon segments around which they construct their myelin sheaths. Some questions relevant to a fuller understanding of myelin formation include the following:

1. How do the neuronal processes that are to be covered with myelin recruit the glial cells that will form myelin around them?
2. Once the appropriate numbers are recruited, how do the glial cells establish the specific territories (internodes) they will myelinate?
3. What role or roles do the axons have in stimulating ensheathment and myelin formation?
4. How do the myelin-forming glial cells in turn influence the development of the axolemma, or axon plasma membrane (e.g., voltage-sensitive ion channels and ion pumps at the nodal regions)?
5. What factors turn on the synthesis of those proteins and lipids used for the construction of the myelin sheath? How are they assembled into the myelin sheath?

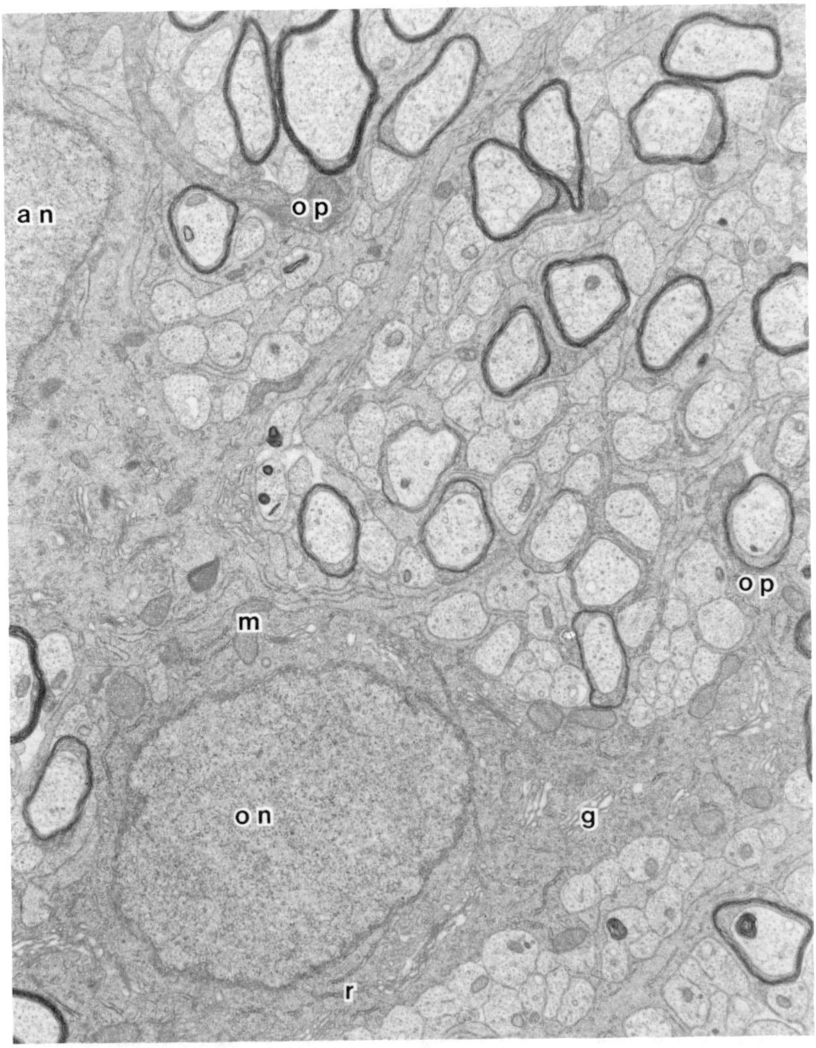

Figure 3. Electron micrograph of a cross section through developing rabbit optic nerve. Both an oligodendroglial cell (*on*, nucleus) and an astrocyte (*an*, nucleus) are visible. Several of the myelin sheaths are seen in close contact with the soma of the oligodendroglia, whereas others are clearly associated with oligodendroglial processes (*op*). The cytoplasm of the oligodendroglial cell is somewhat darker than that of the astrocyte. The cytoplasm of both cells contains distinct strands of rough endoplasmic reticulum (*r*), Golgi apparatus (*g*), and mitochondria (*m*). These organelles in the oligodendroglia are where proteins and lipids needed for myelin sheath formation are made. (Photo courtesy of Drs. Paul Massa and Enrico Mugnaini.)

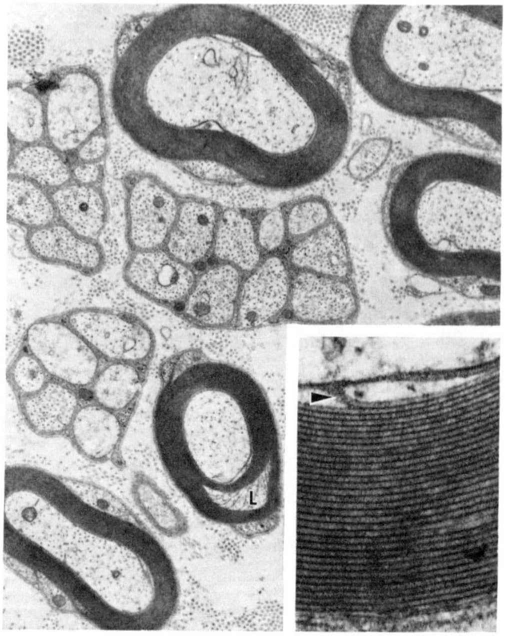

Figure 4. Electron micrograph of a cross section through adult rat sciatic nerve showing the component Schwann cells and axons. Schwann cells develop two distinctive relationships with axons; they form myelin sheaths around the larger axons and enclose the smaller axons within processes in which no myelin is present. This field includes a Schmidt-Lanterman incisure (*L*), characterized by the intercalation of Schwann cell cytoplasm within compact myelin layers. The inset is a high-powered view of a compact myelin section in which the major dense (darker) and intraperiod lines are clearly seen. The internal mesaxon (*arrowhead*) is indicated. Each myelinated nerve and group of unmyelinated fibers are contained within the extracellular basal lamina.

6. During the actual construction of myelin, how are the membrane layers compacted? Another way of asking this question is, How are the glial cytoplasm and extracellular fluids excluded from the major dense and intraperiod lines (fig. 4)? Or, What are the roles of individual myelin lipids and proteins in the sculpturing of individual internodes of myelin?
7. What components make up the nodal and paranodal specialization and Schmidt-Lanterman incisures, and how are they assembled (see "Myelination and Its Effect on Axon Development" below)?
8. How are the rate of myelin formation and its extent (final dimensions) controlled?

Although we cannot answer these questions, it is the search for solutions that makes the study of myelination an exciting area of research.

How Do We Study Myelination?

During the course of normal development myelination occurs. With myelin sheaths comes the more rapid and energy efficient saltatory conduction required for normal behavior. Although abnormal development can be the result of defective myelination, confirmation that behavioral abnormalities are the result of insufficient myelin sheath formation and not other earlier developmental events requires anatomical, physiological, or biochemical evaluation or all three. In the following sections we will learn about myelination, our knowledge coming from studies employing the disciplines of anatomy, physiology, and biochemistry. In this section, I will briefly discuss the differences in approaches that the anatomist, biochemist, and electrophysiologist take in studying myelination.

The anatomist's aim is to obtain a representation of the tissue, usually in the form of a section that can be studied with the naked eye (fig. 1) or with the light or electron microscope (fig. 5). Visual "data" can be used as bases for description. As an example, the micrographs shown in this review are used to describe the characteristics of myelin formation by Schwann cells (figs. 4 and 5) and oligodendroglial cells (figs. 2 and 3). Descriptions such as those in the sections below on PNS and CNS myelin development and axon development are almost always based on sectional views showing limited portions of the cells. It is, however, possible to make three-dimensional reconstructions from a serial sequence of micrographs (e.g., see Webster, 1971; Knobler, Stempak, and Laurencin, 1974, 1976). Because of the skill and effort required to obtain these data, studies are often limited to very few samples, and the generality of the results is brought into question. Finally, quantitative information can be garnered by employing stereological methods to the tissue sections and micrographs (e.g., Weibel and Bolender, 1973; Williams, 1977).

The aim of the biochemist (and the electrophysiologist) is to measure specific properties of the tissue and usually not (because of technical limitations) individual elements such as axons and glia in the tissue. Data, such as the activity of an axonal enzyme or the amount of a specific myelin protein in the sample, represent the sum of enzymes or proteins pooled from all contributing members (axons and myelin sheaths) in the tissue. The results from biochemical and electrophysiological studies that deal with properties of isolated myelin or sodium channels in multifiber nerves are difficult to relate to those of the morphologist, whose descriptions are of individual myelin sheaths and clusters of particles at a single node of Ranvier (fig. 6). I will, therefore, present the

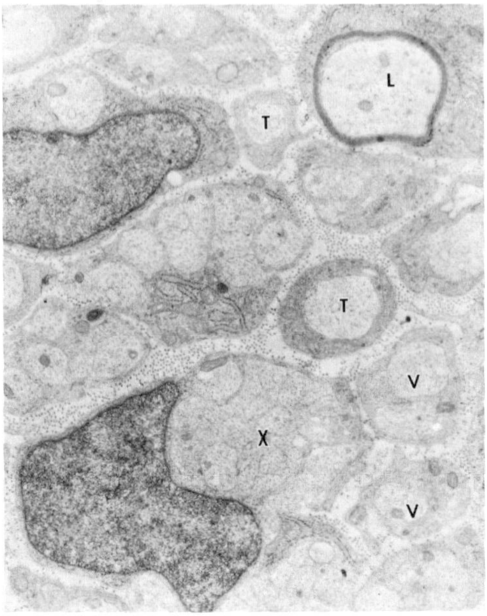

Figure 5. Electron micrograph of a cross section through developing rat trigeminal nerve. At two days, most of the Schwann cells are proliferating, and Schwann cell processes are in contact with bundles of the smaller axons (*X*). The larger axons are isolated within Schwann cell processes, either in small groups (*V*) or alone (*T*). The largest axon (*L*) in this section is not only isolated within the confines of Schwann cell cytoplasm, it has begun to be myelinated.

morphological and biochemical features of myelination in separate sections before in the last two sections making an attempt to weave these viewpoints together and to provide a description of myelination with multidisciplinary implications. Discussion of the electrophysiological studies related to myelination is outside the scope of this review.

Development of PNS Myelin: Morphological Aspects

The morphology and development of PNS myelin have been comprehensively treated in a number of recent reviews that should be consulted for additional reading (Webster, 1975; Asbury, 1975; Landon and Hall, 1976; Peters, Palay, and Webster, 1976; Berthold, 1978; Jacobson, 1978; Spencer and Weinberg, 1978; Hall, 1978; Landon, 1981; Bunge and Bunge, 1981; Raine, 1981; Varon and Manthrope, 1982; Gould, Matsumoto, and Mattingly, 1982; Sears, 1982). In this section, I will briefly discuss the migratory and proliferative properties of Schwann cells and then focus on the process of myelination. This summary includes embryological studies with chicks and quail and nerve development studies mainly with rodents. The generality of these results, scientists hope, reflects developmental patterns in all species. This process can be appreciated by comparing the morphology at the light microscopic level of earlier (fig. 7*A*) and later (figs. 7*B* and 7*C*) stages of developing peripheral nerve with that of the adult stage (fig. 7*D*).

The Schwann cells that give rise to PNS myelin originate in the neural crest and placode regions (see Weston [1970 , 1982] and Le Douarin [1980] for reviews of this area). The progenitor Schwann cells that enter the developing peripheral nerves undergo extensive proliferation in the nerve. The purpose of this proliferation and related migration is to attain sufficient numbers of Schwann cells all along the nerve to completely cover the surfaces of all axons. During these early stages, Schwann cells

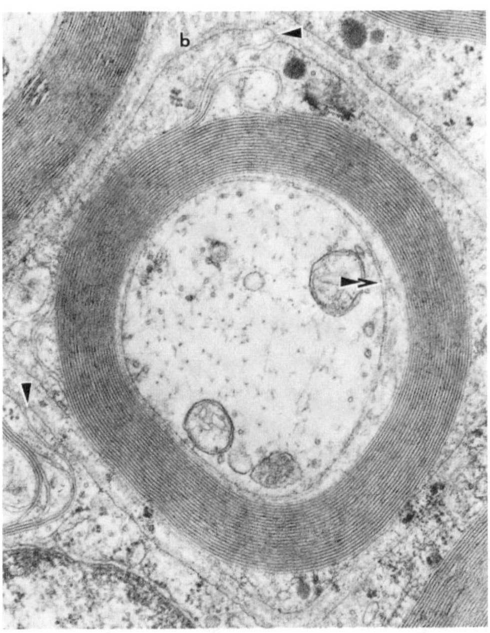

Figure 6. Electron micrograph of an individual myelinated axon from rat trigeminal nerve taken at 14 days after birth. There are already 22 layers of compact myelin. The major dense line is easily visualized. The outer mesaxon of this and another sheath are present (*arrowhead*) as is the site of the inner mesaxon (*double arrowhead*). The basal lamina (*b*) that encloses these nerves is visible adjacent to the Schwann cells' plasma membrane.

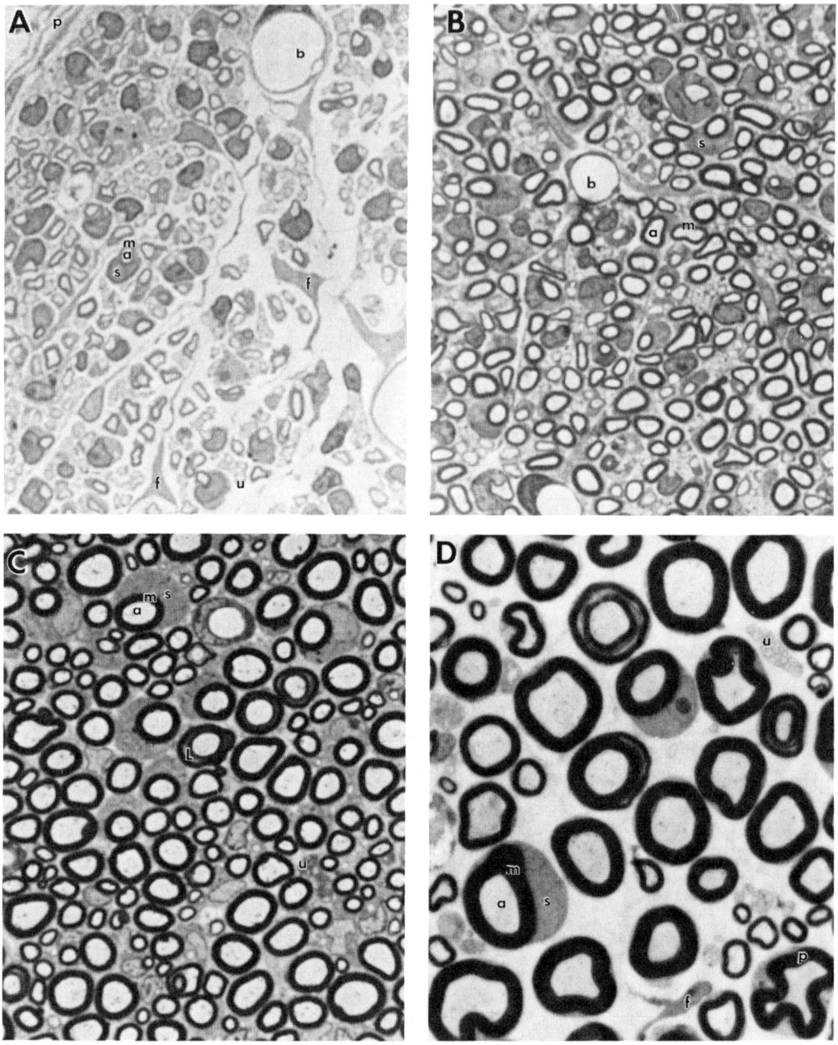

Figure 7. Series of cross-sectional views illustrating different stages of peripheral nerve development at a single magnification (×6,250). Growth can be measured by the increase in the caliber of the axon (*a*) and thickness of its myelin sheath (*m*). The volume of Schwann cell cytoplasm is also increased; this increase is easiest seen in the crescent-shaped perinuclear regions (*s*). *A*, 2-day, *B*, 8-day, *C*, 16-day, and *D*, 65-day sections from rat trigeminal nerve. Fibroblasts (*f*), blood vessels (*b*), perineurial cell layer (*p*, in *A*), bundles of nonmyelinated fibers (*u*), a Schmidt-Lanterman incisure (*L*), and a paranodal region (*p* in *D*), having a crenated axon and myelin sheath, are labeled.

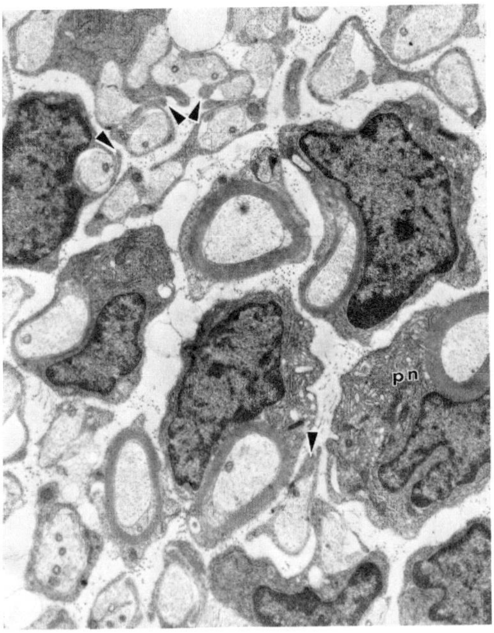

Figure 8. Electron micrograph of developing rat trigeminal nerve at four days after birth. All of the axons are separated from each other by Schwann cell processes. Several fingerlike processes are joined (*arrowheads*) in the initial stages of mesaxon formation. The ensuing myelination will occur as the axons grow. The fluid nature of the compact myelin allows it to expand to accommodate the enlarging axon. The perinuclear regions (*pn*) of the Schwann cells are rich in those organelles involved in the protein and lipid synthesis and the subsequent assembly processes involved in the growth of the myelin sheaths.

surround large bundles of growing axons (figs. 5 and 7*A*; see also Webster, 1975; Peters, Palay, and Webster, 1976; and Landon and Hall, 1976). These "families" each become enclosed by a common basal lamina and are separated by extracellular space containing collagen fibrils. Within the families, the Schwann cells continue dividing so that more and more will form specific interactions with the individual large-caliber axons (fig. 8). It is during this period that the bipotentiality of Schwann cells is expressed. While the Schwann cells that associate with individual large-caliber axons synthesize a myelin sheath around these axons, others, drawn from the same precursor population, develop a stable relationship with smaller caliber axons in which no myelin is formed (figs. 4, 7*C*, and 7*D*). In either case, once the Schwann cells develop their relationship with the axon or axons, they stop proliferating. Schwann cells are surrounded by a basal lamina (figs. 4 and 6), which separates them from all but those associated with the same axon. These resultant tubes of basal lamina form structural pathways from the neuronal perikarya to the target. As the axons continue to grow, the Schwann cells, their myelin sheaths, and their basal lamina must also expand.

For each axon, myelination proceeds in a proximodistal direction, and individual sections of the axon are myelinated once they become separated and enclosed by single Schwann cells. In the longer nerve tracts, myelination is occurring in proximal axons while the distal regions are still growing toward the target sites they will innervate. In cases in which the axons fail to properly innervate their targets and degenerate, the myelination in the proximal regions must be undone (Webster, 1975).

With the electron microscope it is possible to observe the many stages of the interactions between Schwann cells and axons that lead to myelination (figs. 4–6 and 8). The interrelationships seen in thin sections represent only one small portion of the area of contact between these two types of cells. In figures 5 and 8 are shown some of the early stages in the isolation of individual axons from bundles of developing axons. It is clear that the larger caliber axons are individually separated while smaller caliber axons remain together in the confines of shared Schwann cells. In figure 8, the larger caliber axons already have layers of compact myelin around them. Many others are completely separated, and others are in the process of being separated.

The elaboration of the Schwann cell's plasma membrane into myelin is in some way encouraged by the axon (see below). A mesaxon is formed along the regions where the outer surfaces of the Schwann cell join after enclosing the axon. This mesaxon is the origin of the intraperiod line (figs. 5 and 8), the surface where the enclosing folds of Schwann plasma membrane join to separate extracellular fluid from the axon. As compact myelin is formed, the mesaxon is divided into separate inner and outer mesaxons (fig. 6), which are connected to each other by the intraperiod line. The major dense line (fig. 4 inset) is where Schwann cell cytoplasm is excluded, except at the Schmidt-Lanterman incisures (figs. 4 and 7*C*) and paranodal loops.

We do not know the signals that cause the Schwann cells in developing nerve to proliferate and to migrate along and with the growing axons. Nor do we know the cues that cause individual daughter Schwann cells to stop dividing and develop a stable interaction with the section of the axon they will myelinate or the group of small caliber axons they will enclose. We do know that the axons themselves play an important role in all of these events. A part of this knowledge comes from studies performed on animals in which the nerves were surgically manipulated in cross-anastomosis and nerve-grafting experiments (see Spencer and Weinberg, 1978; Spencer, 1979; Aguayo et al., 1979). The result of the manipulations is to have regenerating axons from mylinated nerve grow through regions of degenerated nerve where most of the Schwann cells had previously been associated with unmyelinated axons. The observed myelination of regenerating axons by Schwann cells in the unmyelinated zone is evidence that axons regulate expression of myelin.

In addition to the demonstration that axons influence the behavior of individual Schwann cells, Aguayo and his collaborators (1976) provided evidence that the number of Schwann cells that develop in a nerve is determined by the number of axons present. By experimentally reducing the number of axons present in a developing nerve, they found that the number of Schwann cells was proportionally reduced.

Another approach to studying the influence of the axons on Schwann cells has been through the use of tissue culture. Two excellent reviews dealing with tissue culture approaches to understanding PNS myelination have recently been published by Bunge and Bunge (1981) and Varon and Manthrope (1982). These approaches are centered around the use of populations of Schwann cells that can be isolated from developing nerve (Wood, 1976; Brockes, Fields, and Raff, 1979; Skaper et al., 1980; Kreider et al., 1981; Cassel et al., 1982) and adult (Wrathall, Rigamonti, and Kao, 1981). By growing these cells in isolation or in combination with neurons, fibroblasts, or both, important advances have been made in the following areas:

1. The influence of the axon and axonal contact on the behavior of isolated Schwann cells. Axonal contact was shown not only to influence Schwann cell proliferation and myelin sheath formation (complementing cross-anastomosis and grafting work) but also secretion by the Schwann cells of basal lamina and collagen fibrils.
2. The influence of fibroblasts on the roles of Schwann cells in forming extracellular milieu (i.e., both collagen and perineurium development). Perineurial cells enclose large groups of peripheral nerve fibers, separating them into fascicles (fig. 7*A*). Perineurial fasciculation is part of the sorting mechanism to route bundles of nerves from the main nerve to outlying targets.
3. The role of extracellular matrix in myelination.
4. Mitogenic properties of Schwann cells.

Another question is, How are the final dimensions of each internode of myelin determined? At present, we do not have any information of the biochemistry used in establishing the thicknesses of individual myelin sheaths. Smith and his colleagues (1982) found that there is a morphological relationship between myelin thickness, internodal length, and axon caliber. The size of myelin sheaths, in both normal and remyelinating conditions, is proportional to the surface area of axon that it covers. Do the signals that cue the Schwann cells' synthetic machinery to form so much myelin arise from the axon, axon surface, or both, or are they a consequence of the dimensions that the surface membranes of the Schwann cells attain? Perhaps once the signals are characterized, it will be possible to gain an understanding of the manner in which the myelin-making machinery of the Schwann cell is regulated. In an analogous fashion, the role of the axon in establishing CNS myelin caliber is unknown (see the next section).

Development of CNS Myelin: Morphological Aspects

Information on the morphology and development of CNS myelin is covered in several recent reviews (Privat, 1975; Peters, Palay, and Webster, 1976; Raine, 1977; Privat and Fulcrand, 1978; Skoff, 1980; Hirano, 1981; Paterson, 1981; Raine, 1981; Sears, 1982). The studies came from work on a limited number of species, the rat being the most widely used model. The sequence of events required for myelination of CNS axons shares many features with that of PNS myelination (see previous section). However, unlike the Schwann cells, which are largely of neural crest origin, both the oligodendroglial cells that form CNS myelin and astrocytes (another major type of CNS glial cell) originate in the subependymal zone adjacent to the ventricles of the developing brain and spinal cord. Precursor cells of the myelin-forming oligodendroglia, called *oligodendroblasts*, have been characterized in developing CNS tissue (see Privat and Fulcrand [1978] and Skoff, Price, and Stocks [1976a, 1976b]). Using electron microscope autoradiographic methods, these investigators have shown that, like Schwann cells in the PNS, oligodendroglia proliferate most actively at or around the time that myelination commences. Furthermore, the early stages of oligodendroglial development are dependent on axonal influence. This was shown by Fulcrand and Privat (1977) and Valat et al. (1978) in studies in which axonal contact was removed at various stages of optic nerve development by eye removal (enucleation). Axonal loss at two, five, and eight days after birth (myelination begins on the fourth, fifth, or sixth day) resulted in a reduction in the proliferation and numbers of oligodendroglial cells that were formed within the optic nerve. When enucleation was delayed to day 20, the numbers of oligodendroglia in optic nerve were normal and remained at near normal levels in the nerve. Proliferation was higher than in controls at day 20, indicating that at this stage the axons inhibit cell division. So, like the number of Schwann cells, the number of oligodendrocytes (and processes) that develop in the optic nerve is dependent on the presence of axons. Once formed, mature oligodendroglial cells are retained in the nerve, even when axons are caused to degenerate (see also Bunge, Bunge, and Cochran [1978] and Bunge [1982]).

The way in which oligodendroglial cells distribute themselves among

CNS nerve fibers and claim the territories (axonal internodes) that they will myelinate is not known. There is no clear-cut pattern of myelination either across (Skoff, Price, and Stocks, 1976b) or along (Skoff, 1978; Giorgi and DuBois, 1980) developing optic nerve. This is probably related to the complex interactions between multiple oligodendroglial processes and axons. Furthermore, we as yet have limited understanding on how other factors, such as astrocytes (Skoff, 1976; Bunge, Bunge, and Cochran, 1978; Dupouey et al., 1980; Hatfield and Skoff, 1982) and hormones (Skoff, Skoff, and Katsnelson, 1979) affect CNS myelination. Finally, oligodendroglial cells are probably not a homogeneous population (Mori and Leblond, 1970; Narang and Wisniewski, 1977; Narang, 1977), and it is not known if different populations of CNS glia require different signals to send out processes and myelinate CNS axons. To date, the tissue culture approaches used so successfully to study the cell-cell interactions involved in PNS myelination have hardly been used with CNS tissue. However, it is apparent that such studies as well as those to characterize trophic factors that influence oligodendroglia formation (Bhat, Barbarese, and Pfeiffer, 1981; Wood, Okada, and Bunge, 1981; Raff, Miller, and Noble, 1983) are under way.

Before differentiated oligodendrocytes are formed, the axons are small (fig. 9) and uncovered. Once the contact between the oligodendroglial process and axon is made, the axon becomes enclosed (figs. 3 and 10). Early stages of axon enclosure occur both at the level of the cell perikarya and in processes distant from the oligodendroglial perikarya (figs. 2 and 3). At these stages of development it is obvious that oligodendroglia are contacting many axons and that myelination is occurring in a nonsynchronous fashion; that is, early (formation of mesaxon and first layers) and advanced (many compact layers) are seen in different regions of a single oligodendroglia (fig. 3) (Freidrich and Mugnaini, 1983). The leading "tongues" of cytoplasm enclose the axons (fig. 10) and the junction formed in the region where the external surfaces meet is the origin of the intraperiod line. The collapse of some of the intervening cytoplasm marks the beginnings of the major dense line. As one tongue overlaps the other and they in turn are wrapped around each other, loosely compacted and then tightly compacted myelin is formed (figs. 10 and 11). The early stages of CNS myelination have been studied by Knobler and his colleagues (1974, 1976) using serial reconstruction (see also Peters, Palay, and Webster [1976] for review). Because oligodendroglial cells send their processes in many directions and each of these processes upon contacting segments of axon forms one or several myelin sheaths, there must be a sophisticated organization to supply material for the elaboration of the multiple myelin sheaths (fig. 12). During myelination, columns of associated oligodendroglia are present (fig. 13). These presumably dissociate as the degree of myelination and dimensions of

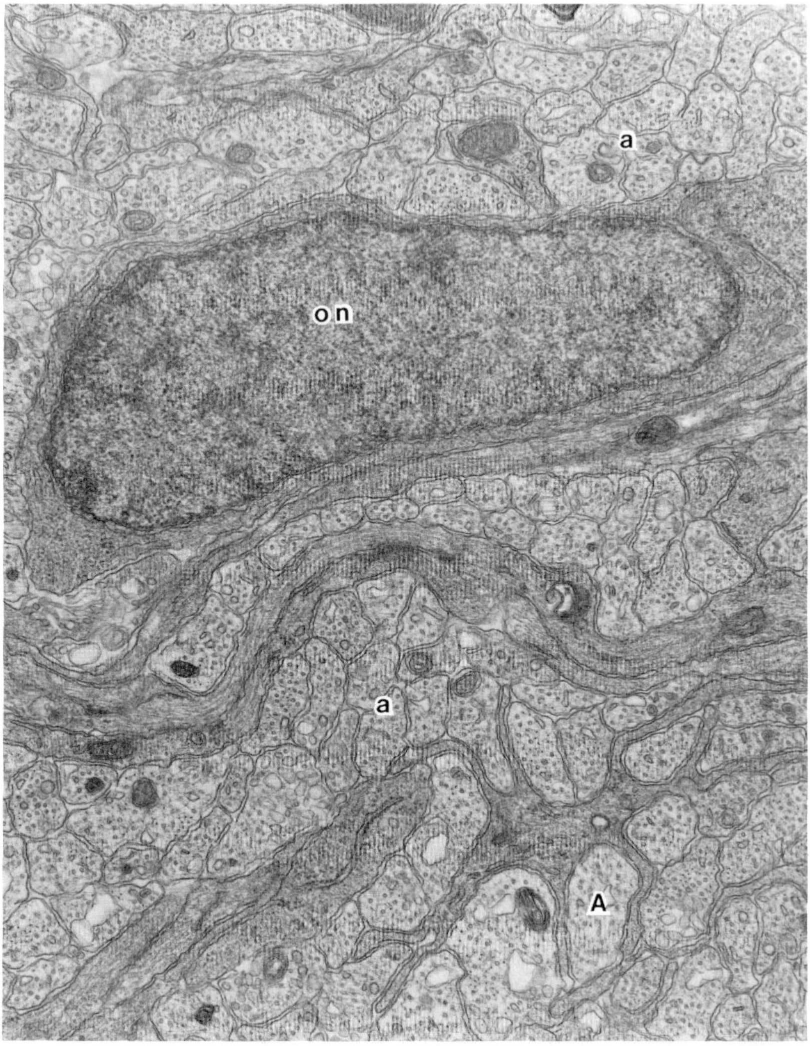

Figure 9. Electron micrograph of developing rabbit optic nerve. At this early stage, no myelin sheaths are seen; however, some of the larger axons (*A*) are being separated from the other axons by glial processes. Presumably, the processes are from nearby oligodendroglial cells. The nucleus of an oligodendroglial cell (*on*) is in the center of the field. Large clusters of the smaller axons (*a*) with no intervening glial processes are common at this stage of development. (Photo courtesy of Drs. Paul Massa and Enrico Mugnaini.)

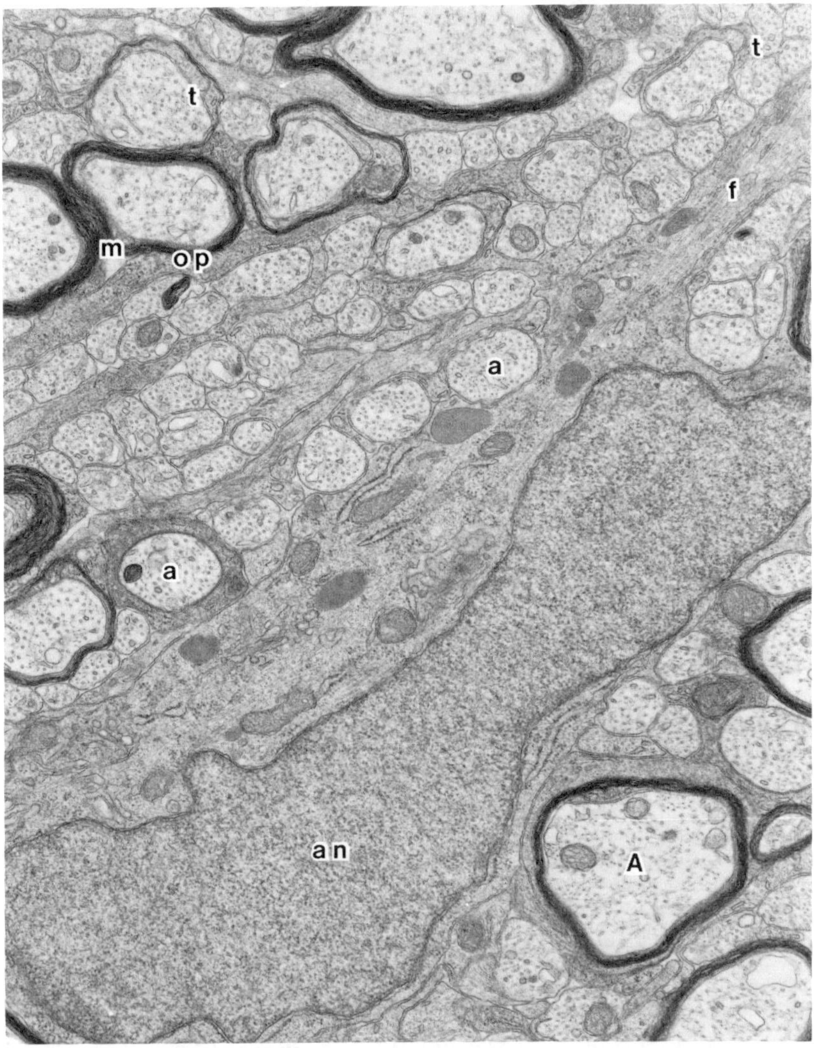

Figure 10. This electron micrograph depicts a stage of rabbit optic nerve development in which compact myelin (*m*) is present around the larger axons (*A*). Other axons (*a*) have been separated from all others by glial processes and the characteristic "tongues" (*t*) of the oligodendroglia indicate that myelination is about to commence. The cell in the center of the field is an astrocyte (*an*), identified by the strands of parallel intermediate-sized filaments (*f*). Oligodendroglial processes (*op*) do not contain these intermediate filaments. (Photo courtesy of Drs. Paul Massa and Enrico Mugnaini.)

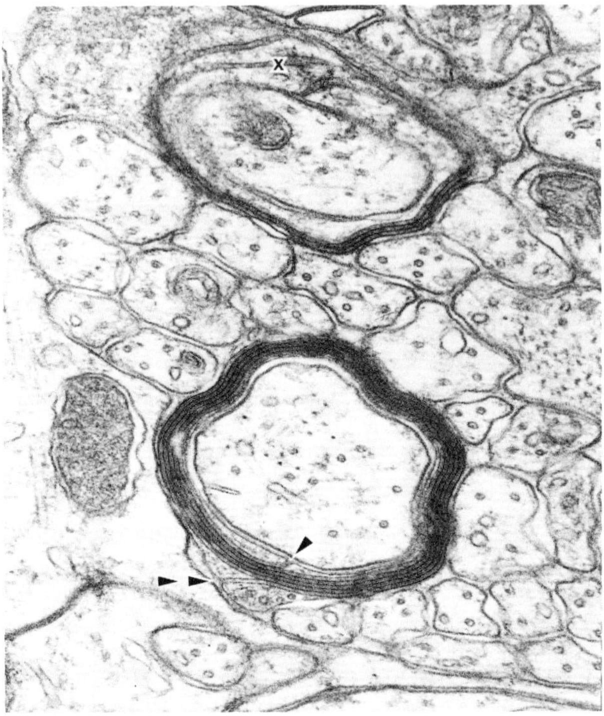

Figure 11. Electron micrograph of developing rat cochlear nucleus. The two largest axons are contained within several layers of myelin. Some layers of one sheath are not yet compacted (*x*). Both the inner (*arrowhead*) and outer (*double arrowhead*) mesaxon are clearly seen as are the alternating major dense (darker) and intraperiod lines (Photo courtesy of Dr. Enrico Mugnaini.)

the tissue are increased. With growth of the tissue the oligodendroglial processes lengthen so that the connections of oligodendroglial cells to axons in adults become impossible to visualize (fig. 14).

Myelination and Its Effect on Axon Development

While oligodendroglia and Schwann cells become committed to making myelin sheaths around axons, the axons go through their own organization to accomodate the insulating myelin. This organization, which establishes saltatory conduction in place of continuous conduction, includes the formation of nodes of Ranvier and the adjacent paranodal specializations. Since the nodes are regularly spaced in relation to the

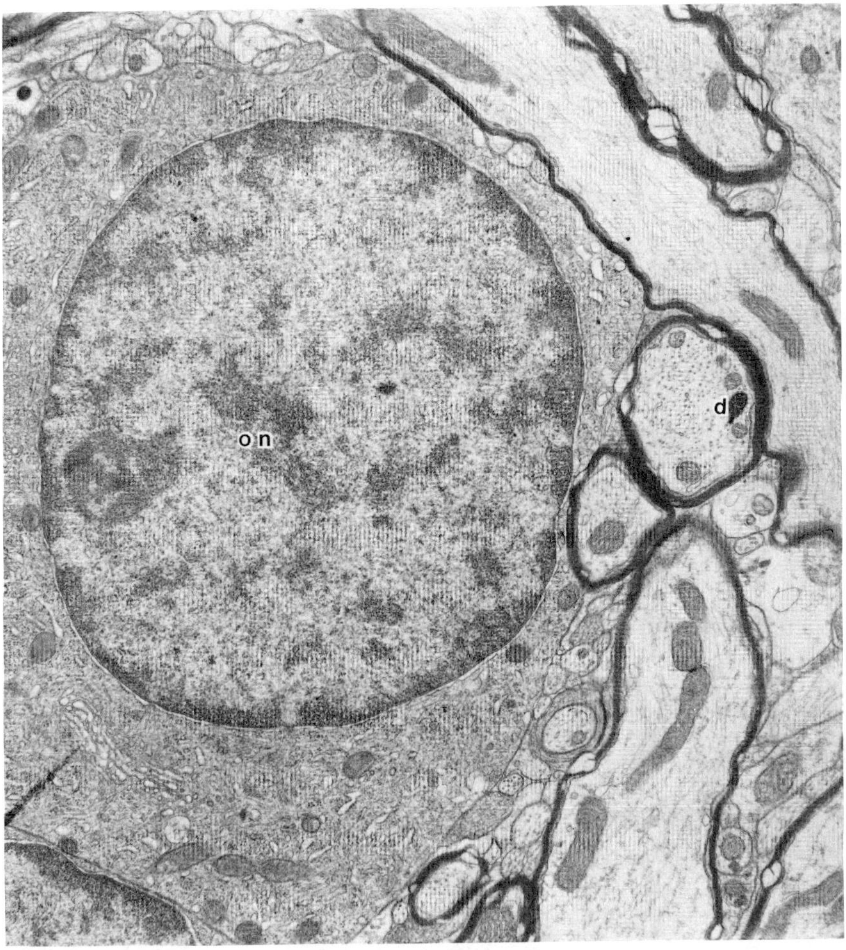

Figure 12. Electron micrograph of developing cochlear nucleus showing two myelinated axons with different orientations. Both are in direct contact with the oligodendroglial cell's plasma membrane. Clusters of polysomes, rough endoplasmic reticulum, Golgi apparatuses, mitochondria, microtubules, and smooth vesicular membrane profiles are all visible in the cytoplasm surrounding the oligodendroglial cell nucleus (*on*). Nuclear pores can also be seen. The richness of this glial cytoplasm should be contrasted with the cytoplasm of the axons, where cytoskeletal elements predominate and membranous organelles include only mitochondria, a few smooth membrane profiles, and dense bodies (*d*). (Photo courtesy of Dr. Enrico Mugnaini.)

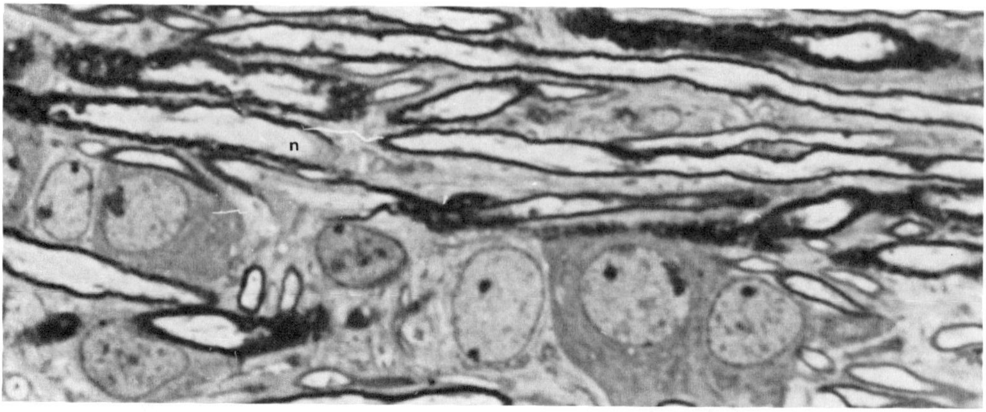

Figure 13. Light micrograph of longitudinal section of developing rat spinal cord at eight days, showing the nuclei of seven stacked oligodendroglial cells. Presumably, these cells were formed by proliferation. This commonly observed arrangement is important for establishing myelin around the axons that pass through this region. Some of the axons are cut close to nodes of Ranvier (*n*), which are commonly seen in longitudinal sections of developing nervous tissue because the internodes are short.

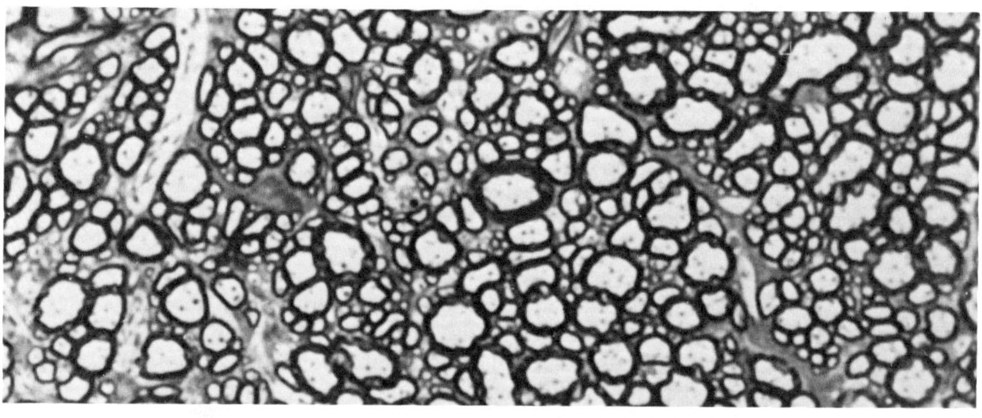

Figure 14. Light micrograph of adult rat spinal cord (cross-sectional view). All the axons in this field appear to be myelinated. The direct relationship between axon caliber and myelin sheath thickness is quite apparent. The developmental growth of axons and myelin sheaths has resulted in a wide spacing of the oligodendroglial cells, and the cells and their lengthened processes are not readily visible.

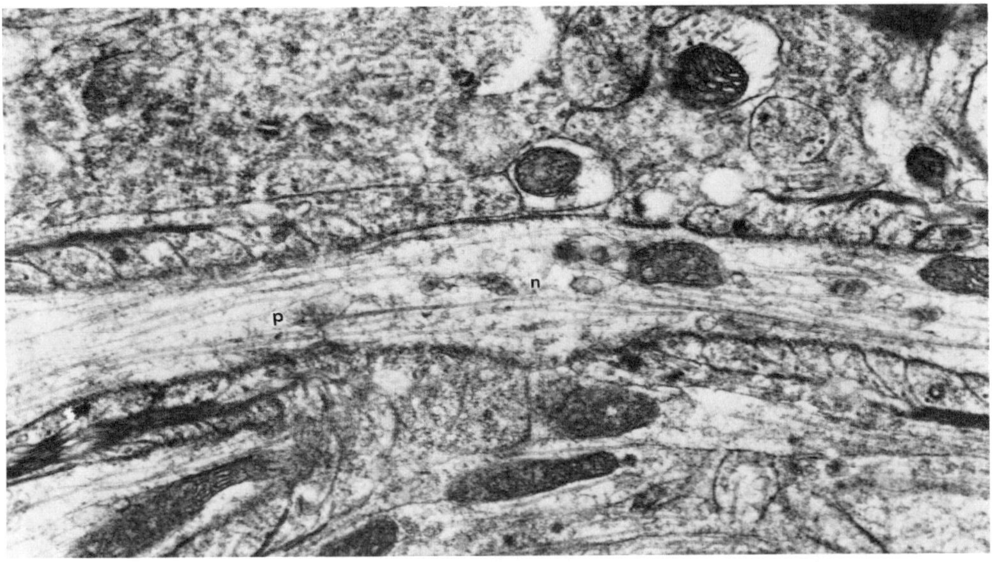

Figure 15. Electron micrograph of longitudinal section through a CNS node of Ranvier (rat spinal cord). The CNS node (*n*) is bared of glial processes and is of slightly greater dimensions than the paranodal axoplasm (*p*). The ends of myelin lamellae are glial loops that make contact with the axon plasma membrane; the outermost loops extend the farthest toward the node. Strands of microtubules are seen coursing through the node of Ranvier.

myelinating glial cell, recent investigations have focused on understanding the relationship between myelination and nodal development.

This subject has been studied largely with conventional electron microscopy and with freeze fracture techniques. In contrast to conventional microscopic techniques, which get a sectional view of membranes, the freeze fracture procedure splits the membrane bilayer, and the samples (replicas) that are prepared from the tissues reveal the inner surfaces of the bilayer, either looking toward the cytoplasm (P face) or toward the cell's exterior (E face). The ultrastructural view of a CNS node and the adjacent paranodal regions is shown in figures 15 and 16. The node itself and the surrounding paranodal axoplasm (fig. 15) are narrower than internodal cytoplasm. The most striking feature of the nodal region is the series of glial loops that are the terminations of the myelin lamellae. Many of these make contact with the axolemmal membrane, and it is their presence that seems to mark and possibly cause the narrowing of the axon. At higher magnification (fig. 16), it is possible to see that there are regularly spaced densities between the glial loops and the

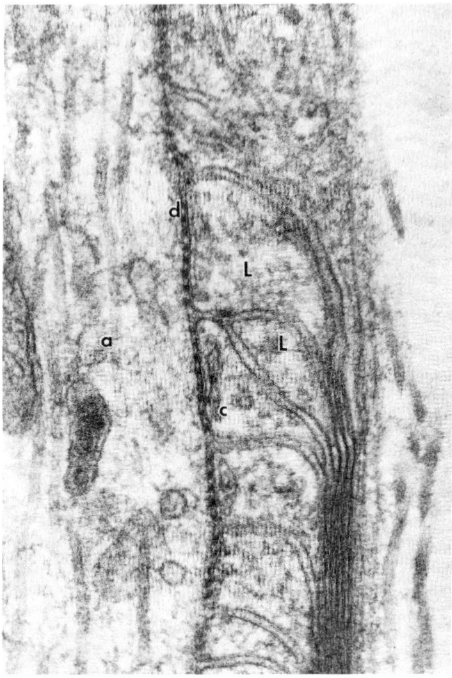

Figure 16. Electron micrograph of the paranode at a high magnification. The regions where the compact myelin layers part to form the loops is seen as is the way in which the loops (*L*) contact the paranodal axon (*a*). Periodic densities (*d*) appear extracellularly between the glial loops and the axon membrane. Within the loops, just adjacent to the axon contact region, are membranous cisternae (*c*). There is no information on the developmental formation of the densities or cisternae or their role in nodal function. (From Rosenbluth, 1981; copyright © 1981 by Raven Press.)

underlying axolemma. These densities (called *transverse bands*) might function to fasten the internodes to the axon, to reduce the flow of materials to the adaxonal fluid space and intraperiod regions of the myelin sheath, or to do both.

A freeze fracture replica (fig. 17) provides a different view of the nodal region. The surface of the axolemma (E face) has numerous particles called *intramembrane particles* (IMPs), which probably represent proteins that pass through the bilayer. It has been argued that these particles represent sodium channel proteins, which control the sodium currents during nerve impulse propagation (Rosenbluth, 1981). The axolemmal membrane (E face) in the adjacent paranodal region shows a pattern of broad stripes and grooves that represents the regions where the glial loops appose the membrane. These regions have far fewer particles than the nodes. There is a paracrystalline diagonal pattern in this region that probably represents the transverse bands seen in thin sections (fig. 16). Seeing this pattern in the interior of the membrane indicates that the transverse bands penetrate deeply into the axolemma. The utility of this structural organization in saltatory conduction is under intense investigation (see Waxman and Ritchie [1981], for review). It is already clear that the differentiation of the nodal axon occurs at or very near the time of glial ensheathment and myelination. It is likely that the formation of the axon's nodal and paranodal specializations require apposition of glial cells or their processes (Rosenbluth, 1981). The origin of

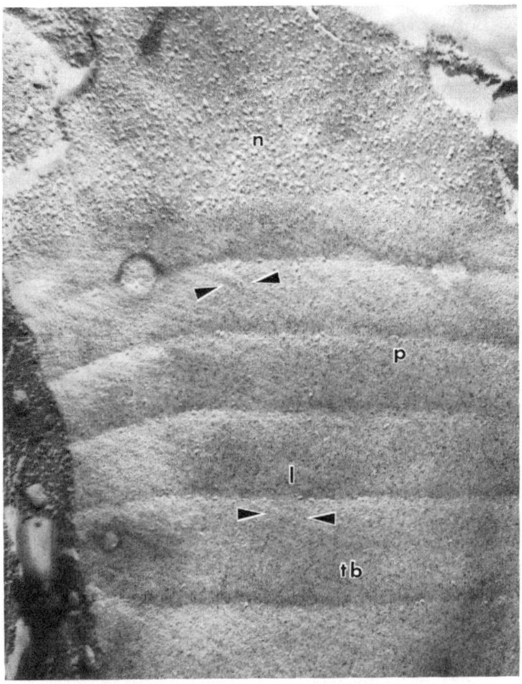

Figure 17. Electron micrograph of a freeze-fractured portion of the E face of a node of Ranvier (*n*) and an adjacent paranode (*p*). The node is covered with intramembrane particles that are believed to represent sodium channel proteins. The replica of the paranode displays broad parallel strips, each corresponding to the region under a single glial loop. There is a pattern of parallel bands over the entire paranode. The spacing (*arrows*) between the diagonal bands is the same as that of the densities seen in ultrathin sections and is a representation on the axolemmal face of the transverse bands (*tb*) that run between glial and axonal membrane. (From Rosenbluth, 1976.)

the axon or glial components of the paranodal specializations (i.e., the proteins, lipids, or both) is not known.

Another influence of glial cells on the axon is felt all along the internode. Two separate and perhaps related effects of the glial cell on the axon are an increase in the size of axons contacted by glial cells and probably the increase in the density of IMPs (presumably proteins) on the P face of the axolemma. Aguayo et al. (1977) found that defective Schwann cells from trembler mice when grafted into normal nerve failed to maintain the same caliber in the regions of axons then covered (and maintained) as was found in proximal and distal regions covered by the animals' own Schwann cells. In this experiment, the axons proximal to the graft grew through trembler Schwann cells in the graft and recruited them to produce myelin. The trembler Schwann cells failed to properly myelinate the normal regenerating axons. The axons underlying the trembler Schwann cells were of smaller diameter than sections of the same axon, both proximal and distal, that were myelinated by normal Schwann cells.

This evidence that Schwann cells influence PNS axon caliber is probably also true of the oligodendroglial cell and CNS axon relationship. In jimpy mice in which CNS myelination is impaired and fewer axons are contacted by oligodendroglial processes (only 1–2 percent of the axons are myelinated), the average axon caliber is reduced (Webster and Sternberger, 1980). Although it is not known how the glial cells influence internodal (and nodal) axon caliber, it has been shown in studies on the squid giant axon (Lasek, Gainer, and Barker, 1977) that the surrounding glial cells provide the giant axon with some of their own proteins. The possibility that glial cells provide the myelinated axons with proteins and other metabolites exists (see Singer, 1968) but has yet to be convincingly demonstrated. One possible way in which the glial cells could influence axonal caliber would require a mechanism to regulate the number of neurofilament proteins maintained in a section of axons, for axon caliber is related to number and possibly transport of neurofilaments (e.g., see Hoffman, Griffin, and Price, 1983).

Black, Foster, and Waxman (1982) used freeze fracture methods to characterize developmental changes in the IMPs on the P face of the axons of the rat optic nerve. They found that the density of IMPs increased as myelination proceeded, and they concluded that the increase was related to ensheathment and myelination. However, in this case it is not known whether glial association was necessary for the particle increase. In this regard, a freeze fracture study of the jimpy mouse optic nerve development might be revealing. Since glial development is abnormal in this mutant, it would be possible to determine whether or not abnormal glial development affects axonal IMP development. These studies must be kept in the context of a study of Pfenninger and Bunge (1974), who found that the density of IMPs on the P face of the axolemma of naked axons in tissue culture increases during development. Whether the internodal distribution of axolemmal proteins (or lipids) is influenced by the glial cells or that they influence the glial cells and cause them to make myelin (Brady and Quarles, 1973; Yates and Wherrett, 1974) merits further study. Furthermore, in considering myelination it is important to realize that this process requires two partners, and that glial ensheathment and the formation of myelin are events that go hand in hand with axonal differentiation.

Myelination: Biochemical Aspects

Studies of the biochemical aspects of myelination have been included in a number of reviews (Davison, 1970; Gregson, 1976; Morell, 1977; Benjamins and Morell, 1978; Baumann, 1980; Norton, 1981a, 1981b; Benjamins and McKhann, 1981; Wiggins, 1982) that should be considered for supplementary reading. An understanding of the biochemistry of myelination requires information of the composition of myelin. This information has come largely from studies of isolated myelin. To a lesser extent, it has come from studies on the changes in composition of tissue during development (myelination), differences in composition of white (myelin-rich) and gray matter, and differences in normal material and pathological material in which myelin is lacking.

Isolated myelin is the product of subcellular fractionation. The starting nervous tissue is usually homogenized in a sucrose solution of low ionic strength so that the myelin sheaths peel away from the axons (Norton, 1981b). The crude myelin collected above a 0.85 M sucrose interface is hypo-osmotically shocked in distilled water to further remove axonal impurities. The multilayered membranes of isolated myelin have the same intraperiod and major dense line spacings as compact myelin fixed in situ. Single-membrane vesicles are also present. This product of fractionation (and its subfractions collected from continuous and discontinuous gradient centrifugation) has been used to quantify both major and minor protein and lipid constituents of myelin. Among the minor components of isolated myelin are several enzymes (Norton, 1981a). There is a good possibility that additional minor components of myelin membranes are lost during the purification procedures (e.g., during hypo-osmotic shock). The following sections will focus on the major components of isolated myelin, because knowledge of their structural and metabolic properties should provide a framework for understanding how myelin sheaths are formed.

Myelin Proteins

Myelin of both CNS and PNS has a relatively simple protein composition (figs. 18 and 19; see also Braun and Brostoff, 1977) compared with

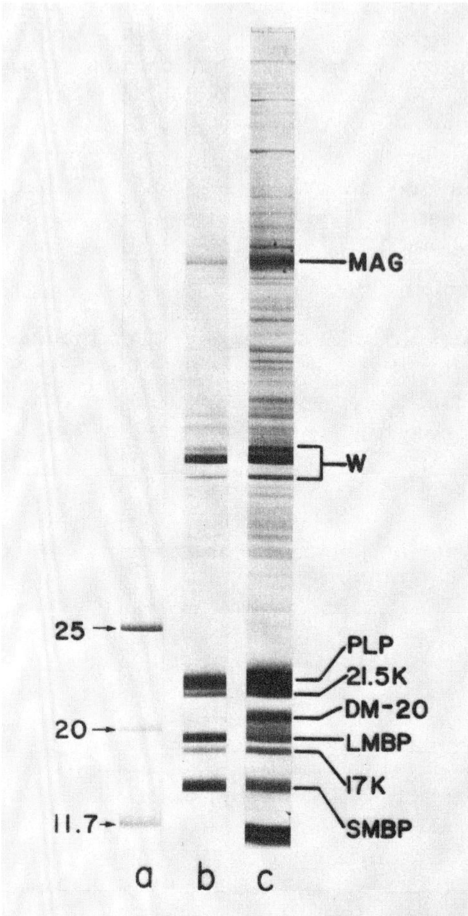

Figure 18. Purified CNS myelin has dissolved in a detergent solution and the proteins separated on polyacrylamide gels by electrophoresis. The proteins migrate according to their molecular weight, and several protein standards are included in the first lane as molecular weight markers. The major proteins of myelin are of low molecular weight and are visualized with Commassie blue dye in the second lane and a silver stain in the third. The designations used here are *LMBP* and *SMBP* for the two main basic proteins of rodent myelin. *PLP* marks the proteolipid protein, and *DM-20* is a hydrophobic protein of lower molecular weight. The Wolfgram proteins (*W*), which include the myelin enzyme marker 2′, 3′-cyclic nucleotide phosphohydrolase and the myelin associated-glycoprotein (*MAG*) are also indicated. With the more sensitive silver stain, a number of minor proteins associated with purified myelin are also visualized. (From Colman, Kreibich, and Sabatini, 1983.)

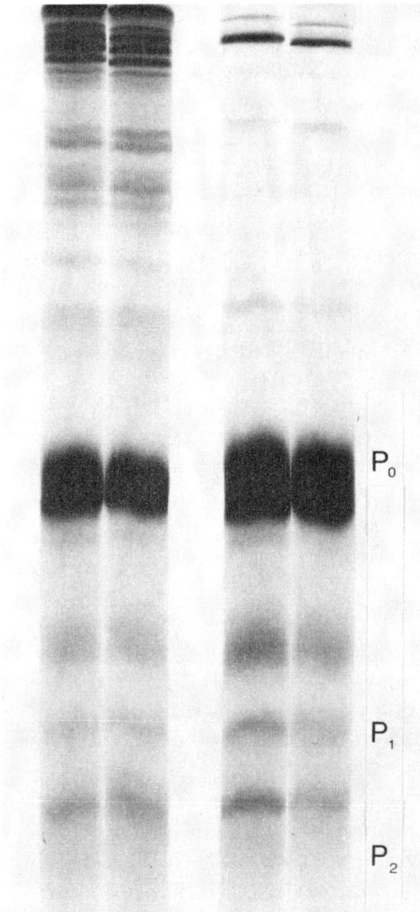

Figure 19. Polyacrylamide gel electrophoretic separation of total mouse sciatic nerve proteins (*lanes on left*) and those of purified PNS myelin (*lanes on right*). The 28,000-dalton P_0 glycoprotein is the dominant band in all of the gels. The large and small myelin basic proteins are the two lower bands. The P_2 basic protein is not a prominent protein in mouse and is not readily visible in these gels. It is apparent that the protein pattern of the total nerve is dominated by the myelin proteins, indicating the role of the myelin-forming Schwann cells in protein formation in this tissue. The proteins of other nonmyelin membranes tend to be of higher molecular weight. (From Wiggins and Morell, 1978; reprinted with permission of Pergamon Press, Ltd.)

other membranes. The proteins of isolated myelin are poorly represented (only 25–30 percent of dry weight, compared with 50 percent or greater in other membranes; see Finean and Michell, 1981). The few proteins that are present in high amounts in compact myelin presumably are in all layers and, therefore, would be instrumental in giving myelin its unique compact multilamellar structure. Like other membrane proteins, those of CNS and PNS myelin fall into the two principal classes—intrinsic and extrinsic (Singer and Nicolson, 1972; Finean and Michell, 1981). The intrinsic proteins, defined as proteins having strong interactions with the hydrophobic region of the membrane, in CNS myelin include a 24,000-dalton (24-kd) protein called the *Folch-Lees proteolipid protein* or *lipophilin* (PLP) and a related protein, the DM-20 protein (Agrawal et al., 1972). These proteins, which make up 40–50 percent of the total CNS myelin protein, are unique to the CNS (Macklin, Braun, and Lees, 1982). PNS myelin contains its own intrinsic protein, a glycoprotein designated P_0 (Greenfield et al., 1973). In addition to this protein, which alone makes up well over 50 percent of the total proteins, there are at least three glycoproteins of lower molecular weight (Kitamura, Suzuki, and Uyemura, 1976; Roomi et al., 1978a; Eylar et al., 1979) that are also intrinsic to PNS myelin.

Extrinsic proteins, defined by a largely hydrophilic association with membranes, include the *myelin basic protein* (MBP; also designated P_1 [PNS] and A_1 [CNS]; see Eylar et al., 1979). This 18-kd protein having approximately 170 amino acids and being present in both CNS and PNS myelin, was one of the first membrane proteins sequenced. It has been widely studied as an agent that causes an autoimmune disease, experimental allergic encephalomyelitis (see Carnegie and Dunkley, 1975, for review). In addition, there is a second related basic protein, called *pre–myelin basic protein* (pre-MBP). This 21-kd protein, described by Barbarese, Braun, and Carson (1977) has an additional chain of approximately 30 amino acids inserted near the amino-terminal portion of the protein. In rodents, there are two additional MBPs in which a section of the above basic proteins (from amino acid residues 120 to 160 of other MBPs) have been deleted (Martenson et al., 1972; Dunkley and Carnegie, 1974). These small MBPs of 14- and 17-kd molecular weight are present in both the CNS and PNS of rat (Greenfield et al., 1982). Studies by Yu and Campagnoni (1982) using protein synthesizing systems with mRNA (messenger ribonucleic acid) from developing rat brain polysomes demonstrated that the four rodent MBPs are synthesized on different mRNAs. The smaller ones are not posttranslational products of the larger ones. Besides these structurally related proteins, there is a distinct basic protein, P_2 (13.5-kd), which was first demonstrated in PNS myelin (Uyemura, Tobari, and Hirano, 1970; London, 1971; and Greenfield et al., 1973) and is also present in CNS myelin of some species (Trapp et al., 1983).

A third class of myelin-specific proteins (of intrinsic nature) are the *Wolfgram proteins*, defined by their solubility, amino acid composition, and migration on polyacrylamide gels (Wolfgram, 1966). These proteins are far more prominent in CNS myelin than in PNS myelin. One of the members of this group of proteins is an enzyme, which catalyzes the breakdown of adenosine 2′,3′-cyclic monophosphate, called *2′,3′-cyclic nucleotide phosphohydrolase* (CNPase), (Drummond and Dean, 1980; Sprinkle et al., 1980). This enzyme is the traditional marker for CNS myelin (Norton, 1981a) but not PNS myelin (see Cammer and Zimmerman, 1982). Roussel and his colleagues (1978) have localized Wolfgram protein with immunocytochemical methods to both oligodendroglial cells and their myelin sheaths.

The *myelin-associated glycoprotein* (MAG) of isolated CNS and PNS myelin is a glycoprotein of high molecular weight (approximately 100 kd), making up less than 1 percent of the total proteins (Quarles, 1980). Although MAG is classified as a myelin protein in fractionation studies, antibodies to purified protein have shown that this protein is not a component of compact myelin. It is located in the myelin-forming glia and in the periaxonal region between the axon and the myelin sheath, a location of potential importance for the initial interaction of glial cells with axons (Sternberger et al., 1979; Quarles, 1980). Some additional properties of the major myelin protein constituents are discussed below.

The Folch-Lees PLP is the main constituent of a heterogeneous mixture of proteins extracted from purified CNS myelin into organic solvents (see Boggs and Moscarello [1978], Lees et al. [1979], and Schlesinger [1981], for reviews). It contains roughly 65 percent nonpolar amino acids, which probably contribute to its solubility in organic solvents and its resistance to proteolysis. In addition, roughly 2 percent of its mass is covalently bound fatty acid (Folch and Stoffyn, 1972; Gagnon et al., 1971). As the main protein of CNS myelin (fig. 18), it is believed to play a structural role in compaction of the external surfaces as suggested from exposure of portions of this protein in the intraperiod line of myelin (Braun, 1977).

The P_0 protein, the dominant protein of myelinated peripheral nerve (fig. 19), accounts for greater than 50 percent of PNS myelin protein (Singh and Spritz, 1974; Wiggins, Benjamins, and Morell, 1975). Unlike the proteolipid protein, it is not soluble in organic solvents and is a glycoprotein (Everly, Brady, and Quarles, 1973; Wood and Dawson, 1973) containing roughly 6 percent carbohydrate (Roomi et al., 1978b). There is ultrastructural evidence that the carbohydrate moiety of this protein is exposed in the intraperiod line (Wood and McLaughlin, 1975; Peterson and Gruener, 1978). This protein probably plays a role in the compaction of both internal and external surfaces of PNS myelin, because the shiverer mutant mouse, which lacks MBP, has structurally normal PNS myelin as determined by X-ray studies and ultrastructural criteria (Privat et al., 1979; Ganser and Kirschner, 1980; Rosenbluth, 1980).

The MBP can be dissociated from isolated myelin by extraction with dilute acid (pH 3) or high salt (Eng et al., 1968; Diebler, Martenson, and Kies, 1972). MBP has been isolated from a number of species, and its amino acid sequence was elucidated by two groups—Carnegie (1971) and Eylar et al. (1971). Most evidence points to a localization of MBP and immunologically related basic proteins at the major dense line of myelin (Braun, 1977; Omlin et al., 1982). This localization may be indicative of a role for these proteins in the exclusion of glial cytoplasm during myelin compaction (at least in the CNS). Unlike MBP, which is a constituent of all CNS and PNS myelin sheaths, the P_2 protein of PNS myelin is present at immunocytochemically detectable levels in a portion but not all myelin sheaths (Trapp et al., 1979; DeArmond et al., 1980; Winter, Mirsky, and Kadlubowski, 1982; Trapp et al., 1983).

Myelin Lipids

Compared with other biologic membranes, myelin is uniquely lipid rich. Because a major responsibility of the myelin-forming oligodendroglia and Schwann cells is to synthesize the lipids needed for the formation and maintenance of their myelin sheaths, an increase in the activities of enzymes of lipid synthesis accompanies myelination. The high proportion of lipids in myelin membranes makes them less dense than other membranes. This property has been used as a basis to isolate myelin fractions from nerve tissue. Lipids are extracted from purified myelin with chloroform : methanol mixtures, and the composition of the lipids determined. (Polyphosphoinositides, highly acidic lipids present in low amounts in myelin, require acidified chloroform : methanol during extraction. They are also rapidly broken down in postmortem tissues, the pool in myelin being somewhat more stable [Eichberg and Hauser, 1973].) Like most other natural membranes, myelin contains phospholipids, glycolipids, and neutral lipids (mainly cholesterol).

Although there are probably no lipids unique to the myelin sheath, there are some lipids that have been found to be enriched in isolated myelin (see tables 1 and 2). This enrichment is a reflection of the commitment of increasing numbers of glial cells to making myelin. In table 1 is shown the percentages of the major lipids recovered in myelin isolated from the cerebral cortex and from total cortex of developing and adult rats.

The following features of myelin lipids are noteworthy. First, lipid compositions of isolated myelin are quantitatively unique (White, 1973). Compared with values for total cortical lipid, myelin is enriched in glycolipid and possibly sterol (mainly cholesterol) at the expense of phospholipid. The more striking differences between the values for myelin and total cortex lipids in the developing rat are probably because myelin is a less dominant component of developing cortex than adult

Table 1. ***Comparison of Brain Lipid Compositions of Cortex and a Purified Myelin Subfraction: Developing and Adult Rat***

Brain Weight and Lipid Compositions	*Developing Rat (15 days old)*		*Adult Rat (115–425 days old)*	
	Total	*Myelin*	*Total*	*Myelin*
Brain weight	1.23	1.21	1.98	1.92
Total phospholipid	74.4	51.4	57.4	44.0
Ethanolamine phosphoglyceride	23.4	18.8	19.5	16.6
Phosphatidylcholine	34.7	16.7	21.2	11.0
Sphingomyelin	3.2	3.5	3.8	3.1
Phosphatidylinositol	2.4	1.4	2.6	1.2
Phosphatidylserine	8.1	7.1	7.2	7.1
Plasmalogen	9.6	13.8	11.9	14.0
Total galactolipid	3.9	21.3	21.4	31.9
Cerebroside	2.8	16.6	14.8	23.3
Sulfatide	1.5	4.4	5.0	7.2
Total sterol	20.4	25.1	23.4	27.7

Note: The results are from a study by Norton and Poduslo (1973). The values for the adults are averages calculated from rats 144, 190, and 425 days old. The brain weight values are in grams, and the lipid values are the percentage of total lipid on a dry weight basis.

Table 2. ***Comparison of Lipid Compositions of CNS and PNS Myelin of Rat and Human***

Lipid Compositions	*CNS Myelin*		*PNS Myelin*	
	Human	*Rat*	*Human*	*Rat*
Total phospholipid	43.1	44.0	54.9	50.6
Ethanolamine phosphoglyceride	14.4	17.6	19.2	19.2
Phosphatidylcholine	10.8	11.0	8.2	9.6
Sphingomyelin	7.8	3.1	18.7	8.6
Phosphatidylinositol	0.4	0.9	9.3*	2.5
Phosphatidylserine	4.7	6.6	—	11.1
Plasmalogen	12.9	15.0	15.9	—
Glycolipid	27.5	31.5	22.1	21.5
Cerebroside	22.7	23.7	—	15.8
Sulfatide	3.8	7.1	—	5.7
Cholesterol	27.7	27.3	23.0	27.2

Note: Lipid compositions of CNS and PNS myelin as calculated from tables I and II of Norton (1981a). Values are expressed as the percentage of total lipid weight.

* Value is for PS plus PI, since the solvent system did not separate these phospholipids.

cortex. Second, a general conclusion from this and other data in the literature is that cerebrosides, sulfatides, and ethanolamine plasmalogen are enriched in myelin, while phosphatidylcholine and probably phosphatidylethanolamine (the plasmalogen is phosphatidalethanolamine) are more highly enriched in nonmyelin membranes. Although cerebroside, sulfatide, and plasmalogen are sometimes referred to as myelin lipids, this reference does not mean that they are found exclusively in myelin. Third, the composition of myelin changes with age, particularly the galactolipids cerebroside and sulfatide.

Comparisons of lipid composition of CNS myelin from different species have been made (e.g., Siakotos, Rouser, Fleischer, 1969; Norton, 1981a). The lipid composition of myelin from all species is quite similar, that is, from frog to human. Some species' differences include a lower value of sphingomyelin in frog and rat myelin (5–7 percent) compared with the bovine and human values (16–18 percent) and the higher amounts of phospholipids in frog (49–53 percent) compared with other species (43–44 percent). The above points that myelin lipids are quantitatively unique and that cerebrosides, sulfatides, and ethanolamine plasmalogen are selectively enriched in myelin appear to be valid for all species studied.

Another interesting feature of myelin lipids is the high proportion of long chain saturated and monounsaturated fatty acids and lower levels of polyunsaturated fatty acids in phospholipids, especially phosphatidylethanolamine, and sphingolipids compared with other membranes and brain subcellular fractions (see Svennerholm, Vanier, and Jungbjer, 1978). In addition, there are several minor lipids that are enriched in isolated myelin compared with other fractions. These include polyphosphoinositides (Hauser, Eichberg, and Gonzalez-Sastre, 1971; Eichberg and Hauser, 1973), sulfogalactosylglycerolipid and galactosyldiglycerides (Inoue, Deshmukh, and Pieringer, 1971; Pieringer et al., 1977), specific gangliosides (Ledeen et al., 1980), and alkanes (Bourre et al., 1977). These minor lipids, accounting for no more than 5 percent of the total, probably are not of major importance in determining the character of compact myelin.

PNS myelin, like CNS myelin, is characteristically lipid rich. It contains the same lipids as CNS myelin but in slightly different proportions (table 2). The major differences include a higher proportion of phospholipid and a lower proportion of glycolipid. The amounts of sphingomyelin present in PNS myelin are greater than those in CNS myelin. Among the minor lipids of peripheral nerve are polyphosphoinositides (Sheltawy and Dawson, 1966; Shaikh and Palmer, 1976), alkanes (Darriet, Cassagne, and Bourre, 1978), and gangliosides (Klein and Mandel, 1975; Fong et al., 1976; Chou, Nolan, and Jungalwala, 1982).

Development of Myelin: Glial Cell Involvement

It is generally accepted that myelin formation and maintenance are largely the responsibility of the myelin-forming glial cells and not that of the neuronal processes covered by the myelin sheaths. This responsibility would include as a major commitment the synthesis of those proteins and lipids present in the myelin sheath (see the preceding section). This commitment, recorded biochemically during development as an increase in myelin constituents, would signify both the growing population of glial cells forming myelin sheaths and the enlargement of individual sheaths.

The times of appearance and rates of increase of myelin proteins, lipids, and enzymes are not all the same; certain constituents appear before others, and their concentrations may increase at different rates (Detering and Wells, 1976; Benjamins and Morell, 1978; Pfeiffer, Barbarese, and Bhat, 1981). This differential accumulation of constituents in isolated myelin is reflected, for example, in table 1, where the report of lipid composition of myelin in the 15-day-old rat is quantitatively different from adult myelin. Two explanations for these findings are (1) that all the glial cells alter the composition of their myelin sheaths during maturation or (2) that individual glial cells make myelin sheaths of different composition. In the latter situation, the glial cells making myelin in later stages of development would form more mature myelin.

At present, we do not have methods to distinguish between these possibilities, that is, determining and comparing the compositions of individual internodes of myelin and seeing if the overall composition changes as more and more layers are added. However, there is some evidence indicating both that the protein composition of myelin sheaths are different from one another and that the composition of the sheath changes with development. Evidence in support of the former notion comes from immunocytochemical studies of P_2 basic protein of PNS myelin. Trapp and two sets of coworkers (1979, 1983), DeArmond et al. (1980), and Winter, Mirsky, and Kadlubowski (1982), all demonstrated that P_2 is present in detectable amounts in some myelin sheaths and not others, although other components (i.e., MBP, PLP, P_0, cerebroside, and sulfatide) are present in all myelin sheaths (Hartman et al., 1979; Trapp et al., 1979; DeArmond et al., 1980, Trapp et al., 1981; Hartman et al., 1982). It is not possible to determine with immunocytochemical technique whether the levels of these constituents are the same in all sheaths and in all layers of each sheath. Hedley-Whyte and Kirschner (1976), in a morphometric study, showed that the periodicity of both CNS and PNS myelin increased in older mice, a finding in agreement with the possibility that the overall composition of myelin changes with maturation.

Another effect of the changing composition of myelin as it matures might be reflected in differences in the composition of inner and outer layers of the same sheath. Gould (1977) showed that the glycoproteins that are laid down early in the formation of individual PNS myelin sheaths are retained for weeks to months in the inner layers of myelin, whereas those formed later and added to outer layers are retained near the sites where they were added. In contrast, phospholipids readily move, probably by lateral diffusion, between inner and outer layers of myelin. If during development, the levels of proteins assembled into inner (earlier) and outer (later) layers are different, these differences could influence the distribution of lipids as a result of their interactions with the proteins (see Rumsby and Crang [1977] and Boggs and Moscarello [1978] for reviews).

Another facet of the autoradiographic studies has been their use in describing the metabolic relationship between myelin-forming glial cells and the myelin sheaths they are elaborating and maintaining. Tennekoon and his colleagues (1977) used [^{35}S]-sulfate to study the synthesis of sulfatide in developing optic nerve. Before differentiated oligodendroglia appeared (two days; see also fig. 19), sulfate was largely incorporated into sulfated glycoprotein and mucopolysaccharide. Once myelination had begun (from postnatal day 6), the incorporation into sulfatide was dramatically increased. This biochemical change was correlated with the localization of radioactivity in thin sections of the tissue to the forming myelin sheaths. The labeling period was 48 hours, and the investigators did not look at earlier times to show that the label originated in synthetic machinery in the oligodendroglial cells. There is evidence that the enzyme involved in sulfatide synthesis (from galactocerebroside) is localized to the Golgi apparatus in brain homogenates (Benjamins, Hadden, and Skoff, 1982) and isolated oligodendroglia to a far higher level than in neurons (Benjamins et al., 1974).

Autoradiographic studies of myelination in the peripheral nerve have been carried out both with developing and mature nerve (for review see Gould, Matsumoto, and Mattingly, 1982). The major sites where tritiated choline and fucose are incorporated into myelin-destined phospholipid and glycoprotein (mainly P_0) are in the Schwann cell cytoplasm near the nucleus, the perinuclear cytoplasm (see fig. 20). With time the label moves from synthetic sites into the myelin, first labeling the outermost layers that are close to the synthetic "machinery." The differences in movement of phospholipid and protein in PNS myelin indicates that myelin behaves as a fluid lipid bilayer (see also Rumsby and Crang, 1977) and that the diffusion of lipids along the concentrically wrapped bilayers is roughly two orders of magnitude faster than that of proteins. This notion that myelin has a fluid character is important for conceptualizing the growth and expansion of myelin sheaths that accompany axonal growth (compare figs. 7A–D).

Immunocytochemical studies, made possible through the purifica-

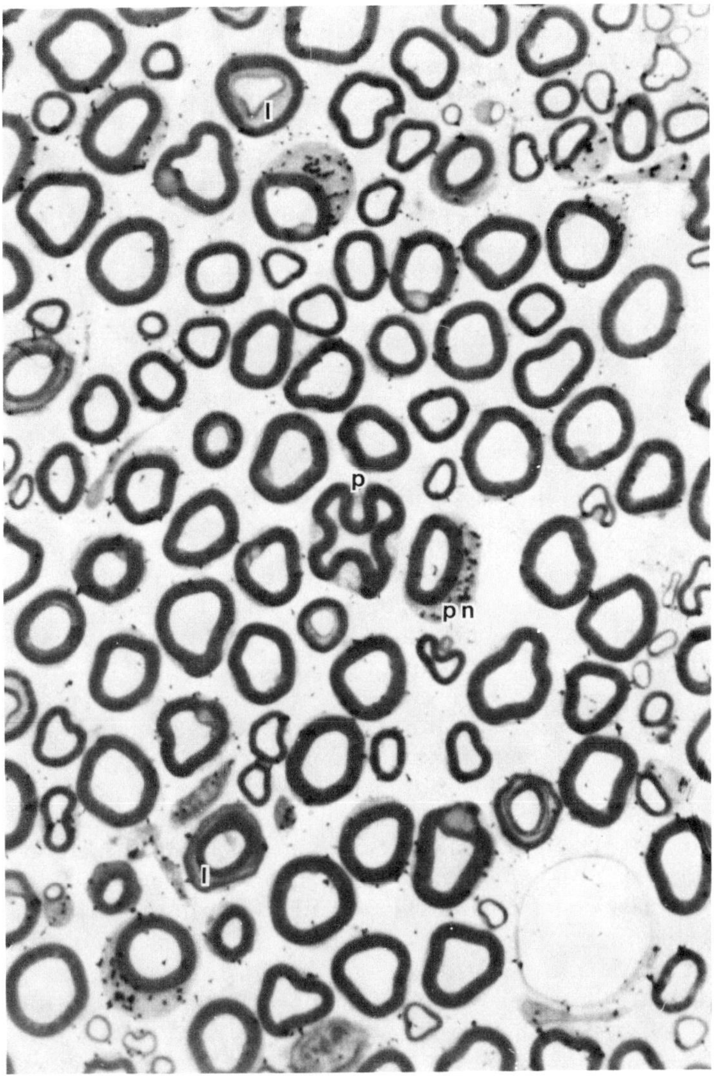

Figure 20. Light microscope autoradiograph of mouse sciatic nerve labeled with tritiated proline. The silver grains represent proline that was incorporated into protein during the two-hour labeling period. The crescent-shaped perinuclear (*pn*) regions of the Schwann cells are the prominent sites of protein synthesis in this tissue. In contrast, Schwann cell cytoplasm at the paranodal (*p*) ends of the internode and in Schmidt-Lanterman incisures (*l*) is relatively unlabeled, indicating that little protein synthesis occurs in these specializations.

tion of myelin proteins and lipids and the raising of antibodies to them, have been used to localize lipids and proteins in both myelin sheaths (Omlin et al., 1982) and in glial cells at and in some cases slightly before the appearance of compact myelin (Sternberger et al., 1978a, 1978b; Hartman et al., 1979; Hartman et al., 1982). During myelin development, the intensity of MBP staining within the glial cells decreases, possibly because MBPs are more efficiently transferred into myelin (i.e., they spend less time in glial cell cytoplasm) as the glial machinery is stabilized, and the formation of compact myelin becomes a more ordered process (Webster and Sternberger, 1980). This immunocytochemical work provides evidence at the level of the myelin-forming cells that their ability to make myelin proteins is correlated in time to the establishment of axon-glial contact. In the case of Schwann cells, but not oligodendroglia, when axonal influence is removed, the ability to "express" myelin constituents diminishes (Mirsky et al., 1980).

The Biochemistry of Myelination: Putting It Together

In the previous sections I discussed the main protein and lipid components of compact myelin and gave evidence that the formation of the myelin sheath is primarily the responsibility of glial cells that have been recruited by axons. In this section I would like to consider the formation of myelin in terms of the biochemistry of the various stages of myelination. Because myelination is not occurring synchronously in any population of glial cells in situ or in culture, we cannot determine in a very straightforward way the order of the biochemical changes in glial cell metabolism that accompany ensheathment and the formation of compact myelin. It is hoped that future advances in such areas as molecular and cell biology and the establishment of microbiochemical methodologies and tissue culture approaches will allow researchers to gain a more comprehensive understanding of how glial cells form and maintain their myelin sheaths.

The earliest stages of myelination include the close apposition of the plasma membrane of glial cells and the larger caliber axons. Following this contact, each glial cell (or process) surrounds the section of axon it will myelinate. All of the specific proteins and lipids that are involved with these initial stages of myelination, with the possible exception of the MAG, are uncharacterized. There is some evidence that gangliosides might be involved (Yates and Wherrett, 1974). The MAG's possible role as a glial surface protein comes from studies that localized it to the myelin axon boundary (Sternberger et al., 1979; Quarles, 1980; Trapp and Quarles, 1982) and identified its early developmental appearance (Johnson, Quarles, and Brady, 1982). There is a need to find other glial and axonal constituents that appear at the area of contact between axons and the myelin-forming glial cells at the time that formation of myelin is starting to occur. Then researchers need to develop approaches to determine how these components would function to influence the formation of myelin, presumably by their ability to effect glial differentiation (i.e., inhibit glial proliferation), on the one hand, or make myelin constituents, on the other. Some of the proteins might be included among the high–molecular weight proteins that are found in diminishing amounts as compact myelin is formed.

Following the formation and interaction of components involved in axon ensheathment would be the transcription of message or activation of transcribed message to allow the formation of the key myelin proteins (and enzymes of myelin lipid formation). Evidence that axonal contact is a prerequisite to the synthesis of PNS myelin proteins has recently been shown by Politis et al. (1982).

The subsequent formation of compact myelin surely must be a process involving interactions among the major protein and lipid components of compact myelin (i.e., the MBP and PLP in the CNS, and P_0 and possibly the MBPs in the PNS). Because of their high levels in myelin and asymmetric distribution in compact myelin, these proteins could potentially function in the processes of compaction that occur at the major dense (MBP and P_0) and intraperiod lines (PLP and P_0). Information on the properties of these proteins, when interacting with lipids, is of potential relevance, although this knowledge must be coupled to an understanding of the intracellular pathways that carry myelin proteins and lipids from sites of synthesis into compact myelin. As an example, it will be difficult to envision an interaction between basic proteins that reside at the internal membrane surfaces with acidic sulfatide and ganglioside. In spite of the fact that they are present at levels sufficient to interact with MBP, the formation in the Golgi would leave them ultimately exposed to the external plasma membrane surfaces. The understanding of how the formation of compact myelin occurs must include both knowledge of the interactions that cause membrane surfaces to come together and in the plane of each membrane to exclude components of the glial plasma membrane from compact myelin. The region in which myelin components might be separated could be the tight junctions that surround compact myelin (Schnapp and Mugnaini, 1975). Recently, Kachar and Reese (1982) gave evidence that these structures might be lipid in nature.

The synthesis of the vast quantities of protein required in the elaboration of multiple layers of myelin membrane takes place in the glial cytoplasm where polyribosomes and rough endoplasmic reticulum are located. In adult and developing Schwann cells these regions are concentrated in the perinuclear areas, adjacent to sites where DNA is transcribed into the mRNAs that would prime the translation of myelin proteins (and enzymes). In autoradiographs of developing and adult peripheral nerve, protein (proline-labeled) and glycoprotein (fucose-labeled) are formed mainly in the perinuclear regions of Schwann cell cytoplasm (fig. 20) (Gould, 1977; Mattingly and Gould, 1982). These areas have an abundance of polyribosomes and cisternae of rER (fig. 8). Although comparable autoradiographic studies have not been carried out in the CNS, the perinuclear regions of developing and mature oligodendroglia appear to contain prominent rER and polyribosomes (fig. 12), but these are usually less evident in long oligodendroglial processes (figs. 3 and 10).

The Golgi apparatus, an organelle functioning in the glycosylation of proteins and in their delivery to cell surfaces (Morre and Ovtracht, 1977; Holtzman and Mercurio, 1980; and Farquhar and Palade, 1981), is also present largely in the perinuclear cytoplasm of Schwann cells and oligodendroglia. This organelle in Schwann cell cytoplasm is the site of fucosylation of protein (predominantly P_0) (Gould, 1977) and sulfation (Young, 1973). Accumulation of P_0 at this site has been demonstrated in Schwann cells of developing peripheral nerve by immunocytochemistry (Trapp et al., 1981). Although the PLP of CNS myelin is not glycosylated, it is fatty acylated. Since other proteins that contain fatty acids are acylated in the Golgi apparatus (Schlesinger, 1981; Magee and Schlesinger, 1982), the oligodendroglial cells' Golgi may be a key site in the transport of PLP into myelin. It is probably an important station for the processing of MAG and other glycoproteins present in the glial plasma membranes, in the compact myelin, or in both.

The endoplasmic reticulum and Golgi apparatus of Schwann cells and oligodendroglia are also the principal sites where myelin phospholipids, cholesterol, and glycoproteins are made. The perinuclear Schwann cell cytoplasm is the site in developing and adult rodent sciatic nerve where phosphatidylcholine and other phospholipids are made (Gould and Dawson, 1976; Dawson and Gould, 1976; Gould and Sinatra, 1981; and Gould, Connell, and Spivack, 1982). These results are in line with those of fractionation studies that demonstrate that the rough and smooth endoplasmic reticulum, purified as "microsomes" and their subfractions, is the main site of phospholipid and sterol formation (for review see Dawson [1973], Parry [1978], and Schroepfer [1981, 1982]). The main site of cerebroside and sulfatide formation is in the Golgi apparatus (Herschkowitz et al., 1969; Benjamins, Hadden, and Skoff, 1982).

It is likely that the sites of synthesis are different from the sites where myelin constituents are added to the forming myelin sheath. In the case of the intrinsic proteins (PLP and P_0) there is a considerable delay between synthesis (measured by incorporation of precursor amino acid) and entry into myelin (measured by accumulation of protein product in a myelin-rich subfraction) (Benjamins and Morell, 1978; Braun, Pereyra, and Greenfield, 1980; Rapaport and Benjamins, 1981; Magee and Schlesinger, 1982). Recent advancements in our understanding of the synthesis and processing of integral membrane proteins (Warren, 1981; Sabatini et al., 1982) should provide a framework for studying the steps required for the incorporation of PLP and P_0 into compact myelin. In contrast, the extrinsic proteins (MBPs) enter myelin soon after synthesis. Furthermore, MBPs are synthesized on free polyribosomes (Campagnoni, Carey, and Yu, 1980; Colman et al., 1982) and inserted without a significant lag into myelin (Benjamins and Morell, 1978; Colman et al., 1982). A model for the formation of myelin must include different pathways for the incorporation of MBP into the major dense region of myelin and PLP into the intraperiod region. It is interesting that Peters, Palay, and Webster (1976) noted that the major dense and intraperiod lines form at different locations (and presumably at different times) in the same sheath.

Conclusion

Myelination requires interactions between myelin-forming glial cells and the axons that are to be myelinated. In addition, there are probably extrinsic factors, including collagen in the peripheral nerve and astrocytes and hormones in the CNS, that influence myelination. One level at which the developmental process of myelination must be understood is at the level of cellular interactions.

We must learn the mechanisms used by axons to influence glial cells and vice versa. In this light it is important to understand the properties of the surfaces of contact between glial cells and axons. We are presently learning about their formation during development through morphological study, but we still know little about the components that make up the structures or about the biochemical information that might be passed at these sites, which may number up to 2,000 in a one-meter nerve.

In terms of extrinsic factors, we have morphological evidence that contact with collagen substratum (and possibly other components) stimulates Schwann cells to myelinate peripheral axons in cultures (Bunge and Bunge, 1981) and that astrocytes form gap junctions with oligodendroglia (Massa and Mugnaini, 1982). We still have no idea about how external material influences Schwann cell differentiation, or how astrocytes influence oligodendroglia and vice versa.

At the subcellular level, we must investigate the regulation of genetic expression. What are the myelination messages and how are they turned on? What is the posttranslational processing that occurs (e.g., phosphorylation of MBP and acylation of PLP), and what role or roles does it have in myelin formation? We must understand the development of the organelles required for the synthesis of myelin constituents, the polysomes, rough and smooth endoplasmic reticulum, the Golgi apparatus, and mitochondria. We must also understand the pathways that get lipids and proteins from synthetic sites to myelin. Are the cytoskeleton, coated vesicles, and lipoproteins involved? If so, at what stages do they fit in?

Finally, we must be careful in considering myelination as too stereotyped a process. Researchers (Knobler, Stempak, and Laurencin, 1974, 1976; Webster, 1971) have shown through serial reconstruction that the initial stages of myelination are quite complex and variable. Is the complexity part of the process of development of a patent system for the synthesis of myelin-destined constituents and the efficient transfer into growing myelin?

It would be foolhardy to say that a comprehensive understanding of myelination is at hand; however, the recent advances that have been made in the area of myelination and in the molecular and cell biology of membrane biogenesis in general may allow us to look forward to the development in the next decade of a clearer understanding of myelination at the cellular and subcellular level.

Acknowledgments

I thank Fred Connell for his help, from preparing the tissues of developing rats to producing figures; Dr. Jack Rosenbluth for discussions related to myelination and its affect on axon development and for figures 16 and 17, and Dr. Raul Rudelli for information on human myelination and for figure 1. I appreciate the help of Helen LaMantia and Lee Antonucci in preparing the manuscript. I also thank Warren Spivack and Gary Mattingly for help and encouragement. Finally, I thank my wife, Marianne, and my children, Milena and Hansko, for giving me the time I needed to put this chapter together.

References

Agrawal, H. C., Burton, R. M., Fishman, M. A., Mitchell, R. F., and Prensky, A. L. (1972). Partial characterization of a new myelin protein component. *J Neurochem* 19, 2083–2089.

Aguayo, A. J., Attiwell, M., Trecarter, J., Perkins, S., and Bray, G. M. (1977). Abnormal myelination in transplanted trembler mouse Schwann cells. *Nature* 265, 73–75.

Aguayo, A. J., Bray, G. M., Perkins, C. S., and Duncan, I. D. (1979). Axon-sheath cell interactions in peripheral and central nervous system transplants. *Soc Neurosci Symp* 4, 361–383.

Aguayo, A. J., Peyronnard, J. M., Terry, L. C., Romine, J. S., and Bray, G. M. (1976). Neonatal neuronal loss in rat superior cervical ganglia: Retrograde effects on developing preganglionic axons and Schwann cells. *J Neurocytol* 5, 137–155.

Asbury, A. K. (1975). The biology of Schwann cells. In *Peripheral neu-*

ropathies. Vol. 1, ed. P. J. Dyck, P. K. Thomas, and E. H. Lambert, pp. 201–212. Philadelphia: W. B. Saunders.

Barbarese, E., Braun, P. E., and Carson, J. H. (1977). Identification of prelarge and presmall basic proteins in mouse myelin and their structural relationship to large and small basic proteins. *Proc Natl Acad Sci USA* 79, 3360–3364.

Baumann, N. (ed.) (1980). *Neurological mutations affecting myelination*. INSERM Symp. 14. Amsterdam: Elsevier/North-Holland.

Benjamins, J. A., Guarnieri, M., Miller, K., Sonneborn, M., and McKhann, G. M. (1974). Sulfatide synthesis in isolated oligodendroglial and neuronal cells. *J Neurochem* 23, 751–757.

Benjamins, J. A., Hadden, T., and Skoff, R. P. (1982). Cerebroside sulfotransferase in Golgi-enriched fractions from rat brain. *J Neurochem* 38, 233–241.

Benjamins, J. A., and McKhann, G. M. (1981). Development, regeneration, and aging of the brain. In *Basic neurochemistry*, ed. G. J. Siegel, R. W. Albers, B. W. Agranoff, and R. Katzman, pp. 445–469. Boston: Little, Brown.

Benjamins, J. A., and Morell, P. (1978). Proteins of myelin and their metabolism. *Neurochem Res* 3, 137–174.

Berthold, C. H. (1978). Morphology of normal peripheral axons. In *Physiology and pathobiology of axons*, ed. S. G. Waxman, pp. 3–63. New York: Raven Press.

Bhat, S., Barbarese, E., and Pfeiffer, S. E. (1981). Requirement for nonoligodendrocyte cell signals for enhanced myelinogenic gene expression in long-term cultures of purified rat oligodendrocytes. *Proc Natl Acad Sci USA* 78, 1283–1287.

Black, J. A., Foster, R. E., and Waxman, S. G. (1982). Rat optic nerve: Freeze-fracture studies during development of myelinated axons. *Brain Res* 250, 1–20.

Boggs, J. M., and Moscarello, M. A. (1978). Structural organization of the human myelin membrane. *Biochim Biophys Acta* 515, 1–21.

Bourre, J. M., Cassagne, C., Larrouquere-Regnier, S., and Darret, D. (1977). Occurrence of alkanes in brain myelin: Comparison between normal and quaking mouse. *J Neurochem* 29, 645–648.

Brady, R. O., and Quarles, R. H. (1973). The enzymology of myelination. *Mol Cell Biochem* 2, 23–29.

Braun, P. E. (1977). Molecular architecture of myelin. In *Myelin*, ed. P. Morell, pp. 201–231. New York: Plenum Press.

Braun, P. E., and Brostoff, S. W. (1977). In *Myelin*, ed. P. Morell, pp. 201–231. New York: Plenum Press.

Braun, P. E., Pereyra, P. M., and Greenfield, S. (1980). Myelin organization and development: A biochemical perspective. In *Myelin chemistry and biology*, ed. G. Hashim, pp. 1–17. New York: Alan R. Liss.

Brockes, J. P., Fields, K. L., and Raff, M. C. (1979). Studies on cultured rat

Schwann cells: I. Establishment of purified populations from cultures of peripheral nerve. *Brain Res* 165, 105–118.

Bunge, R. P. (1982). Cellular and non-cellular influences on myelin-forming cells. In *Neuronal-glial cell interrelationships*, ed. T. A. Sears, pp. 115–130. Berlin: Springer-Verlag.

Bunge, R. P., and Bunge, M. B. (1981). Cues and constraints in Schwann cell development. In *Studies in developmental neurobiology: Essays in honor of Victor Hamberger*, ed. W. M. Cowan, pp. 322–353. New York: Oxford University Press.

Bunge, R. P., Bunge, M. B., and Cochran, M. (1978). Some factors influencing the proliferation and differentiation of myelin-forming cells. *Neurology* 28, 59–67.

Cammer, W., and Zimmerman, T. R., Jr. (1982). 5′-Nucleotidase in sciatic nerves from adult and young rats. *Neurochem Res* 7, 229–233.

Campagnoni, A. T., Carey, G. D., and Yu, Y. T. (1980). In vitro synthesis of the myelin basic proteins: Subcellular site of synthesis. *J Neurochem* 39, 677–686.

Carnegie, P. R. (1971). Amino acid sequence of the encephalitogenic basic protein from human myelin. *Biochem J* 123, 57–67.

Carnegie, P. R., and Dunkley, P. R. (1975). Basic proteins of central and peripheral nervous system myelin. In *Advances in neurochemistry*. Vol. 1, ed. B. W. Agranoff and M. H. Aprison, pp. 95–135. New York: Plenum Press.

Cassel, D., Wood, P. M., Bunge, R. P., and Glaser, L. (1982). Mitogenicity of brain axolemma membranes and soluble factors for dorsal root ganglion Schwann cells. *J Cell Biochem* 18, 433–445.

Chou, K. H., Nolan, C. E., and Jungalwala, F. B. (1982). Composition and metabolism of gangliosides in rat peripheral nervous system during development. *J Neurochem* 39, 1547–1558.

Colman, D. R., Kreibich, G., Frey, A. B., and Sabatini, D. D. (1982). Synthesis and incorporation of myelin polypeptides into CNS myelin. *J Cell Biol* 95, 598–608.

Colman, D. R., Kreibich, G., and Sabatini, D. D. (1983). Biosynthesis of myelin-specific proteins. *Methods Enzymol* 96, 378–385.

Darriet, D., Cassagne, C., and Bourre, J. M. (1978). Sciatic nerve contains alkanes: Comparison between normal mice and neurological mutants—jumpy, quaking and trembler. *J Neurochem* 31, 1541–1543.

Davison, A. N. (1970). The biochemistry of the myelin sheath. In *Myelination*, ed. A. N. Davison and A. Peters, pp. 80–161. Springfield, Ill.: Charles C. Thomas.

Dawson, R. M. C. (1973). The exchange of phospholipids between cell membranes. *Subcell Biochem* 2, 69–89.

Dawson, R. M. C., and Gould, R. M. (1976). Renewal of phospholipids in the myelin sheath. In *Function and metabolism of phospholipids in the central and peripheral nervous system,* ed. G. Porcellati, L. Amaducci, and C. Galli, pp. 95–113. New York: Plenum Press.

DeArmond, S. J., Deibler, G. E., Bacon, M., Kies, M. W., and Eng, L. F. (1980). A neurochemical and immunocytochemical study of P_2 protein in human and bovine nervous systems. *J Histochem Cytochem* 28, 1275–1285.

Detering, N. K., and Wells, M. A. (1976). The non-synchronous synthesis of myelin components during early stages of myelination in the rat optic nerve. *J Neurochem* 26, 253–257.

Diebler, G. E., Martenson, R. E., and Kies, M. W. (1972). Large scale preparation of myelin basic protein from central nervous tissue of several mammalian species. *Prep Biochem* 2, 139–165.

Drummond, R. J., and Dean, G. (1980). Comparison of 2′,3′-cyclic nucleotide 3′-phosphodiesterase and the major component of Wolfgram protein W1. *J Neurochem* 35, 1155–1165.

Dunkley, P. R., and Carnegie, P. R. (1974). Amino acid sequence of the smaller basic protein from rat myelin. *Biochem J* 141, 243–255.

Dupouey, P., Lucas, C. V., Gomes, D., and Jacque, C. (1980). Immunohistochemical localization of the myelin basic protein and of the glial fibrillary acidic protein: Comparative study in normal quaking and jimpy mice. *J Neurosci Res* 5, 387–398.

Eichberg, J., and Hauser, G. (1973). The subcellular distribution of polyphosphoinositides in myelinated and unmyelinated rat brain. *Biochim Biophys Acta* 326, 210–223.

Eng, L. F., Chao, F. C., Gerstl, B., Pratt, D., and Tavastsjerna, M. G. (1968). The maturation of human white matter myelin: Fractionation of the myelin membrane proteins. *Biochemistry* 7, 4455–4465.

Everly, J. L., Brady, R. O., and Quarles, R. H. (1973). Evidence that the major protein in rat sciatic nerve myelin is a glycoprotein. *J Neurochem* 21, 329–334.

Eylar, E. H., Brostoff, S., Hashim, G., Caccam, J., and Burnett, P. (1971). Basic A1 protein of the myelin membrane. *J Biol Chem* 246, 5770–5784.

Eylar, E. H., Uyemura, K., Brostoff, S. W., Kitamura, K., Ishaque, A., and Greenfield, S. (1979). Proposed nomenclature for PNS myelin proteins. *Neurochem Res* 4, 283–293.

Farquhar, M. G., and Palade, G. E. (1981). The Golgi apparatus (complex)—(1954–1981)—from artifact to center stage. *J Cell Biol* 91, 77s–103s.

Finean, J. B., and Michell, R. H. (1981). Isolation, composition and general structure of membranes. In *Membrane structure,* ed. J. B. Finean and R. H. Michell, pp. 1–36. Amsterdam: Elsevier/North-Holland.

Folch, P. J., and Stoffyn, P. J. (1972). Proteolipids from membrane systems. *Ann NY Acad Sci* 195, 86–107.

Fong, J. W., Ledeen, R. W., Kundu, S. K., and Brostoff, S. W. (1976). Gangliosides of peripheral nerve myelin. *J Neurochem* 26, 157–162.

Freidrich, V. L., Jr., and Mugnaini, E. (1983). Myelin sheath thickness in the CNS is regulated near the axon. *Brain Res* 274, 329–331.

Friede, R. L., and Samorajski, T. (1968). Myelin formation in the sciatic nerve of the rat. *J Neuropathol Exp Neurol* 27, 546–570.

Fulcrand, J., and Privat, A. (1977). Neuroglial reactions secondary to wallerian degeneration in the optic nerve of the postnatal rat: Ultrastructural and quantitative study. *J Comp Neurol* 176, 189–224.

Gagnon, J., Finch, P. R., Wood, D. D., and Moscarello, M A. (1971). Isolation of a highly purified myelin protein. *Biochemistry* 10, 4756–4763.

Ganser, A. L., and Kirschner, D. A. (1980). Myelin structure in the absence of basic protein in the shiverer mouse. In *Neurological mutations affecting myelination*, ed. N. Baumann, pp. 171–176. Amsterdam: Elsevier/North-Holland.

Gilles, F. H. (1976). Myelination in the neonatal brain. *Hum Pathol* 7, 244–248.

Giorgi, P. P., and DuBois, H. (1980). Regional differences in thickness and metabolism of the myelin sheath along the optic nerve and tract of rabbit. *Neuroscience* 5, 2013–2022.

Gould, R. M. (1977). Incorporation of glycoproteins in peripheral nerve myelin. *J Cell Biol* 75, 326–339.

Gould, R. M., and Dawson, R. M. C. (1976). Incorporation of newly formed lecithin into peripheral nerve myelin. *J Cell Biol* 68, 480–496.

Gould, R. M., Connell, F., and Spivack, W. D. (1982). Lipid synthesis in mouse sciatic nerve. *J Cell Biol* 95, 257a.

Gould, R. M., Matsumoto, D., and Mattingly, G. (1982). The Schwann cell. In *Handbook of neurochemistry*. 2d ed. Vol. 1, ed. A. Lajtha, pp. 397–414. New York: Plenum Press.

Gould, R. M., and Sinatra, R. S. (1981). Internodal distribution of phosphatidylcholine biosynthesis activity in teased peripheral nerve fibres: An autoradiographic study. *J Neurocytol* 10, 161–167.

Greenfield, S., Brostoff, S., Eylar, E. H., and Morell, P. (1973). Protein composition of myelin of the peripheral nervous system. *J Neurochem* 20, 1207–1216.

Greenfield, S., Weise, M. J., Gantt, G., Hogan, E. L., and Brostoff, S. W. (1982). Basic proteins of rodent peripheral nerve myelin: Immunochemical identification of the 21.5K, 18.5K, 17K, 14K and P_2 proteins. *J Neurochem* 39, 1278–1282.

Gregson, N. A. (1976). The chemistry and structure of myelin. In *The peripheral nerve*, ed. D. N. Landon, pp. 512–604. London: Chapman and Hall.

Hall, S. M. (1978). The Schwann cell: A reappraisal of its role in the peripheral nervous system. *Neuropathol Appl Neurobiol* 4, 165–176.

Hartman, B. K., Agrawal, H. C., Agrawal, D., and Kalmbach, S. (1982). Development and maturation of central nervous system myelin: Comparison of immunohistochemical localization of proteolipid protein and basic protein in myelin and oligodendrocytes. *Proc Natl Acad Sci USA* 79, 4217–4220.

Hartman, B. K., Agrawal, H. C., Kalmbach, S., and Shearer, W. T. (1979). A comparative study of the immunohistochemical localization of basic protein to myelin and oligodendrocytes in rat and chicken brain. *J Comp Neurol* 188, 273–290.

Hatfield, J. S., and Skoff, R. P. (1982). GFAP immunoreactivity reveals astrogliosis in females heterozygous for jimpy. *Brain Res* 250, 123–131.

Hauser, G., Eichberg, J., and Gonzalez-Sastre, F. (1971). Regional distribution of polyphosphoinositides in rat brain. *Biochim Biophys Acta* 248, 87–95.

Hedley-Whyte, E. T., and Kirschner, D. A. (1976). Morphological evidence of alterations in myelin structure with maturation. *Brain Res* 113, 487–497.

Herschkowitz, N., McKhann, G. M., Saxena, S., Shooter, E. M., and Herndon, R. (1969). Synthesis of sulphatide-containing lypoprotein in rat brain. *J Neurochem* 16, 1049–1057.

Hirano, A. (1981). Structure of normal central myelinated fibers. *Adv Neurol* 31, 51–68.

Hodgkin, A. L. (1971). *The conduction of the nervous impulse*. Liverpool, England: Liverpool University Press.

Hoffman, P. N., Griffin, J. W., and Price, D. L. (1983). Role of neurofilaments in regulating axonal caliber. *Trans Am Soc Neurochem* 14, 102.

Holtzman, F., and Mercurio, A. (1980). Membrane circulation in neurons and photoreceptors: Some unresolved issues. *Int Rev Cytol* 67, 1–67.

Inoue, T., Deshmukh, D. S., and Pieringer, R. A. (1971). The association of the galactosyl diglycerides of brain with myelination. *J Biol Chem* 246, 5695–5699.

Jacobson, M. (1978). *Developmental neurobiology*. New York: Plenum Press.

Johnson, D., Quarles, R. H., and Brady, R. O. (1982). Myelin-associated glycoprotein in the CNS and PNS of developing rodents. *Trans Am Soc Neurochem* 13, 213.

Kachar, B., and Reese, T. S. (1982). Evidence for the lipidic nature of tight junction strands. *Nature* 296, 464–466.

Kitamura, K., Suzuki, M., and Uyemura, K. (1976). Purification and partial characterization of two glycoproteins in bovine peripheral nerve myelin membrane. *Biochim Biophys Acta* 455, 806–816.

Klein, F., and Mandel, P. (1975). Gangliosides of the peripheral nervous system. *Life Sci* 16, 751–758.

Knobler, R. L., Stempak, J. G., and Laurencin, M. (1974). Oligodendroglial ensheathment of axons during myelination in the developing rat central nervous system. *J Ultrastruct Res* 49, 34–49.

Knobler, R. L., Stempak, J. G., and Laurencin, M. (1976). Nonuniformity of the oligodendroglial ensheathment of axons during myelination in the developing rat central nervous system: A serial section electron microscopical study. *J Ultrastruct Res* 55, 417–432.

Kreider, B. Q., Messing, A., Doan, H., Kim, S. V., Lisak, R. P., and Pleasure, D. E. (1981). Enrichment of Schwann cell cultures from neonatal rat sciatic nerve by differential adhesion. *Brain Res* 207, 433–444.

Landon, D. N. (1981). Structure of normal peripheral myelinated nerve fibres. *Adv Neurol* 31, 25–49.

Landon, D. N., and Hall, S. (1976). The myelinated nerve fibre. In *The peripheral nerve*, ed. D. N. Landon, pp. 1–105. London: Chapman and Hall.

Lasek, R. J., Gainer, H., and Barker, J. L. (1977). Cell-to-cell transfer of glial proteins to the squid giant axon: The glia-neuron protein transfer hypothesis. *J Cell Biol* 79, 501–523.

Ledeen, R. W., Cochran, F. B., Yu, R. K., Samuels, F. G., and Haley, J. E. (1980). Gangliosides of CNS myelin membrane. *Adv Exp Med Biol* 125, 167–176.

Le Douarin, N. M. (1980). Migration and differentiation of neural crest cells. In *Current topics in developmental biology*. Vol. 16, ed. A. A. Moscona and A. Monroy, pp. 31–85. New York: Academic Press.

Lees, M. B., Sakura, J. D., Sapirstein, V. S., and Curatolo, W. (1979). Structure and function of proteolipids in myelin and non-myelin membranes. *Biochem Biophys Acta* 559, 209–230.

London, Y. (1971). Ox peripheral nerve myelin membrane: Purification and partial characterization of two basic proteins. *Biochim Biophys Acta* 249, 188–196.

Macklin, W. B., Braun, P. E., and Lees, M. B. (1982). Electroblot analysis of the myelin proteolipid protein. *J Neurosci Res* 7, 1–10.

Magee, A. I., and Schlesinger, M. J. (1982). Fatty acid acylation of eucaryotic cell membrane proteins. *Biochim Biophys Acta* 694, 279–289.

Martenson, R. E., Deibler, G. E., Kies, M. W., McKneally, S. S., Shapira, R., and Kibler, R. F. (1972). Differences between the two basic proteins of the rat central nervous system: A deletion in the smaller protein. *Biochim Biophys Acta* 263, 193–203.

Massa, P. T., and Mugnaini, E. (1982). Cell junctions and intramembrane particles of astrocytes and oligodendrocytes: A freeze-fracture study. *Neuroscience* 7, 523–538.

Mattingly, G., and Gould, R. M. (1982). Sites of RNA, protein and phospholipid synthesis in Schwann cells of myelinated axons visualized by teased fiber autoradiography. *J Cell Biol* 95, 399a.

Mirsky, R., Winter, J., Abney, E. K., Pruss, R. M., Garrilovic, J., and Raff,

M. C. (1980). Myelin-specific proteins and glycolipids in myelin-forming cells in culture: Differences between rat Schwann cells and oligodendrocytes. *J Cell Biol* 84, 483–494.

Morell, P. (ed.). (1977). *Myelin*. New York: Plenum Press.

Mori, S., and Leblond, D. S. (1970). Electronmicroscopic identification of 3 classes of oligodendrocytes and a preliminary study of their proliferative activity in the corpus callosum of young rats. *J Comp Neurol* 139, 1–30.

Morre, D. J., and Ovtracht, C. (1977). Dynamics of the Golgi apparatus—Membrane differentiation and membrane flow. *Int Rev Cytol* [Suppl.] 5, 61–188.

Narang, H. K. (1977). Electron microscopic development of neuroglia in epiretinal portion of post-natal rabbits. *J Neurol Sci* 34, 391–406.

Narang, H. K., and Wisniewski, H. M. (1977). The sequence of myelination in the epiretinal portion of the optic nerve in the rabbit. *Neuropathol Appl Neurobiol* 3, 15–27.

Norton, W. T. (1981a). Biochemistry of myelin. *Adv Neurol* 31, 93–121.

Norton, W. T. (1981b). Formation, structure and biochemistry of myelin. In *Basic neurochemistry*, ed. G. J. Siegel, R. W. Albers, B. W. Agranoff, and R. Katzman, pp. 63–92. 3rd ed. Boston: Little, Brown.

Norton, W. T., and Poduslo, S. E. (1973). Myelination in rat brain: Changes in myelin composition during brain maturation. *J Neurochem* 21, 759–773.

Ochoa, J. (1971). The sural nerve of the human foetus: Electron microscope observations and counts of axons. *J Anat* 108, 231–245.

Omlin, F. X., Webster, H. F., Palkovits, C. G., and Cohen, S. R. (1982). Immunocytochemical localization of basic protein in major dense line regions of central and peripheral myelin. *J Cell Biol* 95, 242–248.

Parry, G. (1978). Membrane assembly and turnover. *Subcell Biochem* 5, 261–326.

Paterson, J. A. (1981). Postnatal development of oligodendrocytes. In *Glial and neuronal cell biology*, ed. E. A. Vidrio and S. Federoff, pp. 83–92. New York: Alan R. Liss.

Peters, A., and Vaughn, J. E. (1970). Morphology and development of the myelin sheath. In *Myelination*, ed. A. N. Davison and A. Peters, pp. 3–79. Springfield, Ill.: Charles C. Thomas.

Peters, A., Palay, S. L., and Webster, H. F. (1976). *The fine structure of the nervous system: The neurons and supporting cells*. Philadelphia: W. B. Saunders.

Peterson, R. G., and Gruener, R. W. (1978). Morphological localization of PNS myelin proteins. *Brain Res* 152, 17–29.

Pfeiffer, S. E., Barbarese, E., and Bhat, S. (1981). Non-coordinate regulation of myelinogenic parameters in primary cultures of dissociated fetal rat brain. *J Neurosci Res* 6, 369–380.

Pfenninger, K. H., and Bunge, R. P. (1974). Freeze-fracturing of nerve growth cones and young fibers. *J Cell Biol* 63, 180–196.

Pieringer, J., Rao, G. S., Mandel, P., and Pieringer, R. A. (1977). The association of the sulphogalactosylglycerolipid of rat brain with myelination. *Biochem J* 166, 421–428.

Politis, M. J., Sternberger, N., Ederle, K., and Spencer, P. S. (1982). Studies on the control of myelinogenesis: IV. Neuronal induction of Schwann cell myelin-specific protein synthesis during nerve fiber regeneration. *J Neurosci* 2, 1252–1266.

Privat, A. (1975). Postnatal gliogenesis in the mammalian brain. *Int Rev Cytol* 40, 281–323.

Privat, A., and Fulcrand, J. (1978). Neuroglia from the subventricular precursor to the mature cell. In *Cell, tissue and organ cultures in neurobiology*, ed. S. Federoff and L. Hertz, pp. 11–37. New York: Academic Press.

Privat, A., Jacque, C., Bourre, J. M., Duponey, P., and Baumann, N. (1979). Absence of the major dense line in myelin of the mutant mouse 'shiverer.' *Neurosci Lett* 12, 107–112.

Quarles, R. H. (1980). Glycoproteins from central and peripheral myelin. In *Myelin: Chemistry and biology*, ed. G. Hashim, pp. 55–77. New York: Alan R. Liss.

Raff, M. C., Miller, R. H., and Noble, M. (1983). A glial progenitor cell that develops in vitro into an astrocyte or an oligodendrocyte depending on culture medium. *Nature* 303, 390–396.

Raine, C. S. (1977). Morphological aspects of myelin and myelination. In *Myelin*, ed. P. Morell, pp. 1–49. New York: Plenum Press.

Raine, C. S. (1981). Neurocellular anatomy. In *Basic neurochemistry*, ed. G. J. Siegel, R. W. Albers, B. W. Agranoff, and R. Katzman, pp. 21–47. Boston: Little, Brown.

Rapaport, R. N., and Benjamins, J. A. (1981). Kinetics of entry of P_0 protein into peripheral nerve myelin. *J Neurochem* 37, 164–171.

Richardson, E. P., Jr. (1982). Myelination in the human central nervous system. In *Histology and histopathology of the nervous system*. Vol. 1, ed. W. Haymaker and R. D. Adams, pp. 146–173. Springfield, Ill.: Charles C. Thomas.

Rogart, R. B., and Ritchie, J. M. (1977). Physiological basis of conduction in myelinated nerve fibers. In *Myelin*, ed. P. Morell, pp. 117–160. New York: Plenum Press.

Roomi, M. W., Ishaque, A., Khan, N. R., and Eylar, E. H. (1978a). Glycoproteins and albumin in peripheral nerve myelin. *J Neurochem* 31, 375–379.

Roomi, M. W., Ishaque, A., Khan, N. R., and Eylar, E. H. (1978b). The P0 protein: The major glycoprotein of peripheral nerve myelin. *Biochim Biophys Acta* 536, 112–121.

Rosenbluth, J. (1976). Intramembranous particle distribution at the node of Ranvier and adjacent axolemma in myelinated axons of the frog brain. *J Neurocytol* 5, 731–745.

Rosenbluth, J. (1980). Peripheral myelin in the mouse mutant shiverer. *J Comp Neurol* 193, 729–740.

Rosenbluth, J. (1981). Freeze-fracture approaches to ionophore localization in normal and myelin deficient nerves. *Adv Neurol* 31, 391–418.

Roussel, G., Delaunoy, J. P., Mandel, P., and Nussbaum, J. L. (1978). Ultrastructural localization study of two Wolfgram proteins in rat brain tissue. *J Neurocytol* 7, 155–163.

Rumsby, M. G. (1978). Organization and structure in central-nerve myelin. *Biochem Soc Trans* 6, 448–462.

Rumsby, M. G., and Crang, A. J. (1977). The myelin sheath—A structural examination. In *The synthesis, assembly and turnover of cell surface components*. Vol. 4, ed. G. Poste and G. L. Nicolson, pp. 247–362. Amsterdam: Elsevier/North-Holland.

Sabatini, D. D., Kreibich, G., Morimoto, T., and Adesnik, M. (1982). Mechanisms for the incorporation of proteins in membranes and organelles. *J Cell Biol* 92, 1–22.

Schlesinger, M. J. (1981). Proteolipids. *Annu Rev Biochem* 50, 193–206.

Schnapp, B., and Mugnaini, E. (1975). The myelin sheath: Electron microscopic studies with thin sections and freeze fracture. In *Golgi Centennial Symposium proceedings*, ed. M. Santini, pp. 209–233. New York: Raven Press.

Schroepfer, G. J., Jr. (1981). Sterol biosynthesis. *Annu Rev Biochem* 50, 585–621.

Schroepfer, G. J., Jr. (1982). Sterol biosynthesis. *Annu Rev Biochem* 51, 555–585.

Sears, T. A. (ed.). (1982). *Neuronal-glial cell interrelationships*. Berlin: Springer-Verlag.

Shaikh, N. A., and Palmer, F. B. St. C. (1976). Deposition of lipids in the developing central and peripheral nervous systems of the chicken. *J Neurochem* 26, 597–603.

Sheltawy, A., and Dawson, R. M. C. (1966). The polyphosphoinositides and other lipids of peripheral nerves. *Biochem J* 100, 12–18.

Siakotos, A. N., Rouser, G., and Fleischer, S. (1969). Phospholipid composition of human, bovine and frog myelin isolated on a large scale from brain and spinal cord. *Lipids* 4, 239–242.

Singer, M. (1968). Penetration of labelled amino acids into the peripheral nerve fibre from surrounding body fluids. In *Ciba Foundation symposium on growth of the nervous system*, ed. G. E. W. Wolstensholme and M. O'Connor, pp. 200–215. London: Churchill.

Singer, S. J., and Nicolson, G. L. (1972). The fluid mosaic model of the structure of cell membranes. *Science* 175, 720–731.

Singh, H., and Spritz, N. (1974). Polypeptide components of myelin from rat peripheral nerve. *Biochim Biophys Acta* 351, 379–386.

Skaper, S. D., Manthrope, M., Adler, R., and Varon, S. (1980). Survival, proliferation and morphological specialization of mouse Schwann cells in a serum-free, fully defined medium. *J Neurocytol* 9, 683–697.

Skoff, R. P. (1976). Myelin deficit in the jimpy mouse may be due to cellular abnormalities in astroglia. *Nature* 264, 560–562.

Skoff, R. P. (1978). The pattern of myelination along the developing rat optic nerve. *Neurosci Lett* 7, 191–196.

Skoff, R. P. (1980). Neuroglia: A reevaluation of their origin and development. *Pathol Res Pract* 168, 279–300.

Skoff, R. P., Price, D., and Stocks, A. (1976a). Electron microscopic autoradiographic studies of gliogenesis in rat optic nerve: I. Cell proliferation. *J Comp Neurol* 169, 291–312.

Skoff, R. P., Price, D. L., and Stocks, A. (1976b). Electron microscopic autoradiographic studies of gliogenesis in rat optic nerve: II. Time of origin. *J Comp Neurol* 169, 313–333.

Skoff, R. P., Skoff, A. M., and Katsnelson, I. (1979). Neonatal endocrine abnormalities in myelin deficient jimpy-tabby mice. *Life Sci* 24, 2099–2104.

Skoff, R. P., Toland, D., and Nast, E. (1980). Pattern of myelination and the distribution of neuroglial cells along the developing optic system of the rat and rabbit. *J Comp Neurol* 191, 237–253.

Smith, K. J., Blakemore, W. F., Murray, J. A., and Patterson, R. C. (1982). Internodal myelin volume and axon surface area. *J Neurol Sci* 55, 231–246.

Spencer, P. S. (1979). Neuronal regulation of myelinating cell function. *Soc Neurosci Symp* 4, 275–321.

Spencer, P. S., and Weinberg, H. J. (1978). Axonal specification of Schwann cell expression and myelination. In *The physiology and pathobiology of axons*, ed. S. G. Waxman, pp. 389–405. New York: Raven Press.

Sprinkle, T. J., Wells, M. R., Garver, F. A., and Smith, D. B. (1980). Studies on the Wolfgram high molecular weight CNS myelin protein: Relationship to 2′,3′-cyclic nucleotide-3′-phosphodiesterase. *J Neurochem* 35, 1200–1208.

Stampfli, R. (1981). Overview of studies on the physiology of conduction in myelinated nerve fibers. *Adv Neurol* 31, 11–23.

Sternberger, N. H., Itoyama, Y., Kies, M. W., and Webster, H. F. (1978a). Immunocytochemical method to identify basic protein in myelin forming oligodendrocytes of newborn rat CNS. *J Neurocytol* 7, 251–263.

Sternberger, N. H., Itoyama, Y., Kies, M. W., and Webster, H. F. (1978b). Myelin basic protein demonstrated immunocytochemically in oligodendroglia prior to myelin sheath formation. *Proc Natl Acad Sci USA* 75, 2521–2524.

Sternberger, N. H., Quarles, R. H., Itoyama, Y., and Webster, H. F. (1979). Myelin-associated glycoprotein demonstrated immunocytochemically in myelin and myelin-forming cells of developing rat. *Proc Natl Acad Sci USA* 76, 1510–1514.

Svennerholm, L., Vanier, M. T., and Jungbjer, B. (1978). Changes in fatty acid composition of human brain myelin lipids during maturation. *J Neurochem* 30, 1383–1390.

Tasaki, I. (1982). *Physiology and electrochemistry of nerve fibers*. New York: Academic Press.

Tennekoon, G. I., Cohen, S. R., Price, D. L., and McKhann, G. M. (1977). Myelinogenesis in optic nerve: A morphological, autoradiographic and biochemical analysis. *J Cell Biol* 72, 604–616.

Trapp, B. D., Itoyama, Y., MacIntosh, T. D., and Quarles, R. H. (1983). P_2 protein in oligodendrocytes and myelin of the rabbit central nervous system. *J Neurochem* 40, 47–54.

Trapp, B. D., Itoyama, Y., Sternberger, N. H., Quarles, R. H., and Webster, H. F. (1981). Immunocytochemical localization of P_0 protein in Golgi complex membranes and myelin of developing rat Schwann cells. *J Cell Biol* 90, 1–6.

Trapp, B. D., McIntyre, L. J., Quarles, R. H., Sternberger, N. H., and Webster, H. F. (1979). Immunocytochemical localization of rat peripheral nervous system myelin proteins: P_2 protein is not a component of all peripheral nervous system myelin sheaths. *Proc Natl Acad Sci USA* 76, 3552–3556.

Trapp, B. D., and Quarles, R. H. (1982). Presence of the myelin-associated glycoprotein correlates with alterations in the periodicity of peripheral nerve. *J Cell Biol* 92, 877–882.

Uyemura, K., Tobari, C., and Hirano, S. (1970). Purification and properties of basic proteins in pig spinal cord and peripheral nerve. *Biochim Biophys Acta* 214, 190–197.

Valat, J., Fulcrand, J., Privat, A., and Marty, R. (1978). Radioautographic study of cell proliferation secondary to Wallerian degeneration in the postnatal rat optic nerve. *Acta Neuropathol* 42, 205–209.

Varon, S., and Manthrope, M. (1982). Schwann cells: An in vitro perspective. In *Advances in cellular neurobiology*. Vol. 3, ed. S. Federoff and L. Hertz, pp. 35–95. New York: Academic Press.

Warren, G. (1981). Membrane proteins: Structure and assembly. In *Membrane structure*, ed. J. B. Finean and R. H. Michell, pp. 215–257. Amsterdam: Elsevier/North-Holland.

Waxman, S. G. (ed.) (1978). *Physiology and pathobiology of axons*. New York: Raven Press.

Waxman, S. G., and Ritchie, J. M. (eds.). (1981). *Advances in neurology*. Vol. 31, *Demyelinating diseases*. New York: Raven Press.

Webster, H. F. (1971). The geometry of peripheral myelin sheaths during their formation and growth in rat sciatic nerves. *J Cell Biol* 48, 348–367.

Webster, H. F. (1975). Development of peripheral myelinated and unmyelinated nerve fibers. In *Peripheral neuropathies*. Vol. 1, ed. P. J. Dyck, P. K. Thomas, and E. H. Lambert, pp. 37–61. Philadelphia: W. B. Saunders.

Webster, H. F., and Sternberger, N. H. (1980). Morphological features of myelin formation. In *Neurological mutations affecting myelination*. INSERM Symposium No. 14, ed. N. Baumann, pp. 73–86. Amsterdam: Elsevier/North-Holland.

Weibel, E. R., and Bolender, R. R. (1973). Stereological techniques for electron microscopic morphometry. In *Principles and technique of electron microscopy*. Vol. 3, ed. M. A. Hayat, pp. 237–296. New York: Van Nostrand Rheinhold.

Weston, J. A. (1970). The migration and differentiation of neural crest cells. *Advances in Morphogenesis* 8, 41–114.

Weston, J. A. (1982). Motile and social behavior of neural crest cells. In *Cell behaviour*, ed. R. Bellairs, A. Curtis, and G. Dunn, pp. 429–469. Cambridge: Cambridge University Press.

White, D. A. (1973). The phospholipid composition of mammalian tissues. In *Form and function of phospholipids*. BBA Library, Vol. 3, ed. G. B. Ansell, J. N. Hawthorne, and R. M. C. Dawson, pp. 441–482. Amsterdam: Elsevier/North-Holland.

Wiggins, R. C. (1982). Myelin development and nutritional insufficiency. *Brain Res* 257, 151–175.

Wiggins, R. C., Benjamins, J. A., and Morell, P. (1975). Appearance of myelin proteins in rat sciatic nerve during development. *Brain Res* 89, 99–106.

Williams, M. A. (1977). Quantitative methods in biology. In *Practical methods in electron microscopy*. Vol. 6, ed. A. Glauert. Amsterdam: Elsevier/North-Holland.

Winter, J., Mirsky, R., and Kadlubowski, M. (1982). Immunocytochemical study of the appearance of P_2, in developing rat peripheral nerve: Comparison with other myelin components. *J Neurocytol* 11, 351–362.

Wolfgram, F. (1966). A new proteolipid fraction of the nervous system: I. Isolation and amino acid analysis. *J Neurochem* 13, 461–470.

Wood, J. G., and Dawson, R. M. C. (1973). A major myelin glycoprotein of sciatic nerve. *J Neurochem* 21, 717–719.

Wood, J. G., and McLaughlin, B. J. (1975). The visualization of concanavalin-A binding sites in the interperiod line of rat sciatic nerve myelin. *J Neurochem* 24, 233–235.

Wood, P. (1976). Separation of functional Schwann cells and neurons from normal peripheral nerve tissue. *Brain Res* 115, 361–375.

Wood, P., Okada, E., and Bunge, R. (1981). The use of networks of dissociated rat forsal root ganglion neurons to induce myelination by oligodendrocytes in culture. *Brain Res* 196, 247–252.

Wrathall, J. R., Rigamonti, D. D., and Kao, C. C. (1981). Cultures enriched in Schwann-like cells from dissociated nerve segments of the adult cat. *Brain Res* 224, 218–223.

Yates, A. J., and Wherrett, J. R. (1974). Changes in the sciatic nerve of the rabbit and its tissue constituents during development. *J Neurochem* 23, 993–1004.

Young, R. W. (1973). The role of the Golgi complex in sulfate metabolism. *J Cell Biol* 57, 175–189.

Yu, Y. T., and Campagnoni, A. T. (1982). In vitro synthesis of the four mouse myelin basic proteins: Evidence for the lack of a metabolic relationship. *J Neurochem* 39, 1559–1568.

Ketone Bodies, Lactate, Glucose: Metabolic Fuels for Developing Brain

Jean Holowach Thurston and Richard E. Hauhart

It is established that ketone bodies, as well as glucose, can serve as oxidizable substrates for brain ATP (adenosine triphosphate) formation (Owen et al., 1967; Hawkins, Williamson, and Krebs, 1971). And recently, lactate has been added to the list of potential cerebral respiratory fuels (Vannucci and Duffy, 1976; Levitsky et al., 1977; Hernández et al., 1980). In this presentation we review some of the literature relating to the fact that at particular stages of development and under special circumstances, one or another of these three substrates assumes a predominant role in brain energy production in animals and human beings. Preliminary observations supporting the role of lactate in brain function during hypoglycemia and describing the ameliorating effect of starvation on anoxic survival in developing mice are also presented.

Ketone Bodies

In normal adult animals and man, glucose is the major metabolic fuel for the brain (McIlwain, 1966). However, when ketone bodies (acetoacetate and β-hydroxybutyrate) are available in plasma, for example, during fasting and in diabetes (Owen et al., 1967; Hawkins, Williamson, and Krebs, 1971; Ruderman et al., 1974), the mature brain is enzymatically equipped to oxidize ketone bodies for energy production. In contrast with adults, suckling mammals may demonstrate the physiological finding of ketonemia, particularly when the fat content of the maternal milk is high. Hawkins, Williamson, and Krebs (1971) were the first to demonstrate ketone body utilization by the brain of suckling rats in vivo. In the rat, the activities of the hepatic enzymes of ketogenesis (Lockwood and Bailey, 1971), levels of ketone bodies in plasma (Page, Krebs, and Williamson, 1971), β-hydroxybutyrate transport in brain (Moore et al., 1976), and activities of enzymes of ketone body utilization in brain (Page, Krebs, and Williamson, 1971; Lockwood and Bailey, 1971; Klee and Sokoloff, 1967; Tildon, Cone, and Cornblath, 1971; Middleton, 1973) exhibit an interesting pattern during development: values are low in the immediate postnatal period, rise rapidly during the suckling period, and decline to low adult levels after weaning. In contrast, levels of plasma ketone bodies, their uptake, and their contribution to cerebral oxygen consumption were small to negligible in developing lambs (Jones et al., 1975), infant baboons (Levitsky et al., 1977), and newborn dogs (Gregoire et al., 1978).

While the ketogenic response to fasting is brisk in human infants and older children (Stanley and Baker, 1976; Haymond, Karl, and Pagliara, 1974), the question of plasma ketone bodies as metabolic fuels during the neonatal period has been controversial. Melichar, Drahota, and Hahn (1967) and Persson and Gentz (1966) reported that after a 24-hour lag, ketone body concentrations were in the 1-mM range during the first week of life. But Stanley et al. (1979) reported that in 44 healthy newborns who were denied food for the first eight hours after birth, plasma ketone body concentration remained low despite high levels of plasma free fatty acids and reduced levels of glucose (below 2.2 mM in about one-third of the infants). Others have reported no change in plasma ketone body concentration in human newborns even after the onset of feeding (Aynsley-Green et al., 1977; Anday et al., 1981).

The newborn rat also has a limited capacity for ketogenesis for a brief period immediately after birth. Despite the striking fall in plasma glucose levels and a rise in plasma free fatty acid concentration in newborn rats that are denied food after birth by cesarean section, plasma β-hydroxybutyrate and acetoacetate concentrations fell to undetectable values and remained low for at least two hours (Snell and Walker, 1973). In the suckling rat, blood ketone body concentration increases slowly three to four hours after birth with a sharp rise between 12 and 16 hours (Ferré et al., 1978). Within 24 hours of feeding, blood ketone body levels increased fourfold to tenfold (Snell and Walker, 1973; Yeh and Zee, 1976).

There is considerable evidence that the liver glycogen content may be a regulator of hepatic ketone body formation. Blood ketone body levels increase when liver glycogen is depleted and decrease when the liver glycogen content is high (for a review, see Söling and Seufert, 1978). The limited capacity for ketogenesis at birth (in animals and in man) emphasizes the importance of adequate hepatic glycogen stores (or other sources of metabolic fuel) in the immediate postnatal period. In most mammalian species there is a drop in plasma glucose concentration in the first few hours after birth. The drop is normally short-lived because of the breakdown of the large supply of liver glycogen that is deposited during the last few days of gestation (Shelley, 1961). (Prolonged hypoglycemia, as observed in the prematurely born human infant, could in part reflect the absence or reduction of the preterm hepatic glycogen deposition.) The importance of the hepatic glycogen reserve in maintaining glucose homeostasis after birth is well illustrated by a study of mice with hereditary deficiency of liver phosphorylase "b" kinase. In these mice the failure of liver glycogen levels to fall after birth was associated with a drop in plasma glucose to hypoglycemic levels (plasma glucose levels did not fall in normal control rats) (Gain, Malthus, and Watts, 1981).

Although hepatic ketogenesis appears limited in newborn humans and animals, the capacity for ketone body utilization by brain is not. Activities of the enzymes necessary for ketone body metabolism have been demonstrated even in fetal rat (Dahlquist, Persson, and Persson, 1972; Dierks-Ventling, 1971) and fetal human brain tissue (Page and Williamson, 1971; Patel et al., 1975). In the isolated perfused brains of human fetuses obtained at weeks 12 to 21 of gestation, the uptake and oxidation of β-hydroxybutyrate to CO_2 exceeded that of glucose (Adam et al., 1975). The findings demonstrate the potential use of ketone bodies by fetal brain as alternative substrates for brain oxidative metabolism.

There is an enhanced potential for ketonemia in pregnancy. In the second trimester of pregnancy (weeks 16 to 20 of gestation), levels of β-hydroxybutyrate and acetoacetate in plasma were double the values found in nonpregnant controls during the first 36 hours of fasting (Felig and Lynch, 1970). During the crucial period of myelination and brain development in postnatal rats, ketone bodies were found to be an excellent source (or even preferred to glucose) for the synthesis of brain acetylcholine (Itoh and Quastel, 1970), dicarboxylic amino acids (DeVivo, Fishman, and Agrawal, 1973), and sterols and fatty acids (DeVivo, Leckie, and Agrawal, 1973; Edmond, 1974; Patel and Owen, 1977). Whether ketone bodies could serve similar functions in developing human brain is an interesting question.

Lactate

In 1971 Oldendorf found that lactate was transported into brain by a carrier distinct from that for glucose and amino acids. In later studies, Cremer, Braun, and Oldendorf (1976) showed that the rate of lactate influx into the brains of suckling rats was twice that of recently weaned animals and ten times higher than in adults. Vannucci and Duffy (1976) were the first to suggest that lactate might serve as a metabolic fuel for developing brain. Following intraperitoneal injection of L-[U-^{14}C] lactate in anoxic neonatal rats, there was a progressive increase in brain radioactivity. During recovery from anoxia, pyruvate levels rose promptly, lactate levels remained high, and brain glucose concentration rose above the control value. Based on these findings, the authors postulated a continued influx of lactate from the blood and oxidation of lactate to pyruvate during the recovery period, thereby sparing brain glucose utilization. Since hypoxia greatly lowers the cerebral metabolic rate (Duffy, Nelson, and Lowry, 1972), other explanations are possible for these findings.

Uptake of lactate, a phenomenon not noted in neonates six and twelve weeks of age, was noted in the brains of baboon neonates that were denied food (Levitsky et al., 1977) and in hypoglycemic newborn dogs (Hernández et al., 1980). In both species, cerebral uptake of lactate was directly related to arterial plasma concentration (Levitsky et al., 1977; Hellmann, Vannucci, and Nardis, 1982). At similar arterial concentrations, the cerebral arteriovenous difference for lactate in hypoglycemic puppies was two to three times the value for euglycemic animals (Hellmann, Vannucci, and Nardis, 1982). Together with the greatly decreased brain : blood lactate concentration ratio, the data were taken as evidence of lactate consumption by brain (Hellmann, Vannucci, and Nardis, 1982). However, brain lactate concentration is often significantly reduced during insulin-induced hypoglycemia because of a decreased rate of cerebral glucose utilization (Ferrendelli and Chang, 1973). In insulin-injected weanling mice, brain lactate levels were reduced to one-fourth of the control value (Thurston, Hauhart, and Dirgo, 1976).

It was not possible to evaluate the clinical response of lactate administration to the hypoglycemic puppies because the animals were anesthetized, paralyzed, and artificially ventilated (Hellmann, Vannucci, and Nardis, 1982). It is also likely that hypoglycemic newborn puppies, like hypoglycemic infant mice (Thurston et al., 1972) and newborn infants, may not exhibit the classical neurological dysfunction seen in adults. The hormonal stress of the surgical procedures (Bloom et al., 1973) could also have an influence on the findings. Therefore, we examined the clinical effects of lactate injection in insulin-induced hypoglycemic, but otherwise normal, weanling mice.

Twelve normal suckling 19-day-old mice were grouped by weight into four groups of three. Two groups received subcutaneous injections of 0.9 percent NaCl. The other two groups received 100 units/kg crystalline zinc insulin. When the animals exhibited signs of hypoglycemic stupor or coma, they were given intraperitoneal injections of glucose, lactate, or normal saline. The mice were color coded on their abdomens so that the observers did not know which hypoglycemic animal had received glucose and which had received lactate or saline. Within ten minutes of injection, glucose- and lactate-treated animals were alert and on their feet with normal or near normal activity. There was no response to treatment with normal saline, as expected.

To determine the mechanism of this surprising result, we examined the effect of injection of lactate or insulin in normal weanling mice and of lactate injection in littermates with insulin-induced hypoglycemia. One litter of normal suckling 19-day-old mice was used. Animals were matched by weight and divided into four groups of three. Two groups received subcutaneous injections of 0.9 percent NaCl. The remaining two groups received crystalline zinc insulin, 100 units/kg. When insulin-injected animals exhibited signs of hypoglycemic stupor (paralysis) or coma (loss of righting reflex) (one to two hours), all of the animals were reinjected. One control group received 0.9 percent NaCl intraperitoneally; the other received sodium-L(+)-lactate 18 mmol/kg (pH adjusted to 7.35 with NaOH). One of the two insulin-treated groups of mice received 0.9 percent NaCl intraperitoneally; the other received 18 mmol/kg of lactate. Animals were killed by decapitation ten minutes after the second series of injections. The head was allowed to fall directly into liquid N_2. Blood was collected from the severed neck vessels in heparinized capillary tubes and centrifuged promptly at 4°C. Frozen brains were dissected in a cryostat at −35°C. Perchlorate extracts of plasma and frozen brain tissues were prepared by the method of Lowry and Passonneau (1972). Glycogen was measured by the method of Passonneau and Lauderdale (1974). Remaining metabolites in plasma and brain were measured by the specific enzymic methods of Lowry and Passonneau (1972). Statistical significance among the groups was determined by a one-way analysis of variance, followed by Duncan's multiple comparison test (Duncan, 1955) to identify statistically significant differences between specific pairs of means.

Ten minutes after lactate injection in normal mice, plasma lactate levels increased almost fivefold; the 19 percent increase in plasma glucose concentration was not statistically significant (table 1). The most outstanding finding in the brains of these animals was the greater than twofold increase in glucose concentration (table 2). Concomitantly, there were 18 percent to 20 percent increases in brain glycogen, glucose-6-phosphate, and fructose-1, 6-diphosphate levels. Brain malate and α-ketoglutarate concentrations increased 30 percent and 32 percent, respectively.

The mechanism of the prompt doubling of brain glucose by lactate injection was not evident. It is unlikely that the small increase in plasma glucose could account for the finding (Mayman, Gatfield, and Breckenridge, 1964). With normal or near normal plasma glucose levels, the

Table 1. ***Effect of Lactate, Insulin, and Insulin-plus-lactate Injections on Plasma Metabolites in Weanling Mice***

	Treatment			
Measurement	*NaCl (N = 3)*	*Lactate (N = 3)*	*Insulin (N = 3)*	*Insulin plus Lactate (N = 3)*
Glucose	10.37 ± 0.90	12.24 ± 1.13	0.86 ± 0.44 *	1.51 ± 0.42
Lactate	4.21 ± 0.03	19.52 ± 1.43 *	2.11 ± 0.07 *	16.24 ± 0.93 †

Note: Data, expressed in millimoles per liter, represent the mean plus or minus the standard error of the mean.

* *p* (versus control) <.01.

† *p* (versus insulin) <.01.

Table 2. ***Effect of Lactate, Insulin, and Insulin-plus-lactate Injections on Brain Metabolites in Weanling Mice***

Measurement	*NaCl (N = 3)*	*Treatment* *Lactate (N = 3)*	*Insulin (N = 3)*	*Insulin plus Lactate (N = 3)*
Glycogen	1.49 ±0.05	1.79 ±0.03*	0.51 ±0.13*	0.60 ±0.08
Glucose	1.10 ±0.31	2.46 ±0.21*	0.06 ±0.02*	0.12 ±0.05
Glucose-6-phosphate	0.085±0.004	0.100±0.002*	0.054±0.004*	0.103±0.012†
Fructose-1, 6-diphosphate	0.119±0.006	0.140±0.004*	0.014±0.004*	0.032±0.005†
Lactate	1.98 ±0.28	4.05 ±0.27*	0.57 ±0.07*	3.18 ±0.06†
Malate	0.515±0.061	0.667±0.029	0.272±0.014*	0.481±0.036†
α-Ketoglutarate	0.057±0.008	0.075±0.001	0.040±0.004	0.074±0.014

Note: Data, measured in millimoles per kilogram, represent the mean plus or minus the standard error of the mean.
* *p* (versus control) <.05.
† *p* (versus insulin) <.05.

most common cause of increased brain glucose concentration is a decrease in cerebral metabolic rate for glucose. When this is the case in young mice, we invariably see a crossover at the phosphofructokinase step; that is, increased glucose-6-phosphate and decreased fructose-1, 6-diphosphate levels (Thurston et al., 1975a, 1975b). However, in lactate-injected animals, levels of both phosphorylated intermediates were increased. Increases in α-ketoglutarate and malate were compatible with an increase in the Krebs citric acid cycle flux (Goldberg, Passonneau, and Lowry, 1966).

After insulin injection, plasma glucose levels were less than 1 mM, and plasma lactate levels were reduced 50 percent (table 1). In the brains of the hypoglycemic animals, glucose levels were near zero (table 2). As previously noted (Gorell, Dolkart, and Ferrendelli, 1976; Thurston, Hauhart, and Dirgo, 1976), there was evidence for inhibition of glycolysis at the phosphofructokinase step (Lowry et al., 1964). The fall in fructose-1,6-diphosphate levels (88 percent) was far greater than the fall in glucose-6-phosphate (36 percent) (table 2). The 70 percent decrease in brain lactate levels following insulin injection reflected the decreased glycolytic flux. Decreases in malate (50 percent) and α-ketoglutarate (30 percent) concentrations support a decrease in the metabolic flux through the Krebs citric acid cycle as well (Goldberg, Passonneau, and Lowry, 1966).

In hypoglycemic mice receiving lactate, plasma lactate levels were lower, but not significantly so, than in lactate-injected normal mice (table 1). Lactate injection doubled the level of plasma glucose, but the absolute increase was not large or statistically significant. Clearly, lactate injection increased the cerebral metabolic flux in the glycolytic and the Krebs citric acid cycle pathways (table 2). Concomitant with the recovery from stupor there was a doubling or near doubling of brain glucose, glucose-6-phosphate, fructose-1,6-diphosphate, and malate concentrations.

The preliminary data lend convincing evidence for the role of lactate as a potential oxidizable fuel for immature brain. The biochemical mechanism of the dramatic recovery from hypoglycemic stupor by lactate injection is presently unknown. Nevertheless, it is clear that the restoration of normal central nervous system function was *not* dependent on normal blood or brain glucose levels. Although lactate treatment doubled the brain glucose content, the absolute value was still extremely low (in some animals, brain glucose concentration after lactate treatment was lower than in saline-treated hypoglycemic littermates). Nor is it likely that any change in the brain energy state because of the apparent increased metabolic flux in the citrate cycle (Goldberg, Passonneau, and Lowry, 1966) contributed to the behavioral recovery. It is well known that ATP and phosphocreatine levels are not affected in insulin hypoglycemia. (In fact, in other insulin-induced hypoglycemic mice of this age, phosphocreatine levels increased almost 40 percent [Thurston, Hauhart, and Dirgo, 1976].) The possibility that lactate-induced changes in brain acetylcholine concentration or levels of the neurotransmitter amino acids may contribute to the dramatic recovery following lactate administration is under investigation.

Glucose

After the immediate neonatal period, suckling rats use both glucose and ketone bodies as oxidizable fuels for the brain. Cremer and Heath (1974) found in the brains of 18-day-old rats that "glucose and ketone bodies . . . . were in every way alternative substrates" (p. 527). At high levels of ketonemia, pyruvate and β-hydroxybutyrate contributed equally to the rate of acetyl CoA formation and oxidation (Cremer and Heath, 1974).

After weaning, glucose again becomes the predominant cerebral metabolic fuel in normal mammals. It is interesting that the switch of metabolic fuels is accompanied by a decline in the activities of enzymes of ketogenesis in liver (Lockwood and Bailey, 1971), ketone body utilization in brain (Klee and Sokoloff, 1967; Page, Krebs, and Williamson, 1971; Lockwood and Bailey, 1971; Tildon, Cone, and Cornblath, 1971; Middleton, 1973), and ketone body transport across the blood-brain barrier (Moore et al., 1976). Concomitantly, the rate of glucose transport into brain is doubled (Moore et al., 1971; Cremer et al., 1979).

Glucose, ketone bodies, lactate, or other potential oxidizable substrates (fatty acids, amino acids) generate ATP for normal brain function by oxidative phosphorylation in the Krebs citric acid cycle. However, at any age, only glucose can support brain energy production (by anaerobic substrate phosphorylation) during anoxia or in the presence of drugs that uncouple oxidative phosphorylation, salicylate in particular.

In 1942, Lutwak-Mann reported that 50 percent of adult rats given a toxic dose of salicylate died unless glucose was added to their drinking water for several days prior to salicylate treatment. Developing mice also die in epileptic status after a single toxic dose of salicylate unless they are pretreated with glucose (Thurston et al., 1970). The problem was that within 15–30 minutes after injection, brain glucose levels were near zero and liver glycogen stores were nearly depleted (table 3); how-

Table 3. *Effect of Salicylate Injection on Selected Metabolites in Brain and Liver of Infant Mice*

Measurement	*Control (N = 15)*	*Salicylate (N = 15)*	*p*
		Brain	
Phosphocreatine	3.15 ± 0.19	2.54 ± 0.07	.009
ATP	2.81 ± 0.06	2.70 ± 0.06	NS
Glucose	1.27 ± 0.08	0.44 ± 0.04	<.001
Glucose-6-phosphate	0.116 ± 0.010	0.099 ± 0.006	NS
Glycogen	1.50 ± 0.08	1.24 ± 0.06	.009
Lactate	1.64 ± 0.15	2.67 ± 0.16	<.001
Pyruvate	0.336 ± 0.031	0.382 ± 0.049	NS
		Liver	
ATP	2.24 ± 0.11	2.10 ± 0.11	NS
Glucose	5.60 ± 0.36	5.58 ± 0.51	NS
Glucose-6-phosphate	0.188 ± 0.01	0.190 ± 0.03	NS
Glycogen	62.8 ± 5.4	13.0 ± 3.1	<.001
Lactate	0.78 ± 0.10	4.37 ± 1.07	.003

Source: Data from Thurston et al. (1970).

Note: Data, measured in millimoles per kilogram, represent the mean plus or minus the standard error of the mean. Three litters of 11- to 21-day-old mice were used. Each litter received a dose of salicylate of 700 mg/kg, 750 mg/kg, or 800 mg/kg intraperitoneally. Control animals were given 0.9 percent NaCl. Brain and liver samples were obtained 30 minutes after injection. In this dosage range the biochemical findings were similar; therefore, results were pooled. Statistical significance of the difference between the means of control and experimental values was determined by Student's *t* test. ATP, adenosine triphosphate; NS, not significant ($p > .05$).

Table 4. *Plasma Glucose, Lactate, and β-Hydroxybutyrate Levels after Salicylate Injection in Infant Mice*

	Treatment	
Measurement	*NaCl* *(N = 5)*	*Salicylate* *(N = 6)*
Glucose	6.28 ± 0.46	5.69 ± 0.47
Lactate	3.12 ± 1.93	4.68 ± 1.28
β-Hydroxybutyrate[a]	0.90 ± 0.11	1.74 ± 0.13 *

Source: Data from Thurston et al. (1970; unpublished studies).

Note: Data, expressed in millimoles per liter, represent the mean plus or minus the standard error of the mean. Eleven-day-old suckling littermate mice received 700 mg/kg of salicylate intraperitoneally or 0.9 percent NaCl. Animals were killed 30 minutes after injection. Statistical significance was measured by Student's *t* test.

[a] Plasma β-hydroxybutyrate levels were measured in three controls and three mice injected with salicylate by the method of Williamson, Mellanby, and Krebs (1962).

* p (versus control) = .01.

ever, plasma glucose levels were not significantly different from those of controls, and plasma β-hydroxybutyrate concentration had almost doubled (table 4).

Levels of salicylate in the brains of these animals were in the range known to uncouple oxidative phosphorylation 50 percent to 100 percent in brain homogenates and mitochondria (Brody, 1956; Penniall, Kalnitsky, and Routh, 1956; Penniall, 1958). Therefore, we postulated that the drop in brain glucose reflected increased glycolysis as a result of this phenomenon. Since plasma glucose levels were still in the normal range, apparently the rate of cerebral glucose utilization exceeded the rate of glucose transport from the blood (Pardridge, 1981).

Concurrent administration of glucose with salicylate elevated brain glucose concentration and was associated with striking improvement in the condition of the animals and increased their survival (Thurston et al., 1970). The study producing these results involved two litters of two-week-old mice. The dose of salicylate was 750 mg/kg intraperitoneally. In four animals salicylate was dissolved in 0.5 M NaCL; in five the diluent was 1.0 M glucose. Two other controls were included—four mice received 0.5 M NaCl, and an equal number were injected with 1.0 M glucose. The volume of injections was 30 ml/kg. Animals were killed 15 minutes after injection. Brain glucose measured (mean in mmol/kg plus or minus the standard error of the mean) 1.09 ± 0.07 in mice receiving saline alone, 3.37 ± 0.19 in mice receiving glucose alone, 0.41 ± 0.05 in mice receiving salicylate plus saline, and 1.84 ± 0.22 in those injected with salicylate plus glucose. All values differ significantly from each other ($p < .05$) as determined by a one-way analysis of variance followed by Duncan's multiple comparison test (Duncan, 1955) to identify statistically significant differences between pairs of means. Whether the failure of the brisk salicylate-induced ketogenic response to meet the increased cerebral demands for energy was due to transport limitation or to uncoupling of the link between oxidation and phosphorylation cannot be said. (We did not test the possible beneficial effect of ketone injection on survival of salicylate-treated mice.)

The obligatory role of glucose for brain energy metabolism can also be demonstrated by anoxia. Glucose administration before or at the time of anoxic exposure of newborn and developing mice had a dramatic effect on survival (Holowach-Thurston, Hauhart, and Jones, 1974). Survival after five minutes of anoxic exposure was only 5 percent in control mice eight to twelve days old compared with 82 percent in glucose-treated littermates. Seven litters totaling 78 normal suckling eight- to twelve-day-old mice were used. One half of each litter received 30 ml/kg of 1.0 M glucose subcutaneously; the remainder were injected with an equivalent volume of 0.5 M NaCl. The animals were exposed at 24°C to N_2 gas (O_2 pressure < 5 mm Hg) 15 to 104 minutes after glucose injection (mean 60.7 ± 4.1 minutes). Only one control and one glucose-treated animal were observed at a time. The interval of anoxia ranged from three to seven minutes.

Similar results were seen in newborn animals (Holowach-Thurston, Hauhart, and Jones, 1974). Four litters ($N = 42$) of newborn mice (24–36 hours old) were used. One half of each litter received 1.0 M glucose (30 mmols/kg subcutaneously); the remaining mice were injected with 0.5 M NaCl. The animals were exposed to N_2 gas immediately after injection. The interval of anoxia ranged from 25 to 43 minutes. All 11 glucose-treated mice were alive approximately 40 minutes (mean 39.2 ± 1.1 minutes) after anoxic exposure at 24°C compared with 4 (40 percent) of the ten controls. After 25 minutes of anoxia at 34°C, only 1 (9 percent) of the 11 controls was alive, but all ten of the glucose-treated mice were alive. Even after decapitation, there was a delay in the onset of terminal gasping of the isolated heads of glucose-treated mice, and the last gasp, which signals death of the animal in vivo, occurred much later (table 5). The second finding suggests that glucose had a direct effect in prolonging the primitive respiratory mechanism of the ischemic brain (Selle, 1942; Selle, 1944). The beneficial effect of glucose in anoxic mice was not related to its osmotic activity; mannitol in equal dosage had no effect.

One explanation for the dramatic effect of glucose on anoxic survival was the threefold increase in brain glucose concentration (fig. 1). Levels of ATP, phosphocreatine, and glycogen were unchanged. Glucose administration had no effect on the rate of cerebral glycolysis (Holowach-Thurston, Hauhart, and Jones, 1974), but the continuing supply of glucose for energy use during anoxia had a significant sparing effect on

Table 5. *Effect of Glucose on Gasping of the Isolated Heads of Infant Mice*

Clinical Observation	*Control* (N = 7)	*Glucose-treated* (N = 11)	*p*
Onset of terminal gasping	2.45 ± 0.28	3.50 ± 0.27	.03
Time of last gasp	4.17 ± 0.53	6.72 ± 0.28	<.001

Source: Data from Holowach-Thurston, Hauhart, and Jones (1974).

Note: Data, measured in minutes, represent the mean plus or minus the standard error of the mean. Two litters of ten-day-old mice were used. Animals were treated as were the seven litters of eight- to 12-day-old mice described in the text. Mice were killed by decapitation one hour after glucose or saline injection. The isolated head was observed at 25°C. Statistical significance was determined by Student's *t* test.

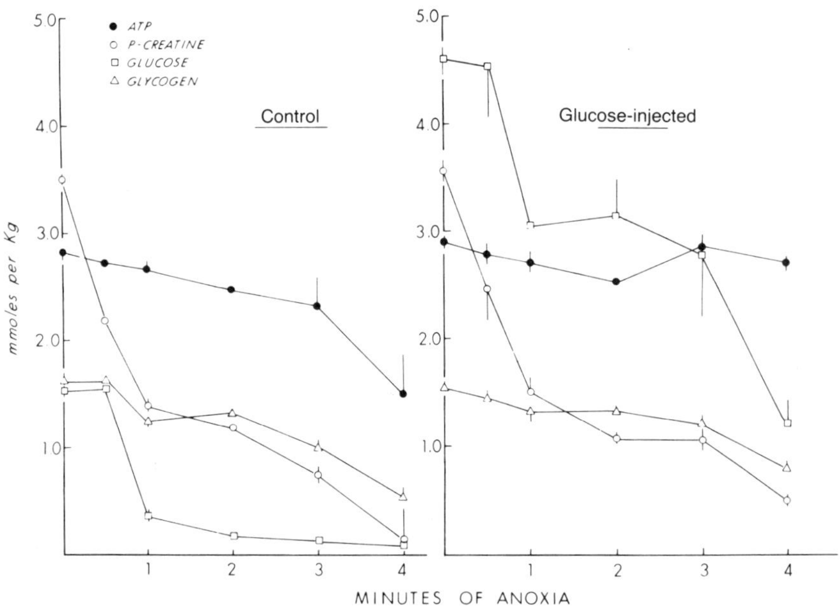

Figure 1. Effect of glucose administration on concentrations of ATP, phosphate-creatine, glucose, and glycogen in the brains of ten-day-old mice at zero time and at intervals after exposure to an atmosphere of 100 percent N_2 gas at 24°C. Each point is the mean value for two to four mice ($N = 42$). Vertical lines represent ± 1 standard error of the mean (three to four samples). Animals were injected with either 30 ml/kg of 0.5 M NaCl (*left*) or 30 ml/kg of 1 M glucose (*right*) one hour before being killed. (Data from Holowach-Thurston, Hauhart, and Jones, 1974.)

the use of the other energy-yielding metabolites (table 6). After four minutes of anoxia, ATP levels were 83 percent higher than in anoxic controls; glycogen levels, 54 percent higher. While the brain glucose concentration in controls was near zero, levels in the glucose-treated mice were still at the preanoxia control value. In terms of anaerobic high-energy phosphate equivalents, the brain energy reserve in glucose-treated animals was now twice as high as in the controls (table 6).

The effect of glucose on anoxic survival in suckling mice appears to contradict the findings of Myers and Yamaguchi (1976) in juvenile monkeys and of Siemkowicz and Hansen (1978) and Siemkowicz and Gjedde (1980) in adult rats. In their studies, glucose treatment prior to the anoxic insult increased mortality and the degree of brain injury in survivors. We believe that the vastly different experimental procedures and the older age of their animals contribute to this discrepancy. In the older animals, cerebral ischemia was produced by occluding the ascending aorta, arresting the heart or by inflating a cup around the neck and dropping the blood pressure. With a lack of blood flow to and from the brain, lactate levels in ischemic monkey brain increased almost threefold after ten minutes (32 mmol/kg compared with 12 mmol/kg in controls). Juvenile monkeys are believed to be two to four years old and not fully grown (Myers, 1973); the adult rats weighed 339 g to 420 g. Therefore, it is likely that the cerebral glycolytic rate (hence lactate production) in these animals was considerably greater than in newborn and infant mice (Thurston and McDougal, 1969; Lowry et al., 1964). With an intact circulation, brain lactate levels in glucose-treated ten-day-old

Table 6. ***Effect of Four Minutes of Anoxia on the Brain Energy Reserve in Infant Mice***

Measurement	*Control (N = 4)*	*Glucose-injected (N = 4)*	*p*
Phosphocreatine	0.15 ± 0.14	0.50 ± 0.05	NS
ATP	1.48 ± 0.39	2.71 ± 0.07	.04
Glucose	0.07 ± 0.01	1.21 ± 0.22	.005
Glycogen	0.52 ± 0.09	0.80 ± 0.03	.05
~ P[a]	4.75 ± 0.19	10.70 ± 0.72	.009

Source: Data from Holowach-Thurston, Hauhart, and Jones (1974).

Note: Data, measured in millimoles per kilogram, represent the mean plus or minus the standard error of the mean. The data are from the study reported in figure 1. Statistical significance measured by Student's *t* test. ATP, adenosine triphosphate; NS, not significant ($p > .05$).

[a]Calculated as described by Lowry et al. (1964): ~ P = (phosphocreatine) + 2 (ATP + glucose) + 2.9 (glycogen).

mice were only 30 percent higher than in control mice after four minutes of complete anoxia (Holowach-Thurston, Hauhart, and Jones, 1974).

There is another possible explanation for the apparent inconsistency of our findings and those reported by Myers and Yamaguchi (1976), Siemkowicz and Hansen (1978), and Siemkowicz and Gjedde (1980). Prior to cerebral ischemia, the monkeys and rats were denied food for 24 hours. In preliminary experiments, we found that 24 hours of fasting produced metabolic changes consistent with a reduced rate of cerebral glucose utilization in seven-to ten-day-old mice (table 7) and increased the brain energy reserve and the energy charge potential (table 8). In plasma, glucose and lactate levels were reduced and β-hydroxybutyrate levels were doubled (table 7). The high energy state of the starved brain had a significant effect on anoxic survival. In starved mice, the time of survival in an atmosphere of pure nitrogen gas was 86 percent longer than in unstarved littermates. It is also noteworthy that administration

Table 7. ***Effect of 24-hour Starvation on Selected Metabolite Levels in Plasma and Brains of Infant Mice***

Measurement	*Control* (N = 6)	*Starved* (N = 6)	*p*
		Plasma	
Glucose	6.56 ± 0.11	4.31 ± 0.14	<.001
β-Hydroxybutyrate	1.11 ± 0.07	2.07 ± 0.08	<.001
Lactate	2.62 ± 0.16	1.59 ± 0.14	<.001
		Brain	
Glycogen	1.73 ± 0.06	1.83 ± 0.08	NS
Glucose	0.60 ± 0.06	0.78 ± 0.07	NS
Glucose-6-phosphate	0.138 ± 0.005	0.165 ± 0.006	.004
Fructose-1,6-diphosphate	0.043 ± 0.003	0.023 ± 0.001	<.001
Pyruvate	0.059 ± 0.005	0.059 ± 0.007	NS
Lactate	1.80 ± 0.12	1.89 ± 0.23	NS
Malate	0.342 ± 0.009	0.308 ± 0.010	.032

Note: Data for plasma metabolites, expressed in millimoles per liter, and data for brain metabolites, measured in millimoles per kilogram, represent the mean plus or minus the standard error of the mean. One litter of normal suckling ten-day-old mice was used. Animals were paired by weight. One-half of the mice were removed from their mothers and placed in a separate cage at room temperature without food or water. Control and starved animals were killed (by decapitation) after 24 hours. Statistical significance was determined by Student's *t* test. NS, not significant ($p > .050$).

Table 8. ***Effect of 24-hour Starvation on Brain Energy Metabolism in Infant Mice***

Measurement	*Control (N = 6)*	*Starved (N = 6)*	*p*
ATP	2.41 ± 0.01	2.70 ± 0.01	<.001
Phosphocreatine	3.13 ± 0.05	3.89 ± 0.02	<.001
ADP	0.398 ± 0.016	0.348 ± 0.005	.013
AMP	0.027 ± 0.001	0.027 ± 0.001	NS
ATP/ADP	6.12 ± 0.24	7.77 ± 0.14	<.001
~P[a]	14.18 ± 0.30	16.14 ± 0.32	.001
ECP[b]	0.921 ± 0.003	0.935 ± 0.001	.001

Note: Data, measured in millimoles per kilogram, represent the mean plus or minus the standard error of the mean. Animals are those reported in table 7. Statistical significance was determined by Student's *t* test. ATP, adenosine triphosphate; ADP, adenosine diphosphate; AMP, adenosine monophosphate; NS, not significant.

[a]Calculated as described in table 6.

[b]The energy charge potential (ECP) was calculated by the formula of Atkinson (1968): ECP = (ATP + 0.5 ADP)/(ATP + ADP + AMP).

of glucose to starved mice prior to anoxic exposure increased anoxic survival compared with that of saline-injected starved littermates. The study producing these results used three litters of eight- to ten-day-old mice. Animals were treated as described in table 7. Starved mice received 30 ml/kg of 1 M glucose subcutaneously 15–20 minutes prior to anoxic exposure (100 percent N_2 gas at 24°C); controls received an equivalent volume of 0.9 percent NaCl. The time to the last gasp for the ten controls (reported in mean minutes plus or minus the standard error of the mean) was 3.12 ± 0.35, 5.83 ± 0.59 for the 14 mice starved 24 hours, and 9.32 ± 0.98 for the four mice starved 24 hours but injected with glucose.

Each mean is significantly different from the other two means ($p < .01$) as determined by a one-way analysis of variance followed by Duncan's multiple comparison test (Duncan, 1955) to identify statistically significant differences between specific pairs of means. These findings offer a biochemical explanation for increased anoxic survival of starved animals and stress the value of glucose in prolonging survival of fed, and starved, developing mice in a model of nonischemic anoxia. In view of the low rate of cerebral glucose utilization in immature brain (Lowry et al., 1964; Thurston and McDougal, 1969), it is also unlikely that sufficient lactate would accumulate to increase anoxic brain damage.

It is relevant that similar effects of starvation were reported by DeVivo, Malas, and Leckie (1975) in adult rats. As in our young mice, biochemical correlates of fasting were mild hypoglycemia, ketonemia, and a rise in the brain phosphorylation potential and the energy charge potential. In animals starved for longer than 24 hours, there was a significant elevation of the electroconvulsive threshold. Thus, starvation may exert a beneficial effect in prolonging anoxic survival or ameliorating hypoxic-ischemic brain damage in adult as well as in developing animals.

The Developmental Pattern of Selected Enzymes of Brain Carbohydrate Metabolism

In contrast with the progressive rise during development and the prompt fall of the high levels of activities of the cerebral enzymes of ketone body utilization following weaning (fig. 2), there are near linear increases of the activities of the key enzymes of brain glucose utilization from birth, through weaning, and into adulthood. In homogenates of rat cerebral cortex, activities of hexokinase, phosphofructokinase, and pyruvate kinase doubled from age five days to 20 days and again from age 20 days to 40 days (fig. 3). Lowry et al. (1964) measured the activities of six

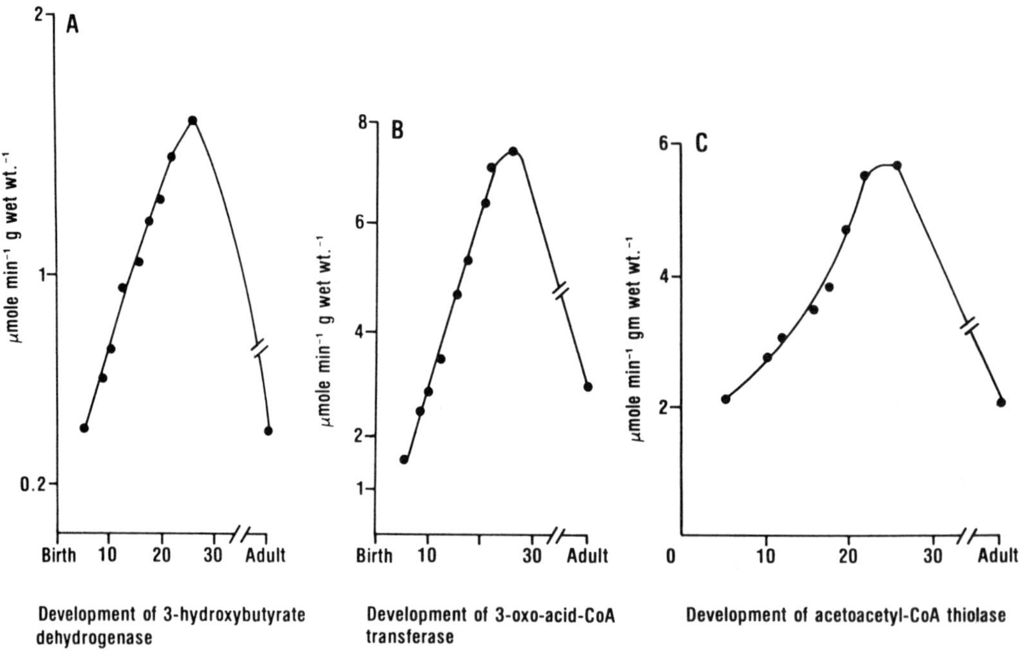

Figure 2. Developmental pattern of (*A*) 3-hydroxybutyrate dehydrogenase, (*B*) 3-oxo-acid-CoA transferase, and (*C*) acetoacetyl-CoA thiolase in purified brain mitochondria prepared from at least eight rats. (Data from Booth, Patel, and Clark, 1980; copyright © 1980 by the International Society for Neurochemistry.)

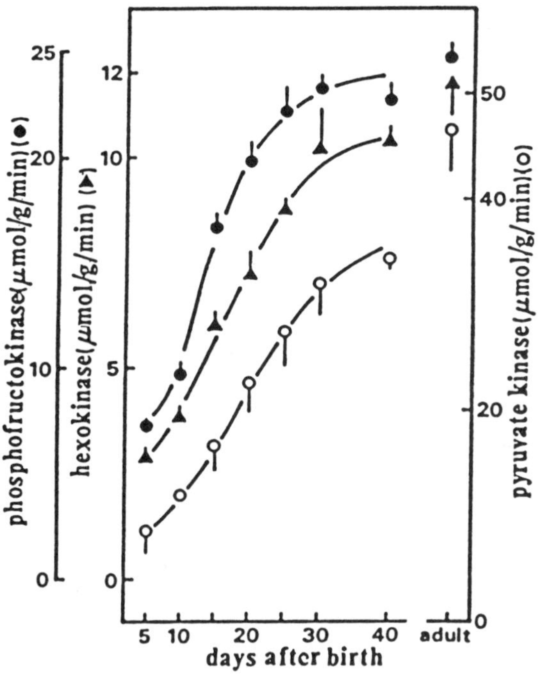

Figure 3. Developmental changes in the activities of hexokinase, phosphofructokinase, and pyruvate kinase in homogenates of rat cerebral cortex. Each value is the mean plus or minus the standard deviation of six experiments. *Solid circles,* phosphofructokinase activity; *solid triangles,* hexokinase activity; *open circles,* pyruvate kinase activity. (Data from Takagaki, 1974; reprinted with permission of Pergamon Press, Ltd.)

other enzymes of glycolysis (phosphoglucoisomerase, aldolase, glyceraldehyde-phosphate dehydrogenase, phosphoglycerate kinase, phosphoglycerate mutase, enolase) in the brains of ten-day-old and adult mice. Although the overall increases during development were smaller, the activities of the six enzymes were 43 percent to 126 percent higher in adult brain. The developmental pattern for brain lactate dehydrogenase is less clear. In one study, the enzyme showed almost a fourfold linear increase in activity in rat brain homogenate between five and 20 days after birth (at which time enzyme activity was 93 percent of the adult value) (Booth, Patel, and Clark, 1980). But in another, there was relatively little difference in the level of lactate dehydrogenase activity in brain homogenates between ten-day-old and adult mice (only 8 percent higher in the latter) (Lowry and Passonneau, 1964). In any event, the

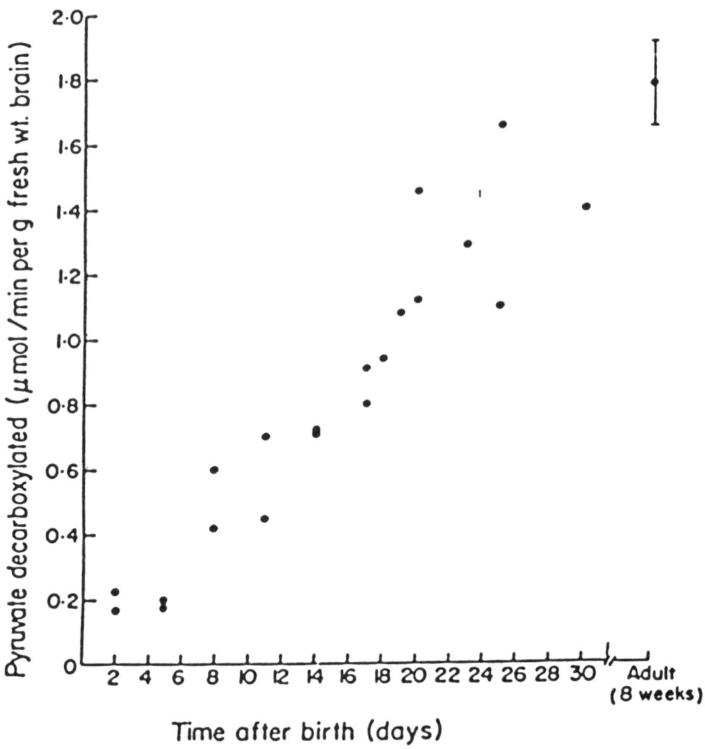

Figure 4. Total activity of pyruvate dehydrogenase in rat brain homogenate during development. All the values are from individual rats except the adult value, which is the mean of values from five animals plus or minus the standard error of the mean. (Data from Cremer and Teal, 1974.)

levels of lactate dehydrogenase activity are so high—five to ten times that of hexokinase (Lowry and Passonneau, 1964; Booth, Patel, and Clark, 1980)—that it seems unlikely that they would be limiting.

The activity of pyruvate dehydrogenase, the prime enzyme in the Krebs citric acid cycle, shows striking alterations during growth and development. Total pyruvate dehydrogenase activity in rat brain homogenate increased fivefold or sixfold from birth to 30 days (fig. 4).

In addition to the use of glucose as metabolic fuel through the Embden-Meyerhof and Krebs citric acid pathways, glucose is a precursor of substrates for the pentose phosphate pathway. It is generally believed that a major function of the pathway during development is to provide sufficient reducing equivalents (nicotinamide-adenine dinucleotide phosphate) for lipid synthesis. The data of Kauffman et al. (1969) suggest that both glucose-6-phosphate dehydrogenase and 6-phosphate-gluconate dehydrogenase may participate in control of the flux through this pathway. We have determined the activities of these two enzymes in the brains of developing mice at five intervals from four to 35 days after birth. Glucose-6-phosphate-dehydrogenase activity increased linearly from 1.36 ± 0.06 μmol $g^{-1}min^{-1}$ (24°C) (N = 4) at four days to 2.42 ± 0.04 μmol $g^{-1}min^{-1}$ (N = 4) at 35 days (considerably less than the many-fold increases in activities of most of the glycolytic enzymes). During the same time span the activity of 6-phosphate-gluconate dehydrogenase fell from 0.447 ± 0.006 μmol $g^{-1}min^{-1}$ (N = 4) to 0.330 ± 0.003 μmol $g^{-1}min^{-1}$ (N = 4). In rat brain homogenates, Kuhlman and Lowry (1956) found an 11 percent drop in glucose-6-phosphate dehydrogenase activity from the newborn to adult. The relatively low level of activity of these two enzymes in the brains of infant mice was somewhat surprising considering the presumed prominent role of the pentose "shunt" during myelination.

Glucose is also the major source of glycerol-phosphate for brain phospholipid synthesis. (The blood-brain barrier is impermeable to glycerol [Thurston, Hauhart, and Dirgo, 1981].) Laatsch (1962) was the first to demonstrate the age dependency of glycerophosphate dehydrogenase activity in whole rat forebrain and spinal cord. We have studied the developmental pattern of the activity of glycerophosphate dehydrogenase in regions of mouse brain (fig. 5). In the first week of life values were low, and there were no regional differences. In the second week, glycerophosphate activity increased sixfold in medulla and cerebellum. Enzyme activity in cerebellum continued to rise, reaching a peak value at about 20 days. Maximum enzyme activity in medulla (also reached at about 20 days) was about one-third less than in cerebellum. Glycerophosphate dehydrogenase activity in the cerebral hemisphere was one-half to one-third of the levels in medulla and cerebellum, respectively, and did not show the sharp developmental increases of activity seen in medulla and cerebellum. Nor did it show the decrease from a peak at 20 to 30 days of age to adulthood—50 percent in medulla, 30 percent in spinal cord (Laatsch, 1962), and 20 percent in cerebellum.

On a quantitative basis, glycogen is an important emergency source of energy for brain. During anoxia the glycogen molecule (as glucose equivalents) generates almost three times as much high-energy phosphate (~P) as phosphocreatine and one-third more than ATP or glucose (Lowry et al., 1964). Yet it is curious that when the blood supply to the brain is cut off, there is a lag in the use of glycogen. This is most apparent in younger animals. While glycogen had virtually disappeared after one to two minutes of ischemia in adult mouse brain, brain glycogen was not depleted until after ten minutes in ten-day old animals (Lowry et al., 1964). In newborn mice, one-half of the brain glycogen reserve was

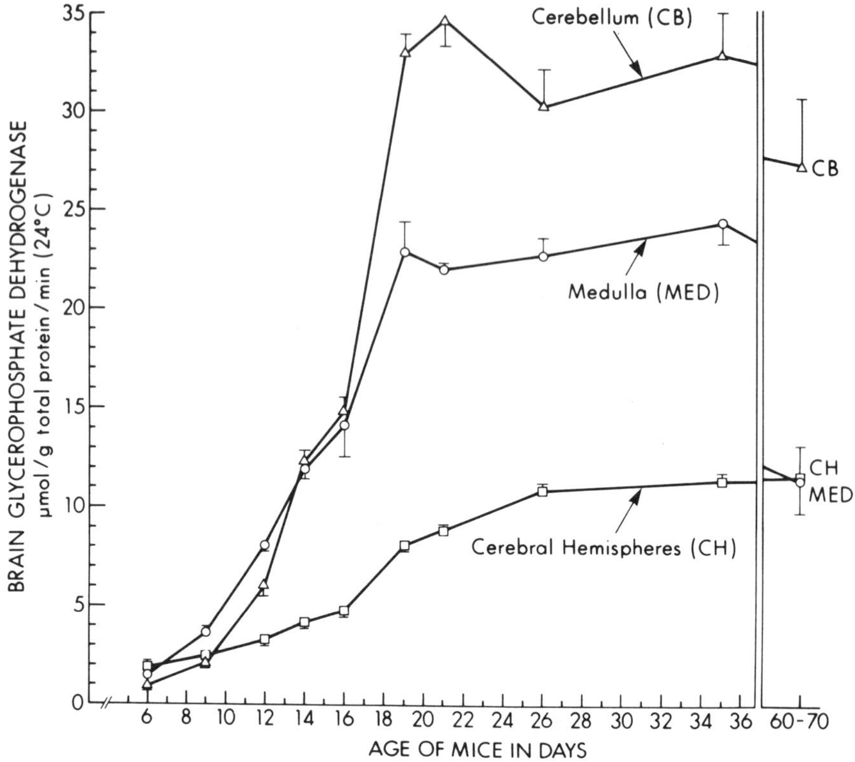

Figure 5. Age-dependent changes in glycerophosphate dehydrogenase activity in homogenates of regions of mouse brain. Glycerophosphate dehydrogenase activity was measured by an adaptation (oxidation of glycerol-1-phosphate in the presence of NAD^+) of the fluorometric procedure of Lowry et al. (1964). Each value is the mean plus or minus the standard error of the mean of levels in four mice.

still present after ten minutes of ischemia (Thurston and McDougal, 1969). The reason for this behavior is not clear. It is not due to exceptionally low levels of phosphorylase activity in immature brain. We have measured total phosphorylase activity in newborn mouse brain. Values in animals less than 24 hours old were more than one-half of those found in adult mouse brain measured under similar conditions (1.70 ± 0.07 μmol $g^{-1}min^{-1}$ [25°C] [N = 10] compared with an adult value of 3.22 μmol $g^{-1}min^{-1}$ [Breckenridge and Norman, 1962]). Nor is the delayed mobilization of brain glycogen in newborn and infant mice during anoxia owed to delayed conversion of phosphorylase "b" (enzyme activity only in the presence of AMP) to phosphorylase "a" (enzyme activity without the addition of AMP). Within two minutes of decapitation, 50 percent of the phosphorylase in newborn mouse brain was converted to the active "a" form (fig. 6).

We could find no studies of glycogen synthetase activity in cerebral tissue early in development. In normal 20-day-old mice total brain glycogen synthetase activity (measured at 37°C in the presence of glucose-6-phosphate) was 0.160 ± 0.009 μmol $g^{-1}min^{-1}$ ($N = 7$) (Thurston et al., 1980). In adult rat brain total glycogen synthetase activity measured by a

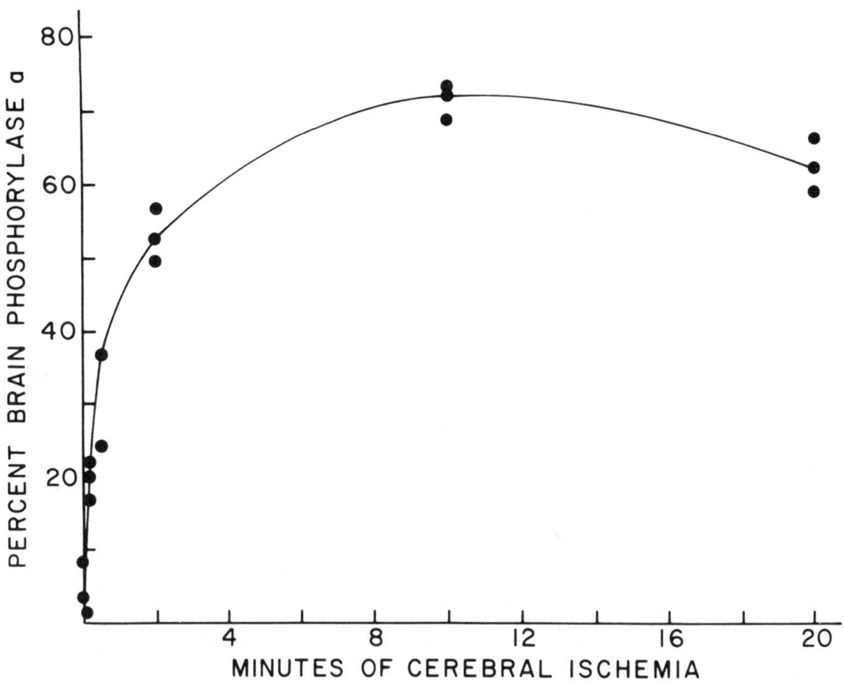

Figure 6. The time course of the conversion of brain glycogen phosphorylase "b" (activity in the presence of AMP) to phosphorylase "a" (activity without AMP) following cerebral ischemia. Enzyme activity was measured in the direction of glycogen breakdown by the fluorometric procedure of Lowry, Schulz, and Passonneau (1967). Animals at zero minutes were frozen whole in liquid N_2. Other animals were decapitated; the isolated heads were kept at 24°C for selected intervals prior to freezing in liquid N_2. Brains were dissected in a cryostat at −35°C, powdered and weighed at −22°C, and homogenates prepared at 4°C. Each value is from one animal.

similar procedure was 0.355 ± 0.027 μmol $g^{-1}min^{-1}$ (N = 8) (Goldberg and O'Toole, 1969). Clearly, at any age, the brain is primarily a glycolytic, not gluconeogenic, organ.

Conclusion

Knowledge of the details of cerebral energy metabolism in developing mouse brain has allowed an effective intervention in some pathological states in infants and young children (Thurston et al., 1970; DeVivo, Keating, and Haymond, 1975; Bray and Gardiner, 1977; Haymond et al., 1978). It is expected that further studies along the same lines will prove equally useful in clinical practice.

Acknowledgments

We thank David B. McDougal, Jr., for critical review of this manuscript and for many helpful discussions. This work was supported in part by U.S. Public Health Service grants NS 06163 and NS 15660 and the Allen P. and Josephine B. Green Foundation of Mexico, Missouri.

References

Adam, P. A. J., Räihä, N., Rahiala, E., and Kekomäki, M. (1975). Oxidation of glucose and D-β-OH-butyrate by the early human fetal brain. *Acta Paediatr Scand* 64, 17–24.

Anday, E. K., Stanley, C. A., Baker, L., and Delivoria-Papadopoulos, M. (1981). Plasma ketones in newborn infants: Absence of suckling ketosis. *J Pediatr* 98, 628–630.

Atkinson, D. E. (1968). The energy charge of the adenylate pool as a regulatory parameter: Interaction with feedback modifiers. *Biochemistry* 7, 4030–4034.

Aynsley-Green, A., Bloom, S. R., Williamson, D. H., and Turner, R. C. (1977). Endocrine and metabolic response in the human newborn to first feed of breast milk. *Arch Dis Child* 52, 291–295.

Bloom, S. R., Daniel, P. M., Johnston, D. I., Ogawa, O., and Pratt, O. E. (1973). Release of glucagon, induced by stress. *Q J Exp Physiol* 58, 99–108.

Booth, R. F. G., Patel, T. B., and Clark, J. B. (1980). The development of enzymes of energy metabolism in the brain of a precocial (guinea pig) and non-precocial (rat) species. *J Neurochem* 34, 17–25.

Bray, P. F., and Gardiner, A. Y. (1977). Salicylism and severe brain edema. *N Engl J Med* 297, 1235.

Breckenridge, B. M., and Norman, J. H. (1962). Glycogen phosphorylase in brain. *J Neurochem* 9, 383–392.

Brody, T. M. (1956). Action of sodium salicylate and related compounds on tissue metabolism in vitro. *J Pharmacol* 117, 39–51.

Cremer, J. E., Braun, L. D., and Oldendorf, W. H. (1976). Changes during development in transport processes of the blood-brain barrier. *Biochim Biophys Acta* 448, 633–637.

Cremer, J. E., Cunningham, V. J., Pardridge, W. M., Braun, L. D., and Oldendorf, W. H. (1979). Kinetics of blood-brain barrier transport of pyruvate, lactate and glucose in suckling, weanling and adult rats. *J Neurochem* 33, 439–445.

Cremer, J. E., and Heath, D. F. (1974). The estimation of rates of utilization of glucose and ketone bodies in the brain of the suckling rat using compartmental analysis of isotopic data. *Biochem J* 142, 527–544.

Cremer, J. E., and Teal, H. M. (1974). The activity of pyruvate dehydrogenase in rat brain during postnatal development. *FEBS Lett* 39, 17–20.

Dahlquist, G., Persson, U., and Persson, B. (1972). The activity of D-β-hydroxybutyrate dehydrogenase in fetal, infant and adult rat brain and the influence of starvation. *Biol Neonate* 20, 40–50.

DeVivo, D. C., Fishman, M. A., and Agrawal, H. C. (1973). Preferential labeling of brain cholesterol by [3-^{14}C]D(-)-3-hydroxybutyrate. *Lipids* 8, 649–651.

DeVivo, D. C., Keating, J. P., and Haymond, M. W. (1975). Reye syndrome: Results of intensive supportive care. *Pediatrics* 87, 875–880.

DeVivo, D. C., Leckie, M. P., and Agrawal, H. C. (1973). The differential incorporation of β-hydroxybutyrate and glucose into brain glutamate in the newborn rat. *Brain Res* 55, 485–490.

DeVivo, D. C., Malas, K. L., and Leckie, M. P. (1975). Starvation and seizures: Observations on the electroconvulsive threshold and cerebral metabolism of the starved adult rat. *Arch Neurol* 32, 755–760.

Dierks-Ventling, C. (1971). Prenatal induction of ketone-body enzymes in the rat. *Biol Neonate* 19, 426–433.

Duffy, T. E., Nelson, S. R., and Lowry, O. H. (1972). Cerebral carbohydrate metabolism during acute hypoxia and recovery. *J Neurochem* 19, 959–977.

Duncan, D. B. (1955). Multiple range and multiple *F* tests. *Biometrics* 11, 1–42.

Edmond, J. (1974). Ketone bodies as precursors of sterols and fatty acids in the developing rat. *J Biol Chem* 249, 72–80.

Felig, P., and Lynch, V. (1970). Starvation in human pregnancy: Hypoglycemia, hypoinsulinemia, and hyperketonemia. *Science* 170, 990–992.

Ferré, P., Pégorier, J., Williamson, D. H., and Girard, J. R. (1978). The development of ketogenesis at birth in the rat. *Biochem J* 176, 759–765.

Ferrendelli, J. A., and Chang, M. M. (1973). Brain metabolism during hypoglycemia. *Arch Neurol* 28, 173–177.

Gain, K. R., Malthus, R., and Watts, C. (1981). Glucose homeostasis during the perinatal period in normal rats and rats with a glycogen storage disorder. *J Clin Invest* 67, 1569–1573.

Goldberg, N. D., and O'Toole, A. G. (1969). The properties of glycogen synthetase and regulation of glycogen biosynthesis in rat brain. *J Biol Chem* 244, 3053–3061.

Goldberg, N. D., Passonneau, J. V., and Lowry, O. H. (1966). Effects of changes in brain metabolism on the levels of citric acid cycle intermediates. *J Biol Chem* 241, 3997–4003.

Gorell, J. M., Dolkart, P. H., and Ferrendelli, J. A. (1976). Regional levels of glucose, amino acids, high energy phosphates, and cyclic nucleotides in the central nervous system during hypoglycemic stupor and behavioral recovery. *J Neurochem* 27, 1043–1049.

Gregoire, N. M., Gjedde, A., Plum, F., and Duffy, T. E. (1978). Cerebral blood flow and cerebral metabolic rates for oxygen, glucose, and ketone bodies in newborn dogs. *J Neurochem* 30, 63–69.

Hawkins, R. A., Williamson, D. H., and Krebs, H. A. (1971). Ketone-body utilization by adult and suckling rat brain in vivo. *Biochem J* 122, 13–18.

Haymond, M. W., Karl, I. E., Keating, J. P., and DeVivo, D. C. (1978). Metabolic response to hypertonic glucose administration in Reye syndrome. *Ann Neurol* 3, 207–215.

Haymond, M. W., Karl, I. E., and Pagliara, A. S. (1974). Increased gluconeogenic substrate in the small-for-gestational-age infant. *N Engl J Med* 291, 322–328.

Hellmann, J., Vannucci, R. C., and Nardis, E. E. (1982). Blood-brain barrier permeability to lactic acid in the newborn dog: Lactate as a cerebral metabolic fuel. *Pediatr Res* 16, 40–44.

Hernández, M. J., Vannucci, R. C., Salcedo, A., and Brennan, R. W. (1980). Cerebral blood flow and metabolism during hypoglycemia in newborn dogs. *J Neurochem* 35, 622–628.

Holowach-Thurston, J., Hauhart, R. E., and Jones, E. M. (1974). Anoxia in mice: Reduced glucose in brain with normal or elevated glucose in plasma and increased survival after glucose treatment. *Pediatr Res* 8, 238–243.

Itoh, T., and Quastel, J. H. (1970). Acetoacetate metabolism in infant and adult rat brain in vitro. *Biochem J* 116, 641–655.

Jones, M. D., Jr., Burd, L. I., Makowski, E. L., Meschia, G., and Battaglia, F. C. (1975). Cerebral metabolism in sheep: A comparative study of the adult, the lamb, and the fetus. *Am J Physiol* 229, 235–239.

Kauffman, F. C., Brown, J. G., Passonneau, J. V., and Lowry, O. H. (1969). Effects of changes in brain metabolism on levels of pentose phosphate pathway intermediates. *J Biol Chem* 244, 3647–3653.

Klee, C. B., and Sokoloff, L. (1967). Changes in D(-)-β-hydroxybutyric dehydrogenase activity during brain maturation in the rat. *J Biol Chem* 242, 3880–3883.

Kuhlman, R. E., and Lowry, O. H. (1956). Quantitative histochemical changes during the development of the rat cerebral cortex. *J Neurochem* 1, 173–180.

Laatsch, R. H. (1962). Glycerol phosphate dehydrogenase activity of developing rat central nervous system. *J Neurochem* 9, 487–492.

Levitsky, L. L., Fisher, D. E., Paton, J. B., and Delannoy, C. W. (1977). Fasting plasma levels of glucose, acetoacetate, D-β-hydroxybutyrate, glycerol, and lactate in the baboon infant: Correlation with cerebral uptake of substrates and oxygen. *Pediatr Res* 11, 298–302.

Lockwood, E. A., and Bailey, E. (1971). The course of ketosis and the activity of key enzymes of ketogenesis and ketone-body utilization during development of the postnatal rat. *Biochem J* 124, 249–254.

Lowry, O. H., and Passonneau, J. V. (1964). The relationship between substrates and enzymes of glycolysis in brain. *J Biol Chem* 239, 31–42.

Lowry, O. H., and Passonneau, J. V. (1972). *A flexible system of enzymatic analysis*, pp. 120–126, 146–218. New York: Academic Press.

Lowry, O. H., Passonneau, J. V., Hasselberger, F. X., and Schulz, D. W. (1964). Effect of ischemia on known substrates and cofactors of the glycolytic pathway in brain. *J Biol Chem* 239, 18–30.

Lowry, O. H., Schulz, D. W., and Passonneau, J. V. (1967). The kinetics of glycogen phosphorylases from brain and muscle. *J Biol Chem* 242, 271–280.

Lutwak-Mann, C. (1942). The effect of salicylate and cinchophen on enzymes and metabolic processes. *Biochem J* 36, 706–728.

McIlwain, H. (1966). *Biochemistry and the central nervous system*, pp. 6–26. Boston: Little, Brown.

Mayman, C. I., Gatfield, P. D., and Breckenridge, B. M. (1964). The glucose content of brain in anaesthesia. *J Neurochem* 11, 483–487.

Melichar, V., Drahota, Z., and Hahn, P. (1967). Ketone bodies in the blood of full term newborns, premature and dysmature infants and infants of diabetic mothers. *Biol Neonate* 11, 23–28.

Middleton, B. (1973). The acetoacetyl-coenzyme A thiolases of rat brain and their relative activities during postnatal development. *Biochem J* 132, 731–737.

Moore, T. J., Lione, A. P., Regen, D. M., Tarpley, H. L., and Raines, P. L. (1971). Brain glucose metabolism in the newborn rat. *Am J Physiol* 221, 1746–1753.

Moore, T. J., Lione, A. P., Sugden, M. C., and Regen, D. M. (1976). β-Hydroxybutyrate transport in rat brain: Developmental and dietary modulations. *Am J Physiol* 230, 619–630.

Myers, R. E. (1973). Two classes of dysergic brain abnormality and their conditions of occurrence. *Arch Neurol* 29, 394–399.

Myers, R. E., and Yamaguchi, M. (1976). Effects of serum glucose concentration on brain response to circulatory arrest. *J Neuropathol Exp Neurol* 35, 301.

Oldendorf, W. H. (1971). Blood brain barrier permeability to lactate. *Eur Neurol* 6, 49–55.

Owen, O. E., Morgan, A. P., Kemp, H. G., Sullivan, J. M., Herrera, M. G., and Cahill, G. F., Jr. (1967). Brain metabolism during fasting. *J Clin Invest* 46, 1589–1595.

Page, M. A., Krebs, H. A., and Williamson, D. H. (1971). Activities of enzymes of ketone-body utilization in brain and other tissues of suckling rats. *Biochem J* 121, 49–53.

Page, M. A., and Williamson, D. H. (1971). Enzymes of ketone-body utilization in human brain. *Lancet* 2, 66–68.

Pardridge, W. M. (1981). Transport of nutrients and hormones through the blood-brain barrier. *Diabetologia* 20, 246–254.

Passonneau, J. V., and Lauderdale, V. R. (1974). A comparison of three methods of glycogen measurement in tissues. *Anal Biochem* 60, 405–412.

Patel, M. S., Johnson, C. A., Rajan, R., and Owen, O. E. (1975). The metabolism of ketone bodies in developing human brain: Development of ketone-body utilizing enzymes and ketone bodies as precursors for lipid synthesis. *J Neurochem* 25, 905–908.

Patel, M. S., and Owen, O. E. (1977). Development and regulation of lipid synthesis from ketone bodies by rat brain. *J Neurochem* 28, 109–114.

Penniall, R. (1958). The effects of salicylic acid on the respiratory activity of mitochondria. *Biochim Biophys Acta* 30, 247–251.

Penniall, R., Kalnitsky, G., and Routh, J. I. (1956). The effects of salicylic acid and related compounds on in vitro rat brain respiration. *Arch Biochem Biophys* 64, 390–400.

Persson, B., and Gentz, J. (1966). The pattern of blood lipids, glycerol and ketone bodies during the neonatal period, infancy and childhood. *Acta Paediatr Scand* 55, 353–362.

Ruderman, N. B., Ross, P. S., Berger, M., and Goodman, M. N. (1974). Regulation of glucose and ketone-body metabolism in brain of anaesthetized rats. *Biochem J* 138, 1–10.

Selle, W. A. (1942). Effects of various chemical agents on survival of primitive respiratory mechanism. *Proc Soc Exp Biol Med* 51, 50–52.

Selle, W. A. (1944). Influence of glucose on the gasping pattern of young animals subjected to acute anoxia. *Am J Physiol* 141, 297–302.

Shelley, H. J. (1961). Glycogen reserves and their changes at birth. *Br Med Bull* 17, 137–143.

Siemkowicz, E., and Gjedde, A. (1980). Post-ischemia coma in rat: Effect of different pre-ischemic blood glucose levels on cerebral metabolic recovery after ischemia. *Acta Physiol Scand* 110, 225–232.

Siemkowicz, E., and Hansen, A. J. (1978). Clinical restitution following cerebral ischemia in hypo-, normo- and hyperglycemic rats. *Acta Neurol Scand* 58, 1–8.

Snell, K., and Walker, D. G. (1973). Glucose metabolism in the newborn rat. *Biochem J* 132, 739–752.

Söling, H., and Seufert, C. (eds.). (1978). *Biochemical and clinical aspects of ketone body metabolism*. Stuttgart: Georg Thieme.

Stanley, C. A., Anday, E. K., Baker, L., and Delivoria-Papadopolous, M. (1979). Metabolic fuel and hormone responses to fasting in newborn infants. *Pediatrics* 64, 613–619.

Stanley, C. A., and Baker, L. (1976). Hyperinsulinism in infancy: Diagnosis by demonstration of abnormal response to fasting hypoglycemia. *Pediatrics* 57, 702–711.

Takagaki, G. (1974). Developmental changes in glycolysis in rat cerebral cortex. *J Neurochem* 23, 479–487.

Thurston, J. H., Hauhart, R. E., and Dirgo, J. A. (1976). Insulin and brain metabolism: Absence of direct action on K^+ and Na^+ transport in mouse brain. *Diabetes* 25, 758–763.

Thurston, J. H., Hauhart, R. E., and Dirgo, J. A. (1981). Effects of a single therapeutic dose of glycerol on cerebral metabolism in the brains of young mice: Possible increase in brain glucose transport and glucose utilization. *J Neurochem* 36, 830–838.

Thurston, J. H., Hauhart, R. E., Dirgo, J. A., and Jones, E. M. (1980). Mechanisms of increased brain glucose and glycogen after hydrocortisone: Possible clinical significance. *Ann Neurol* 7, 515–523.

Thurston, J. H., Hauhart, R. E., Jones, E. M., and Ater, J. L. (1975a). Effects of alloxan diabetes, anti-insulin serum diabetes, and non-diabetic dehydration on brain carbohydrate and energy metabolism in young mice. *J Biol Chem* 250, 1751–1758.

Thurston, J. H., Hauhart, R. E., Jones, E. M., and Ater, J. L. (1975b). Effects of salt and water loading on carbohydrate and energy metabolism and levels of selected amino acids in the brains of young mice. *J Neurochem* 24, 953–957.

Thurston, J. H., Levy, C. A., Warren, S. K., and Jones, E. M. (1972). Permeability of the blood-brain barrier to fructose and the anaerobic use of fructose in the brains of young mice. *J Neurochem* 19, 1685–1696.

Thurston, J. H., and McDougal, D. B., Jr. (1969). Effect of ischemia on metabolism of the brain of the newborn mouse. *Am J Physiol* 216, 348–352.

Thurston, J. H., Pollock, P. G., Warren, S. K., and Jones, E. M. (1970). Reduced brain glucose with normal plasma glucose in salicylate poisoning. *J Clin Invest* 49, 2139–2145.

Tildon, J. T., Cone, A. L., and Cornblath, M. (1971). Coenzyme A transferase activity in rat brain. *Biochem Biophys Res Commun* 43, 225–231.

Vannucci, R. C., and Duffy, T. E. (1976). Carbohydrate metabolism in fetal and neonatal rat brain during anoxia and recovery. *Am J Physiol* 230, 1269–1275.

Williamson, D. H., Mellanby, J., and Krebs, H. A. (1962). Enzymatic determination of D(-)-β-hydroxybutyric acid and acetoacetic acid in blood. *Biochem J* 82, 90–98.

Yeh, Y. Y., and Zee, P. (1976). Insulin, a possible regulator of ketosis in the newborn and suckling rats. *Pediatr Res* 10, 192–197.

The Tricarboxylic Acid Cycle

Alexander L. Miller

Introduction

This chapter deals principally with developmental changes in the tricarboxylic acid (TCA) cycle in brain. The focus is on the changes, over time, in the functioning of the cycle as a whole and of its component reactions and on the varying use of different substrates for acetyl CoA production during the course of development.

In the aggregate, the evidence to be presented on the use of nonglucose fuels by developing brain shows that it selects from an ever-changing metabolic menu that is quite distinct from that used by mature brain. One may ask at the outset, therefore, why this should be so. There is no single answer to such a question, but it is of interest to speculate about how the special circumstances of the developing brain may have dictated these differences.

For most species, including man, one of the most impressive aspects of the development of the central nervous system (CNS) is the degree to which its initial growth exceeds that of the rest of the body. In an adult human, brain is about 2 percent of body weight and accounts for 20 percent of resting oxygen consumption. At birth, the infant brain is more nearly 12 percent of body weight, may account for 50 percent of resting oxygen consumption, and uses glucose at a rate that can totally deplete body glycogen stores in four to six hours (Schwartz and Kalhan, 1975). The muscle mass and liver capacity available for gluconeogenesis are, moreover, much more limited in the infant. Thus, the immature brain is in a far more precarious position with regard to glucose availability than is the adult brain. Given the apparent inability of nonhibernating brain to modify its normal high rate of energy use, the advantage of being able to use multiple energy sources becomes more evident. It is, in fact, a virtual necessity imposed on it by the relatively precocious development of brain relative to the body organs that are responsible for ensuring that an adequate supply of nutrients for brain are continuously present in blood.

An issue that arises repeatedly in neurochemical studies of developing brain is deciding which of many possible bases to use in expressing and interpreting data. The problem occurs because, as discussed in other chapters in this volume, brain weight, protein content, water content, lipid content, DNA content, mitochondrial number, cell number and type, and other like variables all change greatly over the course of brain maturation. In the rat, for instance, whole brain wet weight increases by a factor of six between birth and 30 days, while dry weight, lipid, and protein all almost double as a proportion of wet weight (Pitts and Quick, 1967). Thus an enzyme or metabolite whose total content in brain rises by a factor of 12 over this time span will have doubled on a per-unit-fresh-weight basis and stayed unchanged on a per-unit-protein basis. Arguments can be made for any of a variety of denominators, depending on the purpose of a particular analysis. In this chapter, rates, concentrations, and enzyme activities are expressed, when possible, on a per-unit-weight basis. The factors used for converting dry weight and protein weight to wet weight in the developing rat brain were taken from Pitts and Quick (1967).

Review of the TCA Cycle

This section will briefly review the reaction sequence and functions of the TCA cycle. Subsequent sections will focus on specific developmental changes in substrate use, concentrations of intermediates, and enzyme activities in brain. For purposes of this chapter, the pyruvate dehydrogenase (PDH) reaction is included, though it is not, strictly speaking, a part of the TCA cycle. The sequence of reactions is shown in figure 1.

The principal function of the TCA cycle, regardless of tissue and stage of development, is to provide NADH (reduced nicotinamide-adenine dinucleotide) and FADH (reduced flavin adenine dinucleotide) for oxidation through the electron transport chain, an oxidation that is coupled with the production of three molecules of ATP (adenosine triphosphate) per one NADH oxidized and two molecules of ATP per one FADH oxidized. In addition, one GTP (guanosine triphosphate) is produced in the thiolase reaction. The normal stoichiometry of the cycle is the reduction of three NAD^+ and one FAD^+ per two-carbon fragment of citrate oxidized to two CO_2. When pyruvate is the beginning molecule, its oxidative decarboxylation to acetyl CoA by PDH results in the reduction of NAD^+ to NADH. The reactions of the TCA cycle (including PDH) occur in the mitochondrial matrix, in immediate proximity to the electron transport chain on the mitochondrial cristae. Thirty of the 36 ATP molecules generated by oxidation of one glucose molecule to six CO_2 molecules are generated by oxidation of NADH and FADH produced intramitochondrially by the TCA cycle.

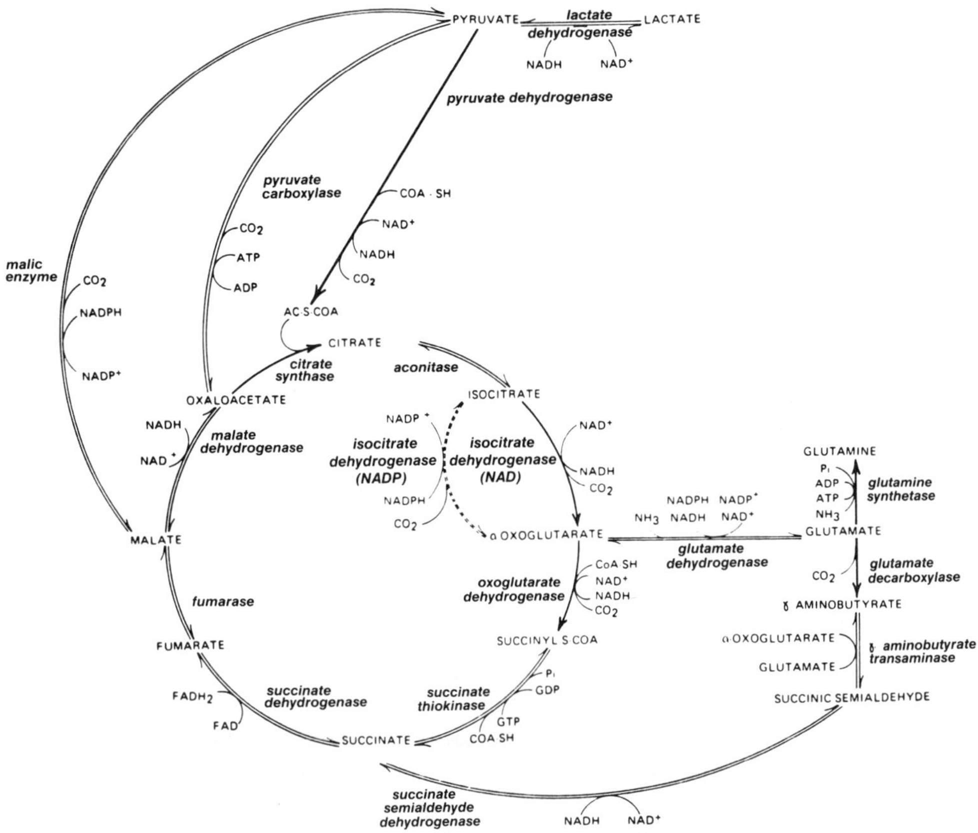

Figure 1. Reactions of the TCA cycle and related pathways.

Aside from ATP production, the TCA cycle has an important role in providing intermediates for various synthetic processes. Thus, acetyl CoA is used for fatty acid and acetylcholine synthesis, and α-ketoglutarate and oxaloacetate are transaminated to glutamate and aspartate, respectively, which are used in protein synthesis and neurotransmission. The blood-brain barrier (BBB) is quite impermeable to all of these compounds, and they must be synthesized in the brain, ultimately being derived from glucose or other substances that do cross the BBB at significant rates. Glutamate is also the precursor for γ-aminobutyrate, a key inhibitory neurotransmitter, and for glutamine, which has a multiplicity of proposed roles in brain. Although there are no gross qualitative differences between immature and mature brain with regard to the operation of these pathways, their rates relative to that of the TCA cycle differ markedly over the course of CNS development. In particular, the period of rapid brain growth requires that relatively large amounts of acetyl CoA, α-ketoglutarate, and oxaloacetate be used for lipid and protein synthesis. Since cycle functioning is dependent on equal amounts of acetyl CoA and oxaloacetate condensing to form citrate, in a steady state the rates of production of each of these intermediates must increase when their rates of utilization through noncycle pathways differ. Increased rates of acetyl CoA production can be achieved by increased rates of pyruvate oxidation or ketone body utilization. Oxaloacetate is replenished by the ATP-dependent anaplerotic reaction, pyruvate carboxylase, in which carbon dioxide is fixed. Developmental changes in the in vitro activity of this enzyme are discussed below, but it is clear from the preceding that its functional activity must increase in conjunction with increased net rates of protein synthesis.

Oxygen Consumption and Mitochondrial Function

Since the principal function of the TCA cycle is to provide substrates for the respiratory chain, by which their oxidation is linked to ATP production, rates of oxygen consumption are normally closely correlated with rates of ATP production and with rates of turning of the TCA cycle. ATP production, electrical activity, and functional activity of cerebral tissue are all interdependent, so determinations of cerebral oxygen consumption in vivo provide valuable information about these parameters.

The development of the Kety-Schmidt nitrous oxide method for measuring cerebral blood flow (CBF) allowed direct in vivo measurements of cerebral metabolic rates of oxygen consumption (CMR_{O_2}) in animals in which arteriovenous differences across the brain could be obtained (Kety and Schmidt, 1948). These determinations have generally been done in man and other relatively large mammals because of the technical difficulties of studying small animals.

The most extensive data on developing brain are from children (table 1). Settergren, Lindblad, and Persson (1980) studied 65 lightly anesthetized children, aged 11 days to 15 years. There was no significant correlation between age and CMR_{O_2} in their subjects. CMR_{O_2} was the same, whereas CBF was higher in children than in comparably treated adults. The arteriovenous differences of oxygen across the brain increased with age, while those for lactate decreased. Kennedy and Sokoloff (1957) studied nine awake children, aged three to 11 years, and 12 awake young adults and found both CBF and CMR_{O_2} to be higher in the children. There was no discernible trend for either CBF or CMR_{O_2} to decline with age in the children. These human studies are in agreement that CBF is higher in children than in young adults by 44–77 percent. They are not consistent in their findings with regard to CMR_{O_2}, though the 24 percent higher rate found by Kennedy and Sokoloff in children is small compared to the 77 percent higher CBF. The discrepant findings with regard

Table 1. *Cerebral Blood Flow and Cerebral Metabolic Oxygen Consumption Rates in Children and Adults*

Study	*Mean Age (Yrs)*	*State*	*CBF (ml g^{-1} min^{-1})*	*CMR_{O_2} (μmol g^{-1} min^{-1})*
Settergren, Lindblad, and Persson (1980)	8.3	Anesthetized	0.65	1.35
Alexander et al. (1968)	22.0	Anesthetized	0.45	1.35
Kennedy and Sokoloff (1957)	6.1	Awake	1.06	2.31
Kennedy and Sokoloff (1957)	24.5	Awake	0.60	1.86

Note: CBF, cerebral blood flow; CMR_{O_2}, cerebral metabolic rate of oxygen consumption.

to $CMRO_2$ could be attributable to the differential effects of anesthesia on $CMRO_2$ of adults and children. It will be of interest, as positron-emitting transverse tomography techniques are further developed and applied to evaluate children, to see if the higher rate in awake children is confirmed and, if so, whether it is uniformly or heterogeneously distributed among brain regions. It should be noted that many review articles juxtapose the $CMRO_2$ for children found by Kennedy and Sokoloff (1957) with that for adults found by Kety and Schmidt (1948), implying a 50 percent higher rate in the children. In fact, the difference between adults and children studied in the same way was only 24 percent (Kennedy and Sokoloff, 1957).

Baboons, dogs, sheep, and rats have all been studied in vivo for their brain rates of oxygen consumption at various stages of development. Baboons showed little change in $CMRO_2$ between birth and 12 weeks of age, but the numbers of animals studied were very small (Levitsky et al., 1977). Gregoire et al. (1978) and Hernández et al. (1978) reported that CBF and $CMRO_2$ in newborn dogs were only about half the rates found in adult dogs. The sheep fetus has a $CMRO_2$ about 70 percent that of the lamb (Jones, 1979). Twenty-day-old rats had $CMRO_2$ values about 80 percent those found in adult rats (Dahlquist and Persson, 1976). This difference was not statistically significant. Direct measurements of oxygen consumption in younger rats have not been reported, but estimates of rates of high-energy phosphate use by the "closed head" technique (Lowry et al., 1964) have concluded that rates at birth and seven days afterward are lower than those of the adult by a factor of five to ten (Duffy, Kohle, and Vannucci, 1975). In vitro studies of rat brain slices tend to confirm the in vivo observations. The rates of cortical O_2 consumption, on a wet-weight basis, approximately double between the fifth postnatal day and adulthood, with most of this increase occurring by day 20 (Mourek, 1970).

The finding of quite large postnatal increases in rates of oxygen consumption in some species and little change in others is most easily explained by the degree of maturity of the central nervous system at birth. Rats are particularly immature when born, and dogs somewhat less so; however, sheep, humans, and baboons are comparatively mature at birth. Thus, the period of increase in the rate of oxygen consumption on a wet-weight basis seems to be during myelination and development of more adult patterns of electrical activity. This, of course, is reasonable, in light of the fact that the principal energy-requiring function of brain is neurotransmission.

A question that has received some attention is whether the mitochondria in developing brain are qualitatively different from those in mature brain. Of particular note is the finding by two laboratories of ratios of ADP (adenosine diphosphate) to oxygen that are lower than adult ratios in isolated brain mitochondria from suckling rats. The normal stoichiometry when respiration is supported by NAD-linked substrate pairs is for each atom of oxygen consumed, three ADP and phosphate atoms form three ATP. With glutamate and malate as substrates, the ADP:O ratio in brain mitochondria from early postnatal rats is in the range 1.5–2.0, rather than 3.0 (Milstein, White, and Swaiman, 1968; Holtzman and Moore, 1973). Such marked changes were not found in rabbit brain by Pigareva (1968) but were reported by Swaiman (1970) in this species. The possibility that the relatively inefficient coupling of oxidation and phosphorylation by isolated mitochondria of immature brain is an artifact of preparation must, of course, be considered. These mitochondria might, for instance, be more susceptible to damage in the isolation process, or they might be contaminated with a kinase that would regenerate ADP from ATP, yielding falsely low ADP:O ratios. To date, however, no evidence that these findings are artifactual has been presented, and they remain of considerable theoretical and practical interest.

Data on respiratory rates from isolated rat brain mitochondria are in agreement with the finding in brain slices of progressive increases postnatally (Holtzman and Moore, 1975; Land et al., 1977; Patel and Clark, 1981). Maximum rates are reached by 20–30 days of age. When β-hydroxybutyrate is used as the substrate instead of pyruvate, brain mitochondrial respiratory rates decline after age 20 days (Land et al., 1977; Patel and Clark, 1981).

Use of Fuels Other Than Glucose

The first in vivo demonstration of a substantial reliance of mammalian brain on a fuel other than glucose was in 1967 when Owen et al. showed that ketone body use could account for over 50 percent of brain oxygen consumption. The discovery arose from the observation that normal cerebral functioning was maintained long beyond the time when all body glucose stores and precursors would have been consumed if the brain had continued to be solely reliant on glucose for energy in persons undergoing prolonged fasting. As discussed above, in some species, including man, developing brain can account for 50 percent or more of total body energy use during a period when glucose stores and precursors are at relatively low levels. Significant use of fuels other than glucose by brain is likely, therefore, to be a more frequent occurrence in these species when they are immature than when they are adult.

Ketone Bodies

When fatty acids are degraded in the liver, some of the flow of carbon atoms is into the ketone bodies, D-β-hydroxybutyrate (BHB), acetoacetate (AcAc), and acetone. BHB and AcAc can be directly taken up from blood and metabolized in brain. Acetone can contribute to glucose synthesis and, thereby, indirectly to brain nutrition (Reichard et al., 1979). Since 1967 a large amount of information on ketone body metabolism by adult and developing brain has accumulated, though some important questions remain unanswered.

There are distinct species differences in the degree to which blood ketone bodies constitute an energy source for the brain in the normal course of development. Rat milk, for instance, has a high lipid content, and metabolism of ketone bodies can account for half or more of brain oxygen use in the late suckling period (Hawkins, Williamson, and Krebs, 1971). Sheep, guinea pigs, and dogs, on the other hand, normally seem to derive very little of their cerebral energy needs from metabolism of ketone bodies (Weng, Nakamura, and Spitzer, 1973; Jones et al., 1975; Gregoire et al., 1978; Booth, Patel, and Clark, 1980). In the series of children studied by Settergren, Lindblad, and Persson (1980), BHB and AcAc use accounted for, on average, about 20 percent of brain oxygen consumption. This figure is almost certainly at the high end of the normal range, because the patients were all undergoing surgical procedures and would have been denied food for at least 12 hours prior to the determinations. In newborns and infants (one to seven months) denied food for six to nine hours, AcAc and BHB accounted for 11–13 percent of cerebral O_2 use (Kraus, Schlenker, and Schwedesky, 1974). The conclusion to be drawn from these human data is that ketone bodies are a significant fuel for developing postnatal human brain after relatively brief periods of fasting but ordinarily are not the dominant fuel, unlike the case in prolonged fasting of adults.

Because of the rapid rates at which developing, postnatal rat brain can use ketone bodies, much of the literature cited below on control and regulation of ketone body use and on interactions with use of other fuels by developing brain was written about research on this species. Although the theoretical possibility exists that rat brain is intrinsically different from that of human and other mammals, all of the evidence indicates that species differences are quantitative rather than qualitative and that conclusions based on studies of the rat are generally applicable to mammalian brain.

As with any blood-borne metabolic fuel, the possible rate-limiting steps in utilization can be at the level of the blood-brain barrier or at subsequent intracellular degradative reactions. Under no conditions have brain ketone body concentrations been found to exceed those expected from blood and cerebrospinal fluid contamination. It seems, therefore, that rates of utilization are controlled by rates of entry into brain. As indicated below, however, the activities of the enzymes that convert ketone bodies to acetyl CoA in brain do rise and fall in a pattern that follows the developmental pattern of ketone body use by brain.

Unlike the case with glucose, which is metabolized by brain at the same rate over a wide range of blood concentrations, brain uptake and use of ketone bodies is clearly dependent on their concentrations in blood. This has been shown by a number of investigators using human subjects (Persson, Settergren, and Dahlquist, 1972; Kraus, Schlenker, and Schwedesky, 1974; Settergren, Lindblad, and Persson, 1980) and rats (Hawkins, Williamson, and Krebs, 1971; Ruderman et al., 1974; Gjedde and Crone, 1975; Daniel et al., 1977; Pollay and Stevens, 1980). It is also well established that unidirectional transport and net uptake of ketone bodies by brain are age dependent. In postnatal humans the percentage of ketone bodies extracted from blood is highest in newborns and lowest in adults, differing by a factor of seven (Kraus, Schlenker, and Schwedesky, 1974). In rats, brain uptake and net extraction of ketone bodies from blood rise postnatally till age two to three weeks and decrease thereafter by a factor of at least three (Hawkins, Williamson, and Krebs, 1971; Daniel et al., 1971; Cremer, Braun, and Oldendorf, 1976; Dahlquist and Persson, 1976; Moore et al., 1976; Daniel et al., 1977; Cremer, 1981). The major questions addressed by more recent studies of ketone body transport from blood into brain have concerned (1) the nature of the transport process, and (2) the modifiability of the transport system by experimental conditions that produce prolonged hyperketonemia. Present knowledge in these areas is summarized below.

There is general agreement that movement of ketone bodies from blood into brain is, at least in part, carrier mediated, and that AcAc is transported somewhat more rapidly than BHB. (Since BHB concentrations are normally two to five times higher than those of AcAc, however, BHB utilization by brain usually exceeds that of AcAc.) The characteristics of the carrier system have been investigated in several laboratories. The D isomer is transported at several times the rate of the L isomer of

Table 2. ***Kinetic Constants for BHB Transport into Brain of Fed and Starved Rats***

Constant	*Study*	*Fed*	*Starved*
K_m (mM)	Gjedde and Crone (1975)	2.5	10.1
	Pollay and Stevens (1980)	13.9	8.53
V_{max} (μmol g^{-1} min^{-1})	Gjedde and Crone (1975)	0.16	0.22
	Pollay and Stevens (1980)	1.25	0.70

Note: BHB, D-β-hydroxyglutarate.

BHB (Moore et al., 1976; Daniel et al., 1977). The L-lactate and D-BHB inhibit one another's unidirectional transport into brain, and their transport is highest in the late suckling period (Cremer, Braun, and Oldendorf, 1976), suggesting that both share a monocarboxylic acid transport system. The kinetic constants K_m and V_{max} have been determined with BHB as substrate in fed and starved rats. The values reported by Gjedde and Crone (1975) and by Pollay and Stevens (1980) are shown in table 2. The discrepancies between these studies may be due in part to the use of tritiated water as the reference molecule in the Gjedde and Crone study (Pollay and Stevens, 1980). The contribution of a diffusionary component that is not carrier mediated was found by Pollay and Stevens to be large in starved animals. Further work defining the mechanism or mechanisms of ketone body entry into brain is clearly indicated, including comparisons between the kinetic constants of BHB entry into brains of developing and adult rats.

The observation of high rates of ketone body use by rat brain during the suckling period and prolonged starvation suggests that continued exposure to high blood levels of ketone bodies increases the permeability of the blood-brain barrier to them. Some support for this conclusion comes from the studies by Moore et al. (1976) of postweanling rats given high-fat diets. Rat pups made hyperketonemic both prenatally and postnatally did not differ from controls in their brain BHB dehydrogenase activities until after weaning (Sherman and Wilson, 1978). More extreme prolonged ketonemia caused a variety of differences in enzymes and substrates related to glutamate metabolism in brains of rat pups (Tildon, Ozand, and Cornblath, 1975). Whether these changes were attributable to increased brain permeability to ketone bodies is unknown but seems worthy of investigation.

It is apparent that, at all ages, the concentration dependence of ketone body use by brain requires that rates of glucose oxidation be adjusted if brain energy production is to remain constant. A number of investigators have noted that high rates of ketone body use by brain are associated with release of lactate from brain into blood (Owen et al., 1967; Hawkins, Williamson, and Krebs, 1971; Cremer and Heath, 1974; Ruderman et al., 1974; Settergren, Lindblad, and Persson, 1980). Since one-eighteenth as much ATP is generated by metabolism of glucose to lactate instead of oxidation to CO_2, release of lactate by brain has the effect of diminishing ATP production from glucose metabolism. This effect could be achieved by inhibition of brain PDH, leading to accumulation of pyruvate and lactate in brain. Indeed, hyperketonemia led to modest increases in brain pyruvate and lactate concentrations in suckling and adult rats in vivo (DeVivo et al., 1978; Miller et al., 1982), BHB slowed the rates of pyruvate decarboxylation by brain mitochondria and synaptosomes from eight- and 21-day-old rats (but not adults—Booth and Clark, 1981), and BHB increased rates of pyruvate carboxylation by rat brain mitochondria (Patel, 1974a). It has been stated that inhibition of PDH and consequent lactate release may be the major regulatory site of glucose metabolism in brain when ketone bodies are used (Robinson and Williamson, 1980), but other evidence, considered below, indicates an important effect of ketone bodies on brain glycolytic rate in developing rats. As a practical matter, moreover, sole reliance on lactate release into blood as a means of regulating glucose oxidation would seem to have at least one significant drawback. Blood lactate levels are quite variable and, with exercise, can temporarily exceed those of brain by a wide margin. Unless lactate transport from brain to blood can occur against a concentration gradient, fluctuations in blood lactate levels could result in rather pronounced changes in brain lactate and pyruvate concentrations, if the glycolytic rate were unchanged, thereby disturbing brain-homeostasis. The question of how blood lactate and ketone body levels interact in their effects on brain glucose metabolism is an important one that has yet to be addressed experimentally.

Brain metabolite concentrations during hyperketonemia in adult rats change in patterns that are consistent with inhibition of phosphofructokinase and PDH and with stimulation of glycogen synthase (Ruderman et al., 1974; DeVivo et al., 1978). Brain glycogen levels are higher in offspring of rat mothers fed high-fat diets (Wapnir, Tildon, and Cornblath, 1973), and changes in brain metabolite levels during acute hyperketonemia in 20-day-old rats were also consistent with inhibition of phosphofructokinase and PDH (Miller et al., 1982). There is, therefore, evidence for effects of hyperketonemia on glucose metabolism at sites other than the PDH reaction. In recent work, Miller et al. (1982) found that brain glucose utilization in 20-day-old rats was reduced during acute hyperketonemia, whether measured with [2-^{14}C]glucose or [^{3}H]deoxyglucose as tracer. If increased release of lactate had been the only adaptation to increased use of ketone bodies, then the rate of use of [^{3}H]deoxyglucose would not have fallen, because this compound is phosphorylated but not metabolized further by brain. Thus, hyperketonemia resulted in a slowing of the whole glycolytic pathway, including the hexokinase reaction.

Lactate

Though brain use of lactate has not been studied to the same extent as ketone bodies, an accumulating body of evidence indicates that blood lactate can function as a fuel for immature brain. These data come from (1) studies of the blood-brain transport system for lactate, (2) arterio-venous differences across neonatal dog brain during hypoglycemia, and (3) observation of the effects of hyperlactatemia on brain glucose metabolism.

The effort to identify and characterize the system transporting lactate from blood into brain was greatly forwarded by the application of the intracarotid injection technique to the study of this process (Oldendorf, 1972; Oldendorf, 1973). A carrier system for L-lactate in adult rat brain was found that exhibited cross-inhibition with pyruvate, acetate, and propionate, but not with D-lactate. Nemoto and Severinghaus (1974) confirmed the stereospecificity of the carrier and showed that in adult rats net movement of lactate into brain was slight compared to movement of glucose. Subsequent studies using the intracarotid injection technique in developing rats have demonstrated that both carrier-mediated influx and nonsaturable influx of L-lactate into brain are many-fold higher at 15–22 days of age than in adults, that pyruvate is transported similarly, and that the V_{max} and K_m for transport of lactate into brain are both considerably higher in suckling rats than in adult rats (V_{max} of 1.95 versus 0.023 μmol. g^{-1} min^{-1} and K_m of 9.70 versus 0.89 mM) (Cremer, Braun, and Oldendorf, 1976; Cremer et al., 1979; Cremer, 1981). It is possible that even higher rates of transport of lactate occur earlier in the suckling periods. Spatz et al. (1978) found that uptake of lactate by isolated rat cerebral capillaries was about fourfold greater in the immediate neonatal period than 16–21 days after birth. A recent report on lactate uptake by developing rabbit brain found that the brain uptake index was greatest seven to 14 days after birth and declined thereafter (Cornford, Braun, and Oldendorf, 1982).

The observation that increased influx of ketone bodies into brain is associated with increased efflux of lactate has led to speculation that the same carrier system is used for both molecules. In support of this hypothesis are the findings that (1) developmental changes in the lactate and BHB transport systems in rat brain parallel one another; (2) BHB and lactate can each partially inhibit the transport of the other; and (3) in starved adult rats the brain uptake for lactate increases, as it does for BHB (Cremer, Braun, and Oldendorf, 1976). On the other hand, the evidence that capillary lactate uptake is highest in the neonatal period

(Spatz et al., 1978) while BHB permeability rises continuously till age three weeks (Moore et al., 1976) suggests the existence of two different carriers. It is apparent that further studies to define more clearly the relationships between lactate and the ketone bodies as fuels for developing brain are needed.

The work on arteriovenous differences of lactate across the developing brain has, for technical reasons, been done on dogs and other animals larger than developing rats. Thus, the kinetic data from unidirectional transport studies in the rat are not directly applicable to an analysis of the net fluxes found in other species. Nemoto, Hoff, and Severinghaus (1974) found that in hypoglycemic adult dogs infused with lactate (blood level = 8 mM) about 25 percent of the CMR_{O_2} was accounted for by lactate uptake with the rest coming from glucose. In the newborn dog, however, there is apparently much greater permeability of the BBB to lactate. During hypoglycemia net uptake of lactate into brain increased to the point that lactate became more important than glucose as a fuel for cerebral oxidative metabolism (Hernández et al., 1980; Vannucci et al., 1981). It is important to note that blood lactate levels rose by 50–350 percent in the hypoglycemic animals (Hernández et al., 1980) while brain lactate levels fell by 33–75 percent (Vannucci et al., 1981). Thus, a substantial blood-brain gradient was created, whereas this gradient was negligible in the normoglycemic state. In later work it was found that creation of a gradient by lactate infusion did lead to uptake by newborn dog brain in direct proportion to the arterial concentration (Hellmann, Vannucci, and Nardis, 1982). The same investigators noted that during hypoglycemia, arteriovenous differences of lactate were twofold to threefold greater than during normoglycemia at similar arterial lactate concentrations. During hypoglycemic hyperlactatemia the lactate taken up by brain must have been largely metabolized, because the increases in brain lactate concentrations were small relative to the amounts taken up. The information about the normoglycemic hyperlactatemic dogs was insufficient to conclude with certainty that significant amounts of the lactate taken up by brain were metabolized.

Studies of arteriovenous differences across the brains of baboons, calves, and humans have also found evidence for a contribution of blood lactate to brain oxidative metabolism under some conditions. In the neonatal baboon, lactate uptake accounted for about 10 percent of brain oxygen consumption, whereas no net uptake was found at six or twelve weeks of age (Levitsky et al., 1977). In hypoglycemic newborn calves, blood lactate levels rose, and net uptake (assuming oxidation) was equivalent to 21 percent of brain oxygen consumption (and almost 50 percent of glucose utilization—Gardiner, 1980). In a four-year-old boy with glucose 6-phosphatase deficiency, blood lactate levels were quite high (4.8–8.8 mM), and uptake by brain was sufficient to account for 30–40 percent of cerebral oxygen consumption (Fernandes, Berger, and Smit, 1982).

The general observation in developing humans, as noted above, is that lactate is released from brain when ketone bodies are used. Under these conditions the concentration gradient is from brain to blood. Studies of brain lactate uptake by normal developing humans with hyperlactatemia (produced by infusion or exercise) have not been done.

Recent work in this laboratory has examined the effects of hyperlactatemia on brain glucose metabolism in vivo in 20-day-old rats (Miller, Kiney, and Staton, 1984). When blood lactate levels were raised to 4–6 mM by intraperitoneal administration of Na-lactate, brain rates of glucose utilization fell by 30–50 percent during the 15 minutes following injection. These decreases were found with both [2-^{14}C]glucose and [^{3}H]deoxyglucose as tracers. Since deoxyglucose is not metabolized further after phosphorylation, these results showed that the entire glycolytic pathway was less active in brain in the hyperlactatemic animals. Despite the decrease in glucose utilization, brain concentrations of TCA cycle intermediates and associated amino acids either rose or were unchanged, indicating that influx and metabolism of blood lactate compensated for the decrease in glucose use.

In summary, there is a growing body of evidence that blood lactate can enter and nourish developing brain under both normoglycemic and hypoglycemic conditions. There are many similarities between lactate and the ketone bodies as fuels for brain: (1) they enter developing brain much more rapidly than they do adult brain; (2) they appear to enter brain in proportion to their blood concentrations by both carrier-mediated and simple diffusion; and (3) their increased use by brain is accompanied by decreases in glycolytic rate in normoglycemic animals. In fact, the same transport system may be used by lactate and the ketone bodies. A number of questions about lactate use by brain remain unanswered: Is it more rapid during hypoglycemia or is its increased uptake just a function of a larger concentration gradient? When blood lactate and ketone body levels are both high, at what rates are they used and how do they compete with one another for entry into the brain? Are there adaptational increases in lactate uptake by brain with starvation or with conditions, such as glucose 6-phosphatase deficiency, which result in chronic hyperlactatemia?

Other Nonglucose Fuels

Some substances, such as mannose and medium-chain fatty acids, can enter and be metabolized by brain at fairly rapid rates but are not present in blood in sufficient concentrations to constitute a significant source of fuel for brain at any age. Other blood-borne substances, such as amino acids, are used by other organs, but do not cross the blood-brain barrier at sufficient rates to function as an energy supply for brain. Long-chain fatty acids, such as palmitate, have not, in general, been found to function as an exogenous energy source for brain. In one study of starved newborn puppies, however, net uptake of free fatty acids into brain sufficient to account for 25 percent of brain oxygen consumption was found (Weng, Nakamura, and Spitzer, 1973). This was more than twice the oxygen consumption attributable to oxidation of ketone bodies by the brains of these animals. Based on work with brain homogenates, Warshaw and Terry (1976) postulated that rat brain oxidizes fatty acids in a developmental pattern similar to that found for ketone bodies. Demonstration that oxidation of fatty acids from blood contributes measurably to the oxygen consumption of developing rat or human brain in vivo has yet to be reported.

Metabolites

This section presents data on developmental changes in brain levels of TCA cycle intermediates and related amino acids in normal animals in vivo. A later section deals with changes found when subjects' physiological conditions are altered or when they are treated with various drugs and toxins.

In the past decade, considerable effort has gone into devising methods that stop brain metabolism as instantaneously as possible in order to examine in vivo levels of compounds whose levels change rapidly postmortem. Most of this work has been on rodents; there are no data from humans. The relatively small size and slow metabolic rate of the neonatal rat and mouse brain are advantageous in that freezing of the whole brain from the surface occurs more quickly and postmortem artifacts develop less rapidly than in adults. The problem that arises in studies of rodent brain at different stages of development is that the contribution of postmortem changes to the results can be different at each age. Thus, as the size of the brain and the thickness of the skull, hair, and skin increase, the time to freeze deeper parts of the brain increases, using methods such as immersion or decapitation into liquid nitrogen. The changing geometry of the head has a different effect on the speed of heating by microwave irradiation. For five- and ten-day-old rats a 2,450-MHz apparatus was slower than for 15- and 20-day-olds, whereas a 915-MHz device is better for adult rats (Miller and Shamban, 1977; Medina, Deam, and Stavinoha, 1980). Thus, comparing data from animals of widely variant head sizes killed in a microwave device of a single wavelength may be invalid.

Using a freeze-blowing apparatus to obtain brains (Veech et al., 1973), researchers found that the delay between the initial disturbance of brain metabolism and freezing is less than one second and varies little, if at all, between rodents of different sizes. When the technique was modified to study developing rat brain five, ten, 15, and 20 days old, some, but not all, TCA cycle intermediates and related amino acids were found to increase markedly over this time span (Miller and Shamban, 1977)

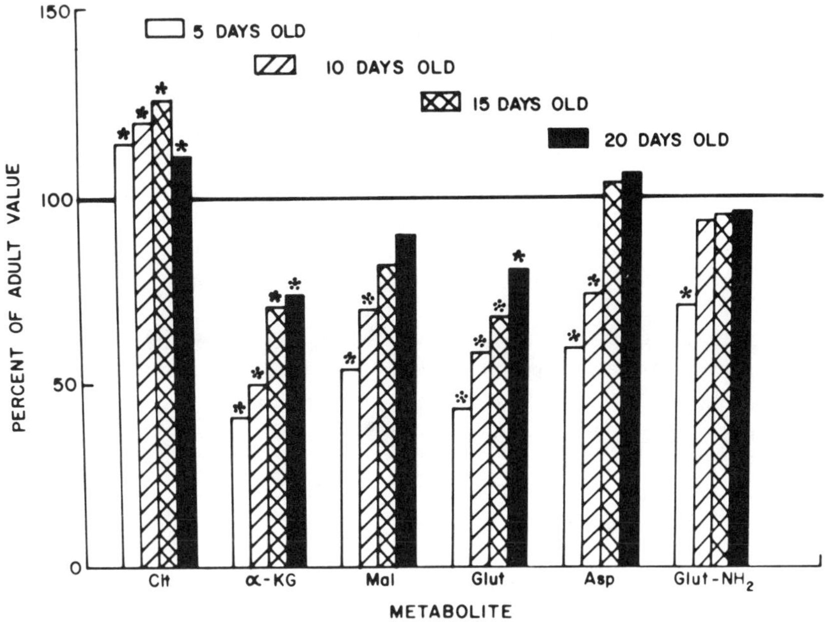

Figure 2. Concentrations of TCA cycle intermediates and related amino acids in developing rat brain compared with adult values (*Cit*, citrate; *α-KG*, α-ketoglutarate; *Mal*, malate; *Glut*, glutamate; *Asp*, aspartate; and *Glut-NH$_2$*, glutamine). The asterisks indicate statistical significance ($p < .05$) in the comparison with adult levels. (From Miller and Shamban, 1977; reproduced with permission of Pergamon Press, Ltd.)

(fig. 2). Steady-state metabolite concentrations do not yield direct information about flux through metabolic pathways, but the observed changes in developing brain are generally consistent with available data on metabolic rates. Thus, the increases in α-ketoglutarate, malate, and glutamate concentrations correlate with rising activity of the TCA cycle (see above and Miller and Shamban [1977] for discussion) and with the development of compartmentation of glutamate metabolism in rat brain (Patel and Balazs, 1975). Interpretation of the higher-than-adult citrate levels in developing rat brain is difficult, depending importantly on the intracellular locations of this metabolite. It would be teleologically reasonable that cytosolic citrate levels should be high in suckling rat brain, both to inhibit phosphofructokinase activity when ketone bodies are being rapidly used and to serve as the source of acetyl CoA for fatty acid synthesis.

The freeze-blowing apparatus is not suitable for regional studies of brain metabolite levels. With the recent development of radioisotopic methods for quantitating regional glucose utilization in brain (Sokoloff et al., 1977; Hawkins, Hass, and Ransohoff, 1979), the value of being able to measure regional metabolite concentrations as an aid to interpreting changes in metabolic rates is apparent. In the study comparing freeze-blowing and microwave irradiation in developing rats (Miller and Shamban, 1977) an apparatus with 1.3-kW output was used, and a number of differences in whole brain metabolite levels were found. More recently, a 2,450-MHz device has been used with 6-kW output (Medina et al., 1975) to stop brain metabolism of 20-day-old rats; results in whole brain have been very similar to those obtained with the freeze-blowing apparatus (A. L. Miller, M. D.; C. A. Kiney, B.S.; and D. Zirkman, B. S., unpublished data, 1981). Thus, it seems likely that given microwave devices of sufficient power and appropriate wavelength, studies of regional brain metabolite levels in developing rodents of different ages and under different conditions will be possible in the future.

Enzymes

Developmental changes in brain enzyme activities can provide considerable information about changes in metabolic flux in vivo. This has proved to be the case for intermediary metabolism in the brain, even in instances in which the enzyme rates found in vitro clearly exceed in vivo fluxes by a wide margin. This section will summarize data from studies of normal brain, and information about effects of drugs, diseases, toxins, for example, will be presented in the next section.

As with brain metabolite concentrations, more brain enzyme determinations have been done on rats than on any other species. Table 3 contains representative studies from the literature on the rat and is organized according to pathway and ages studied. For purposes of comparison, all rates are expressed in the same units and adjusted to compensate for differences in assay temperature. (See table 3 note.)

The activities of the three enzymes that metabolize BHB and AcAc to acetyl CoA follow a developmental pattern similar to that noted above for ketone body transport. Peak activities are in the late suckling period, with the exception of total acetoacetyl CoA thiolase. When this enzyme has been fractionated into its cytoplasmic and mitochondrial components, however, it has been found that the mitochondrial activity changes in the same time course as the other enzymes. In the late suckling period, flux through these reactions has been estimated to be about 0.3 μmol g^{-1} min^{-1} (Cremer and Heath, 1974), which is well below the enzyme activity levels listed in table 3 and is in agreement with the conclusion that transport into brain is the rate-limiting step in ketone body metabolism.

The normal rate of turning of the TCA cycle is about 1–1.5 μmol g^{-1} min^{-1}, and flux through the PDH reaction about 0.7–1.2 μmol g^{-1}

min^{-1} in rat brain in the 16–20-day-old period (Cremer and Heath, 1974; Miller and Corddry, 1981). The reported activities of the TCA cycle enzymes are generally in excess of these rates, with the exceptions of PDH, aconitase, and α-ketoglutarate dehydrogenase (α-KGDH). Of these, PDH and α-KGDH are thought to be key regulatory enzymes in many tissues, including brain (Stryer, 1981), and particular attention has focused on developmental changes in and abnormalities of PDH. Regulation of citrate synthase and isocitrate dehydrogenase (ICDH) (NAD) activities can also influence the rate of the TCA cycle (Stryer, 1981), and the fact that their in vitro activities considerably exceed normal net flux through these reactions does not mean that their in vivo regulation is not a factor in adjusting the rate of the TCA cycle in brain. Nevertheless, the regulatory aspects of these enzymes in developing brain have not been extensively studied. Diseases in which PDH inhibition or deficiency is thought to play an etiologic role in causing brain damage are discussed in the next section. The points to be made here are that PDH and α-KGDH seem to normally operate near the upper limits of their activity in brain and that increasing rates of glucose oxidation by developing brain correlate well with increases in PDH activity (Cremer and Teal, 1974).

As might be predicted from rising rates of protein synthesis in brain during the suckling period of the rat, the activity of the anaplerotic reaction, pyruvate carboxylase, increases in brain during this time. The data from the two different laboratories in which these studies were performed are slightly discrepant in that Land and Clark (1975) found declining activities after 22 days of age, whereas they were reported to increase by Patel (1974a) and by Wilbur and Patel (1974).

A discussion of glutamate and γ-aminobutyrate metabolism in developing brain is beyond the scope of this chapter. Activities of the relevant enzymes are listed for the interested reader, because some of the reactions occur partly or completely within mitochondria.

Data about developing brain in humans are much less extensive than those from the rat. Activities of ketone body–metabolizing enzymes, except total acetoacetyl CoA thiolase, increase postnatally and are well above net rates of utilization (Page and Williamson, 1971; Patel et al., 1975). In the earlier study, rates declined in adults, but they were at their highest in adult brain in the work of Patel et al. (1975). This remains an unresolved issue, as does the question whether prolonged ketonemia in humans leads to adaptive increases in BBB permeability, in activities of ketone body–metabolizing enzymes in brain, or increases in both.

Other human studies have included assays of succinate dehydrogenase (SDH). Contradictory results have been reported. Diebler, Farkas-Bargeton, and Wehrle (1979) found SDH activity to increase by a factor of ten in the late fetal months in cortex, but Mahajan et al. (1982) found the highest rate in cortex of small fetuses (0–100 g), twice the rate in cortex during the late fetal period (700–1,000 g body weight). Given the in

Table 3. *Enzyme Activity Levels in Developing Rat Brain*

Enzyme	Reference	Age (Days) 0–5	6–10	11–15	16–20	Adult
HBDH	2	0.7	1.2	1.9	2.6	0.7
HBDH	13	1	1.5	2.5	3.5	1.5
HBDH	15	—	0.4	0.9	1.4	0.7
CoA transferase	2	3.6	5.3	7.7	11	7.2
CoA transferase	13	3.5	6	8.5	12	6
CoA transferase	15	—	17	26	48	19
AcAc CoA thiolase (M)	2	4.8	5.5	7.2	9.2	4.8
AcAc CoA thiolase (C)	2	5.5	5.2	4.9	4.6	3.1
AcAc CoA thiolase (M)	11	3.6	5.3	7.7	10.6	5.3
AcAc CoA thiolase (C).	11	6	5.5	5.5	5.3	3
AcAc CoA thiolase (T)	13	9.5	12	9.5	9.5	6
AcAc CoA thiolase (M)	15	—	—	4.6	5	2
PDH	2	0.2	0.5	1	1.9	4.8
PDH	4	0.2	0.35	0.55	0.85	1.8
PDH	9	0.2	0.5	1.2	2.0	4.9
PDH	24	0.56	—	—	—	1.44
Citrate synthase	2	19	26	48	67	72
Citrate synthase	9	22	36	50	72	71
Citrate synthase	21	7	12	21	27	33
Aconitase	21	0.5	0.7	0.9	1.5	2.2
ICDH (NADP) (M)	3	—	2.7	—	2.0	—
ICDH (NADP) (C)	3	—	2.5	—	1.6	—
ICDH (NADP)	18	11	9	9	7.6	—
ICDH (NADP)	22	9.4	9.6	9.2	6.9	—
ICDH (NADP)	17	9.7	—	7.7	—	5.2
ICDH (NAD)	17	4.1	—	8.8	—	18
α-KGDH	14	0.25	0.4	0.55	0.75	0.85

Note: Values are in μmol min^{-1} g^{-1} fresh weight of whole brain at 37°C. Many rates are quite approximate, having been extracted from small graphs. Most of the enzymes are, at least in part, mitochondrial in location. A few, such as glutamic acid decarboxylase, are cytoplasmic but are listed because they are part of a pathway in which some of the reactions are intramitochondrial. Where values were reported in other units, the developmental changes in rat brain protein and dry weight reported by Pitts and Quick (1967) were used to convert to activities per unit fresh weight. Where assays were run at temperatures other than 37°C, the rates have been scaled up or down by a factor of 2 for each 10°C temperature difference. The enzyme abbreviations are as follows (abbreviation, trivial name, Enzyme Commission number): HBDH, β-hydroxybutyrate dehydrogenase, 1.1.1.30; CoA transferase, 3-oxo-acid CoA transferase, 2.8.3.5; AcAc CoA thiolase, acetoacetyl CoA thiolase, 2.3.1.9; PDH, pyruvate dehydrogenase, 1.2.4.1; citrate synthase, 4.1.3.7; aconitase, 4.2.1.3; ICDH (NADP or NAD), isocitrate dehydrogenase (NADP or NAD), 1.1.1.42 or 1.1.1.41; α-KGDH, α-ketoglutarate dehydrogenase, 1.2.4.2; SDH, succinate dehydrogenase, 1.2.99.1; MDH, malate dehydrogenase, 1.1.1.37; PC, pyruvate carboxylase, 6.4.1.1; PEPCK,

Table 3. (Continued)

Enzyme	*Reference*	*0–5*	*6–10*	*11–15*	*16–20*	*Adult*
		Age (Days)				
SDH	1	—	—	4.3	7.3	8.0
SDH	5	0.3	—	—	—	1.4
SDH	6	1.7	—	3	5	6
MDH	7	53	67	108	153	230
MDH	18	174	222	310	408	578
MDH	21	55	85	110	175	240
MDH	24	480	580	1,000		1,426
PC	8	0	0.1	0.4	0.97	0.65
PC	15	—	0.4	0.55	0.8	1.1
PC	21	0.1	0.17	0.27	0.35	0.80
PEPCK	21	0.02	0.02	0.025	0.036	0.055
GOT	10	31	43	84	125	218
GOT(P)	21	7	13	20	28	35
GOT	24	12	14	17		18
GlDH	3	—	2.4	—	5.1	—
GlDH	7	3	5.3	6.7	9.7	14.3
GlDH	10	4.8	7.2	10	16	19
GlDH	12	0.3	0.5	1	—	—
GlDH	13	12.5		32		40
GS	12	0.2	0.5	0.8	1	1
GS	23	0.2	0.4	0.8	—	—
Glutaminase	12	3.1	5.1	9.3	—	—
GAD	19	0.09	0.18	0.36	0.52	0.77
GABA transaminase	20	0.27	0.35	0.68	1.10	2,107
SSADH	16	1.67	2.24	3.06	4.18	5.39

phosphoenolpyruvate carboxykinase, 4.1.1.32; GOT, glutamate oxaloacetate transaminase, 2.6.1.1; GlDH, glutamate dehydrogenase (NAD or NADP), 1.4.1.2. or 1.4.1.4; GS, glutamine synthetase, 6.3.1.2; glutaminase, 3.5.1.2; GAD, glutamic acid decarboxylase, 4.1.1.15; GABA transaminase, γ-aminobutyrate transaminase, 2.6.1.19; SSADH, succinic semialdehyde dehydrogenase, 1.2.1.16. The letters *C*, *M*, *P*, and *T* in parentheses indicate cytoplasmic, mitochondrial, particulate, and total activities, respectively. The reference numbers indicate the following reports: (1) Banik and Davison (1969); (2) Booth, Patel, and Clark (1980); (3) Chase et al. (1976); (4) Cremer and Teal (1974); (5) Gregson and Williams (1969); (6) Jilek (1970); (7) Kuhlman and Lowry (1956); (8) Land and Clark (1975); (9) Land et al. (1977); (10) MacDonnell and Greengard (1974); (11) Middleton (1973); (12) Ozand et al. (1975); (13) Page et al. (1971); (14) Patel (1974b); (15) Patel (1979); (16) Pitts and Quick (1967); (17) Reijnierse, Veldstra, and Van den Berg (1975); (18) Robins and Lowe (1961); (19) Sims and Pitts (1970); (20) Sims et al. (1968); (21) Wilbur and Patel (1974); (22) Wilson (1972); (23) Wu (1964); (24) Zubairu et al. (1982).

vitro data suggesting that human brain oxidative metabolism increases throughout intrauterine life (Himwich et al., 1959), it seems more probable that SDH activity would follow a similar pattern, rather than declining. Malate dehydrogenase (MDH) activities increase prenatally in most areas of human brain, whereas ICDH (NADP) stays level or decreases (Sabster et al., 1979).

Overall, data from humans are scanty, difficult to obtain, and subject to considerable underestimation because of autolysis of the more labile enzymes. One conclusion from animal studies that would be especially worthwhile to confirm in humans is that PDH activity is a key regulator of the rate of glucose oxidation by developing brain.

Abnormal States

Changes in TCA cycle functioning have been reported under a number of conditions. These can be broadly subdivided into (1) those in which a primary defect in a cycle enzyme has been proposed, (2) those in which the effect or effects of a drug, disease, or toxin (exogenous or endogenous) on cycle functioning has been implicated as etiologic in the production of CNS pathology, and (3) those in which a change in cycle functioning is an attempt to compensate for decreased availability of glucose as a metabolic fuel. Each of these areas will be discussed separately.

Since the first report of a child with an intermittent movement disorder and a pyruvate decarboxylase deficiency (Blass, Avigan, and Ulbendorf, 1970), a burgeoning literature on the association between abnormalities of the PDH complex and various neurological syndromes has developed. Though there is considerable controversy over clinical and biochemical aspects of some of the cases, particularly with regard to the role of PDH deficiency in Friedreich's ataxia (Barbeau et al., 1978; Kark, Rodriquez-Budelli, and Blass, 1978; Kark, Rosenberg, and Schut, 1978; Stumpf and Parks, 1978; Melancon et al., 1979; Robinson et al., 1981; and see "News and Comment," 1982), there seems little reason to doubt that childhood neurological syndromes caused by abnormalities of structure or activation (or both) of the brain PDH complex do exist (Blass et al., 1978; Koster, Slee, and Fernandes, 1978; DeVivo et al., 1979; Robinson, Taylor, and Sherwood, 1980). The PDH complex is a large, five-unit, membrane-bound complex that is regulated both allosterically and by phosphorylation-dephosphorylation (Stryer, 1981). Its in vitro and in vivo properties in brain and other tissues are active areas of investigation in a number of laboratories, and one may doubtless expect further reports on PDH abnormalities in various neurological diseases as these properties are elucidated. Among the questions that remain unanswered are (1) How low must activity be to be of clinical significance? (2) What are the effects of nutrition, concurrent illnesses, agonal status, medications, time from death to sampling, and other similar variables on postmortem human brain PDH activity? (3) To what extent and under what conditions do peripheral tissues (platelets, white cells, liver, cultured fibroblasts) that can be obtained before death reflect changes in brain PDH? (4) Does brain PDH differ in any significant ways from PDH in other tissues? (No major differences have been found thus far.)

Defects in other TCA cycle enzymes in developing have not been reported, with the exception that a minority of the patients with low PDH activity also have low activity of the structurally similar α-KGDH complex. Whether other defects have gone unrecognized is, of course, an open question. Certainly, the findings with PDH and α-KGDH support the conclusion that even quite marked deficiencies in TCA cycle enzymes are not necessarily lethal *in utero*.

A second area of investigation has been on the effects of various toxins and disease states on brain TCA cycle functioning. In at least some instances the argument has been advanced that the effects on the TCA cycle are of pathogenetic significance in causing brain damage. Two genetically determined disorders of amino acid metabolism in which keto acids and hydroxy acids accumulate to high levels are phenylketonuria and branched-chain ketoaciduria (maple syrup urine disease). Because dietary therapy that prevents the increases in metabolites of phenylalanine or branched-chain amino acids also prevents CNS damage, the hypothesis that these metabolites are toxic to the CNS has arisen. They have been tested for their ability to inhibit mitochondrial functions by several investigators. Bowden, McArthur, and Fried (1971) reported that pyruvate decarboxylation by rat brain homogenates was inhibited by α-ketoisocaproic acid (KIC). Patel et al. (1973) confirmed this finding in both rat and human brain homogenates by using KIC concentrations of 5–10 mM. They found no effect of α-keto-β-methylvalerate or α-ketoisovalerate (KIV). All of these keto acids were competitive inhibitors of α-ketoglutarate decarboxylation by rat brain mitochondria and human brain homogenates (Patel, 1974b). On the other hand, KIC and KIV did not affect rates of oxidation of [U-^{14}C]glucose by slices of developing rabbit brain (Swaiman and Milstein, 1965). Benavides et al. (1976) found that phenylalanine metabolites, especially phenylpyruvate, inhibited the activities of the enzymes of ketone body utilization in homogenates of developing rat brain. Others have noted that phenylpyruvate inhibits pyruvate carboxylation and decarboxylation by human brain homogenates (Patel, Grover, and Wilbur, 1973). Evidence from studies of isolated mitochondria has supported the hypotheses that both KIC and phenylpyruvate interfere with transport of pyruvate, BHB, and AcAc (Clark and Land, 1975; Patel, Booth, and Clark, 1977; Patel and Clark, 1981). KIC and KIV both inhibited passage of BHB and lactate across the BBB of suckling rats studied by the intracarotid injection technique (Cremer, 1981).

There is no single unifying hypothesis that brings together all of the above observations, and even if there were, the relationship between the disturbance in energy metabolism and the neuropathology of phenylketonuria and maple syrup urine disease would still need to be shown. While there is certainly a need for more animal experiments to further explore these findings, other approaches to understanding the brain biochemical abnormalities of patients with these diseases are now possible. Thus, positron emission transverse tomography, with labeled glucose, ketone bodies, and oxygen, can be used to test the hypothesis that uptake and metabolism of these substrates by brain regions is altered in untreated but not in adequately treated patients. Nuclear magnetic resonance imaging of high-energy phosphates adds the potential for examining the energy state of the brain under these conditions.

Phenylketonuria and maple syrup urine disease are examples of relatively rare CNS diseases in which specific effects of endogenous toxins on specific mitochondrial functions have been postulated. Much more common are clinical states that have deleterious effects on the development of all tissues, including brain. Among these are several that represent major public health problems and connected with which there is some evidence that there are mitochondrial dysfunctions in brain that are associated with impaired brain development. Included in these conditions are malnutrition, hypothyroidism, and lead intoxication. Some of the abnormalities which have been described are (1) decreased brain oxygen consumption in conjunction with increased brain glucose uptake by young protein-calorie malnourished monkeys (Mehta, Nain, and Kalsi, 1980); (2) delayed increases and decreases of the enzymes of ketone body metabolism in brains of rats with neonatal hypothyroidism (Patel, 1979); (3) decreased incorporation of ^{14}C BHB into brain protein and lipid in protein-calorie malnourished rats from birth to 15 days of age without decreased incorporation into brain amino acids or lower activities of the enzymes of ketone body metabolism in brain (Swiatek et al., 1979); (4) decreased brain glutamate dehydrogenase activity in undernourished rat pups (Chase et al., 1976); (5) decreased brain glutamate levels in undernourished nine-day-old rats (Thurston et al., 1971); (6) increased dependence on glucose as opposed to ketone bodies as cerebral substrate during starvation of 20-day-old pups suffering from intrauterine growth retardation (Dahlquist, 1976); (7) decreased oxidation of the glutamate-malate substrate pair and decreased cytochromc oxidase activity in brain mitochondria from suckling rats on a high-lead diet (Holtzman and Hsu, 1976); (8) delayed development of the mature pattern of labeling of TCA cycle amino acids by [^{14}C]glucose and delayed development of glutamate compartmentation in brains of undernourished, hypothyroid, and lead-intoxicated rat pups (see Patel and Balazs [1975] for review). The precise nature of these abnormalities and their causes remain to be defined, but these observations make it clear that mitochondrial and TCA cycle abnormalities are part of the biochemical sequelae of these clinical conditions. Whether these abnormalities play a role in the pathogenesis of the permanent brain damage caused by these conditions remains to be firmly established.

The final area for discussion of TCA cycle functioning in developing brain is glucose utilization decreased by hypercapnia or hypoglycemia. Under these circumstances the brain must use nonglucose fuels to a greater extent if normal or near-normal rates of energy production are to be maintained.

Hypercapnic acidosis inhibits glycolysis in adult and developing brain (Miller, Hawkins, and Veech, 1975; Borgstrom, Norberg, and Siesjo, 1976; Miller and Corddry, 1981). Acutely, at least, brain oxygen consumption is maintained at normal or near normal rates, indicating increased reliance on nonglucose fuels. The principal energy sources identified in mature brain are the pools of endogenous intermediates of glycolysis and the TCA cycle and associated amino acids, especially glutamate (Folbergrova et al., 1975; Miller, Hawkins, and Veech, 1975). Recent work on developing rat brain has established that the metabolic responses to hypercapnic acidosis are qualitatively the same as in adult brain, but quantitatively lower, in proportion to the lower rates of brain glucose utilization in the younger animals (Miller and Corddry, 1981). Because of the greater capacity of developing rat brain to take up and metabolize ketone bodies and lactate from blood, it might be anticipated that high blood levels of these substrates would have a sparing effect on the consumption of glutamate and other endogenous fuels during hypercapnia. The evidence to date, however, does not support this hypothesis. When 20-day-old rats were made hyperketonemic or hyperlactatemic as well as hypercapnic, they had even greater decreases in rates of brain glucose utilization than during hypercapnia alone (Miller et al., 1982; Miller, Kiney, and Staton, 1984). Thus, the inhibitory effects of ketone bodies and lactate noted above were additive with that of hypercapnia, and no significant sparing of the consumption of endogenous intermediates was found. However, these experiments examined only the acute situation (15 minutes), and it remains possible that during longer term hypercapnia there could be compensatory increases in ketone body and lactate uptake or diminution of the inhibitory effect of these substrates on glycolytic activity.

During hypoglycemia, brain glucose utilization falls because of the failure of transport of glucose into brain to keep pace with utilization. Brain oxygen consumption is maintained at near-normal rates in adult humans, rabbits, sheep, and rats (Kety et al., 1948; Eisenberg and Seltzer, 1962; Della Porta et al., 1964; Pappenheimer and Setchell, 1973; Norberg and Siesjo, 1976). Oxygen consumption is also maintained in developing brain during hypoglycemia in newborn dogs (Hernández et al., 1980) and fetal sheep (Mann et al., 1973). Indirect assessment of cerebral energy use by the "closed head" technique (Lowry et al., 1964) indicates that brain energy use in newborn rats is also unchanged by hypoglycemia (Vannucci and Vannucci, 1978). Therefore, the brain must oxidize non-glucose fuels during hypoglycemia. As is the case with hypercapnia, consumption of endogenous intermediates and amino acids can suffice initially, and indeed, lactate and glutamate levels decrease in the brains of hypoglycemic developing rats (Davis, Himwich, and Pederson, 1970; Vannucci and Vannucci, 1978). During periods of more than a few minutes, however, other substrates must be used. Endogenous fatty acids already present in the brain have been suggested but not proved to be this energy source (Pappenheimer and Setchell, 1973; Agardh et al., 1981). As discussed above, blood lactate has been found to be oxidized at rapid rates by brain of hypoglycemic newborn dogs. Whether lactate is also an oxidative substrate for brain of other species is an important but unanswered question.

References

Agardh, C. D., Chapman, A. G., Nilsson, B., and Siesjo, B. K. (1981). Endogenous substrates utilized by rat brain in severe insulin-induced hypoglycemia. *J Neurochem* 36, 490–500.

Alexander, S. C., Smith, T. C., Strobel, G., Stephen, G. W., and Wollman, H. (1968). Cerebral carbohydrate metabolism of man during respiratory and metabolic alkalosis. *J Appl Physiol* 24, 66–72.

Banik, N. L., and Davison, A. N. (1969). Enzyme activity and composition of myelin and subcellular fractions in the developing rat brain. *Biochem J* 115, 1051–1062.

Barbeau, A., Melancon, S., Butterworth, R. F., Filla, A., Izumi, K., and Ngo, T. T. (1978). Pyruvate dehydrogenase complex in Friedreich's ataxia. *Adv Neurol* 21, 203–217.

Benavides, J., Gimenez, C., Valdivieso, F., and Mayor, F. (1976). Effect of phenylalanine metabolites on the activities of enzymes of ketone-body utilization in brain of suckling rats. *Biochem J* 160, 217–222.

Blass, J. P., Avigan, J., and Ulbendorf, B. W. (1970). A defect in pyruvate decarboxylase in a child with an intermittent movement disorder. *J Clin Invest* 49, 423–432.

Blass, J. P., Cederbaum, S. D., Kark, R. A. P., and Rodriquez-Budelli, M. (1978). Pyruvate dehydrogenase deficiency in 35 patients. *Monographs in Human Genetics* 9, 12–15.

Booth, R. F. G., and Clark, J. B. (1981). Energy metabolism in rat brain: Inhibition of pyruvate decarboxylation by 3-hydroxybutyrate in neonatal mitochondria. *J Neurochem* 37, 179–185.

Booth, R. F. G., Patel, T. B., and Clark, J. B. (1980). The development of enzymes of energy metabolism in the brain of a precocial (guinea pig) and non-precocial (rat) species. *J Neurochem* 34, 17–25.

Borgstrom, L., Norberg, K., and Siesjo, B. K. (1976). Glucose consumption in rat cerebral cortex in normoxia, hypoxia and hypercapnia. *Acta Physiol Scand* 96, 569–574.

Bowden, J. A., McArthur, C. L., III, and Fried, M. (1971). The inhibition of pyruvate decarboxylation in rat brain by α-ketoisocaproic acid. *Biochem Med* 5, 101–108.

Chase, H. P., Rodgerson, D. O., Lindsley, W., Jr., Thorne, T., and Cheung, G. (1976). Brain glucose utilization in undernourished rats. *Pediatr Res* 10, 102–107.

Clark, J. B., and Land, J. M. (1975). Phenylketonuria and maple syrup urine disease and their association with brain mitochondrial substrate utilization. In *Normal and pathological development of energy metabolism*, ed. F. A. Hommes and C. J. Van den Berg, pp. 177–192. New York: Academic Press.

Cornford, E. M., Braun, L. D., and Oldendorf, W. H. (1982). Developmental modulations of blood-brain barrier permeability as an indicator of changing nutritional requirements in the brain. *Pediatr Res* 16, 324–328.

Cremer, J. E. (1981). Nutrients for the brain: Problems in supply. *Early Hum Dev* 5, 117–132.

Cremer, J. E., Braun, L. D., and Oldendorf, W. H. (1976). Changes during development in transport processes of the blood-brain barrier. *Biochim Biophys Acta* 448, 633–637.

Cremer, J. E., Cunningham, V. J., Pardridge, W. M., Braun, L. D., and Oldendorf, W. H. (1979). Kinetics of blood-brain barrier transport of pyruvate, lactate and glucose in suckling, weanling and adult rats. *J Neurochem* 33, 439–445.

Cremer, J. E., and Heath, D. F. (1974). The estimation of rates of utilization of glucose and ketone bodies in the brain of the suckling rat using compartmental analysis of isotopic data. *Biochem J* 142, 527–544.

Cremer, J. E., and Teal, H. M. (1974). The activity of pyruvate dehydrogenase in rat brain during postnatal development. *FEBS Lett* 39, 17–20.

Dahlquist, G. (1976). Cerebral utilization of glucose, ketone bodies and oxygen in starving infant rats and the effect of intrauterine growth retardation. *Acta Physiol Scand* 98, 237–247.

Dahlquist, G., and Persson, B. (1976). The rate of cerebral utilization of glucose, ketone bodies, and oxygen: A comparative *in vivo* study of infant and adult rats. *Pediatr Res* 10, 910–917.

Daniel, P. M., Love, E. R., Moorehouse, S. R., and Pratt, O. E. (1977). The transport of ketone bodies into the brain of the rat (in vivo). *J Neurol Sci* 34, 1–13.

Daniel, P. M., Love, E. R., Moorehouse, S. R., Pratt, O. E., and Wilson, P. (1971). Factors influencing utilization of ketone-bodies by brain in normal rats and rats with ketoacidosis. *Lancet* 2, 637–638.

Davis, J. M., Himwich, W. A., and Pederson, V. C. (1970). Hypoglycemia and developmental changes in free amino acids of rat brain. *J Appl Physiol* 29, 219–222.

Della Porta, P., Maiolo, A. T., Negri, V. U., and Rossella, E. (1964). Cerebral blood flow and metabolism in therapeutic insulin coma. *Metabolism* 13, 131–140.

DeVivo, D. C., Haymond, M. W., Obert, K. A., Nelson, J. S., and Pagliara, A. S. (1979). Defective activation of the pyruvate dehydrogenase complex in subacute necrotizing encephalomyelopathy (Leigh disease). *Ann Neurol* 6, 483–494.

DeVivo, D. C., Leckie, M. P., Ferrendelli, J. S., and McDougal, D. B., Jr. (1978). Chronic ketosis and cerebral metabolism. *Ann Neurol* 3, 331–337.

Diebler, M. F., Farkas-Bargeton, E., and Wehrle, R. (1979). Developmental changes of enzymes associated with energy metabolism and the synthesis of some neurotransmitters in discrete areas of human neocortex. *J Neurochem* 32, 429–435.

Duffy, T. E., Kohle, S. J., and Vannucci, R. C. (1975). Carbohydrate and energy metabolism in perinatal rat brain: Relation to survival in anoxia. *J Neurochem* 24, 271–276.

Eisenberg, S., and Seltzer, H. (1962). The cerebral metabolic effects of acutely induced hypoglycemia in human subjects. *Metabolism* 11, 1162–1168.

Fernandes, J., Berger, R., and Smit, G. P. A. (1982). Lactate as energy source for brain in glucose-6-phosphatase deficient child. *Lancet* 1, 113.

Folbergrova, J., Norberg, K., Quistorff, B., and Siesjo, B. K. (1975). Carbohydrate and amino acid metabolism in rat cerebral cortex in moderate and extreme hypercapnia. *J Neurochem* 25, 457–462.

Gardiner, R. M. (1980). The effects of hypoglycemia on cerebral blood flow and metabolism in the new-born calf. *J Physiol* 298, 37–51.

Gjedde, A., and Crone, C. (1975). Induction processes in blood-brain transfer of ketone bodies in starvation. *Am J Physiol* 229, 1165–1169.

Gregoire, N. M., Gjedde, A., Plum, F., and Duffy, T. E. (1978). Cerebral blood flow and cerebral metabolic rates for oxygen, glucose, and ketone bodies in newborn dogs. *J Neurochem* 30, 63–69.

Gregson, N. A., and Williams, P. L. (1969). A comparative study of brain and liver mitochondria from new-born and adult rats. *J Neurochem* 16, 617–626.

Hawkins, R. A., Hass, W. K., and Ransohoff, J. (1979). The measurement of regional brain glucose utilization *in vivo* using [2-^{14}C]glucose. *Stroke* 10, 690–703.

Hawkins, R. A., Williamson, D. H., and Krebs, H. A. (1971). Ketone-body utilization by adult and suckling rat brain *in vivo*. *Biochem J* 122, 13–18.

Hellmann, J., Vannucci, R. C., and Nardis, E. E. (1982). Blood-brain barrier permeability to lactic acid in the newborn dog: Lactate as a cerebral metabolic fuel. *Pediatr Res* 16, 40–44.

Hernández, M. J., Brennan, R. W., Vannucci, R. C., and Bowman, G. S. (1978). Cerebral blood flow and oxygen consumption in the newborn dog. *Am J Physiol* 234, R209–R215.

Hernández, M. J., Vannucci, R. C., Salcedo, A., and Brennan, R. W. (1980). Cerebral blood flow and metabolism during hypoglycemia in newborn dogs. *J Neurochem* 35, 622–628.

Himwich, W. A., Benaron, H. B. W., Tucker, B. E., Babuna, C., and Stripe, M. (1959). Metabolic studies on perinatal human brain. *J Appl Physiol* 14, 873–877.

Holtzman, D., and Hsu, J. S. (1976). Early effects of inorganic lead on immature rat brain mitochondrial respiration. *Pediatr Res* 10, 70–75.

Holtzman, D., and Moore, C. L. (1973). Oxidative phosphorylation in immature rat brain mitochondria. *Biol Neonate* 22, 230–242.

Holtzman, D., and Moore, C. L. (1975). Respiration in immature rat brain mitochondria. *J Neurochem* 24, 1011–1015.

Jilek, L. (1970). The reaction and adaptation of the central nervous system to stagnant hypoxia and anoxia during ontogeny. In *Developmental neurobiology*, ed. W. A. Himwich, pp. 331–369. Springfield, Ill.: Charles C. Thomas.

Jones, M. D., Jr. (1979). Energy metabolism in the developing brain. *Semin Perinatol* 3, 121–129.

Jones, M. D., Jr., Burd, L. I., Makowski, E. L., Meschia, G., and Battaglia, F. C. (1975). Cerebral metabolism in sheep: A comparative study of the adult, the lamb, and the fetus. *Am J Physiol* 229, 235–239.

Kark, R. A. P., Rodriquez-Budelli, M., and Blass, J. P. (1978). Evidence for a primary defect of lipoamide dehydrogenase in Friedreich's ataxia. *Adv Neurol* 21, 163–180.

Kark, R. A. P., Rosenberg, R. N., and Schut, L. J., eds. (1978). Discussion: Oxidative disorders. *Adv Neurol* 21, 245–256.

Kennedy, C., and Sokoloff, L. (1957). An adaptation of the nitrous oxide method to the study of the cerebral circulation in children; Normal values for cerebral blood flow and cerebral metabolic rate in childhood. *J Clin Invest* 36, 1130–1137.

Kety, S. S., and Schmidt, C. F. (1948). The nitrous oxide method for the quantitative determination of cerebral blood flow in man: Theory, procedure and normal values. *J Clin Invest* 27, 476–483.

Kety, S. S., Woodford, R. B., Harmel, M. H., Freyhan, F. A., Appel, K. E., and Schmidt, C. F. (1948). Cerebral blood flow and metabolism in schizophrenia: The effects of barbiturate semi-narcosis, insulin coma and electroshock. *Am J Psychiatry* 104, 765–770.

Koster, J. F., Slee, R. G., and Fernandes, J. (1978). Lactic acidosis due to a defect in the pyruvate dehydrogenase complex: A possible brain pyruvate dehydrogenase phosphatase deficiency. *Monographs in Human Genetics* 9, 7–11.

Kraus, H., Schlenker, S., and Schwedesky, D. (1974). Developmental changes of cerebral ketone body utilization in human infants. *Hoppe Seylers Z Physiol Chem* 355, 164–170.

Kuhlman, R. E., and Lowry, O. H. (1956). Quantitative histochemical changes during the development of the rat cerebral cortex. *J Neurochem* 1, 173–180.

Land, J. M., Booth, R. F. G., Berger, R., and Clark, J. B. (1977). Development of mitochondrial energy metabolism in rat brain. *Biochem J* 164, 339–348.

Land, J. M., and Clark, J. B. (1975). The changing pattern of brain mitochondrial substrate utilization during development. In *Normal and pathological development of energy metabolism*, ed. F. A. Hommes and C. J. Van den Berg, pp. 155–167. New York: Academic Press.

Levitsky, L. L., Fisher, D. E., Paton, J. B., and Delannoy, C. W. (1977). Fasting plasma levels of glucose, acetoacetate, D-β-hydroxybutyrate, and lactate in the baboon infant: Correlation with cerebral uptake of substrates and oxygen. *Pediatr Res* 11, 298–302.

Lowry, O. H., Passonneau, J. V., Hasselberger, F. X., and Schulz, D. W. (1964). Effect of ischemia on known substrates and cofactors of the glycolytic pathway in brain. *J Biol Chem* 239, 18–30.

MacDonnell, P. C., and Greengard, O. (1974). Enzymes in intracellular organelles of adult and developing rat brain. *Arch Biochem Biophys* 163, 644–655.

Mahajan, R. G., Mandal, S., Sinha, A. K., and Mukherjee, K. L. (1982). Succinate dehydrogenase, monoamine oxidase and glutamine synthetase in developing human foetal brain regions. *J Neurochem* 38, 356–359.

Mann, L. I., Duchin, S., Halverstram, J., Mastrantonio, J., Weiss, R., and Schulman, J. (1973). The effect of hypoglycemia on fetal brain function and metabolism. *Am J Obstet Gynecol* 117, 45–50.

Medina, M. A., Deam, A. P., and Stavinoha, W. B. (1980). Inactivation of brain tissue by microwave irradiation. In *Cerebral metabolism and neural function*, ed. J. V. Passonneau, R. A. Hawkins, W. D. Lust, and F. A. Welsh, pp. 56–69. Baltimore, Md.: Williams & Wilkins.

Medina, M. A., Jones, D. J., Stavinoha, W. B., and Ross, D. H. (1975). The levels of labile intermediary metabolites in mouse brain following rapid tissue fixation with microwave irradiation. *J Neurochem* 24, 223–227.

Mehta, S., Nain, C. K., and Kalsi, H. K. (1980). Energy metabolism of the brain in malnutrition: An experimental study in young rhesus monkeys. *J Med Primatol* 9, 335–342.

Melancon, S. B., Potier, M., Dallaire, L., Rollin, P., Fontaine, G., and Grenier, B. (1979). Pyruvate dehydrogenase, lipoamide dehydrogenase and citrate synthase activity in fibroblasts from patients with Friedreich's and Charlevoix-Saguenay ataxia. *Can J Neurol Sci* 6, 241–242.

Middleton, B. (1973). The acetoacetyl-coenzyme A thiolases of rat brain and their relative activities during postnatal development. *Biochem J* 132, 731–737.

Miller, A. L., and Corddry, D. H. (1981). Brain carbohydrate metabolism in developing rats during hypercapnia. *J Neurochem* 36, 1202–1210.

Miller, A. L., Hawkins, R. A. and Veech, R. L. (1975). Decreased rate of glucose utilization by rat brain *in vivo* after exposure to atmospheres containing high concentrations of CO_2. *J Neurochem* 25, 553–558.

Miller, A. L., Kiney, C. A., Corddry, D. H., and Staton, D. M. (1982). Interactions between glucose and ketone body use by developing brain. *Developmental Brain Research* 256, 443–450.

Miller, A. L., Kiney, C. A., and Staton, D. M. (1984). Effects of lactate on glucose metabolism of developing rat brain. *Developmental Brain Research* 14, 33–40.

Miller, A. L., and Shamban, A. (1977). A comparison of methods for stopping intermediary metabolism of developing rat brain. *J Neurochem* 28, 1327–1334.

Milstein, J. M., White, J. G., and Swaiman, K. F. (1968). Oxidative phosphorylation in mitochondria of developing rat brain. *J Neurochem* 15, 411–415.

Moore, T. J., Lione, A. P., Sugden, M. C., and Regen, D. M. (1976). β-Hydroxybutyrate transport in rat brain: Developmental and dietary modulations. *Am J Physiol* 230, 619–630.

Mourek, J. (1970). Oxidative metabolism of nervous tissue during ontogeny in rat. In *Developmental neurobiology*, ed. W. A. Himwich, pp. 370–390. Springfield, Ill.: Charles C. Thomas.

Nemoto, E. M., Hoff, J. T., and Severinghaus, J. W. (1974). Lactate uptake and metabolism by brain during hyperlactatemia and hypoglycemia. *Stroke* 5, 48–53.

Nemoto, E. M., and Severinghaus, J. W. (1974). Stereospecific permeability of rat blood-brain barrier to lactic acid. *Stroke* 5, 81–84.

"News and Comment." (1982). *Science* 216, 711–712.

Norberg, K., and Siesjo, B. K. (1976). Oxidative metabolism of the cerebral cortex of the rat in severe insulin-induced hypoglycemia. *J Neurochem* 26, 345–352.

Oldendorf, W. H. (1972). Blood brain barrier permeability to lactate. *Eur Neurol* 6, 49–55.

Oldendorf, W. H. (1973). Carrier-mediated blood-brain barrier transport of short-chain monocarboxylic organic acids. *Am J Physiol* 224, 1450–1453.

Owen, O. E., Morgan, A. P., Kemp, H. G., Sullivan, J. M., Herrera, M. G., and Cahill, G. F., Jr. (1967). Brain metabolism during fasting. *J Clin Invest* 46, 1589–1595.

Ozand, P. T., Stevenson, J. H., Tildon, J. T., and Cornblath, M. (1975). The effects of hyperketonemia on glutamate and glutamine metabolism in developing rat brain. *J Neurochem* 25, 67–71.

Page, M. A., Krebs, H. A., and Williamson, D. H. (1971). Activities of enzymes of ketone-body utilization in brain and other tissues of suckling rats. *Biochem J* 121, 49–53.

Page, M. A., and Williamson, D. H. (1971). Enzymes of ketone-body utilization in human brain. *Lancet* 2, 66–68.

Pappenheimer, J. R., and Setchell, B. P. (1973). Cerebral glucose transport and oxygen consumption in sheep and rabbits. *J Physiol* 233, 529–551.

Patel, A. J., and Balazs, R. (1975). Factors affecting the development of metabolic compartmentation in the brain. In *Metabolic compartmentation and neurotransmission*, ed. S. Berl, D. D. Clarke, and D. Schneider, pp. 363–383. New York: Plenum Press.

Patel, M. S. (1974a). The effect of ketone bodies on pyruvate carboxylation by rat brain mitochondria. *J Neurochem* 23, 865–867.

Patel, M. S. (1974b). Inhibition by the branched-chain 2-oxo acids of the 2-oxoglutarate dehydrogenase complex in developing rat and human brain. *Biochem J* 144, 91–97.

Patel, M. S. (1979). Influence of neonatal hypothyroidism on the development of ketone-body-metabolizing enzymes in rat brain. *Biochem J* 184, 169–172.

Patel, M. S., Auerbach, V. H., Grover, W. D., and Wilbur, D. O. (1973). Effect of the branched-chain α-keto acids on pyruvate metabolism by homogenates of human brain. *J Neurochem* 20, 1793–1796.

Patel, M. S., Grover, W. D., and Wilbur, D. O. (1973). Pyruvate metabolism by homogenates of human brain: Effects of phenylpyruvate and implications for the etiology of the mental retardation in phenylketonuria. *J Neurochem* 20, 289–296.

Patel, M. S., Johnson, C. A., Rajan, R., and Owen, O. E. (1975). The metabolism of ketone bodies in developing human brain: Development of ketone-body-utilizing enzymes and ketone bodies as precursors for lipid synthesis. *J Neurochem* 25, 905–908.

Patel, T. B., Booth, R. F. G., and Clark, J. B. (1977). Inhibition of acetoacetate oxidation by brain mitochondria from the suckling rat by phenylpyruvate and α-ketoisocaproate. *J Neurochem* 29, 1151–1153.

Patel, T. B., and Clark, J. B. (1981). Mitochondrial/cytosolic carbon transfer in the developing rat brain. *Biochim Biophys Acta* 677, 373–380.

Persson, B., Settergren, G., and Dahlquist, G. (1972). Cerebral arteriovenous difference of acetoacetate and D-β-hydroxybutyrate in children. *Acta Paediatr Scand* 61, 273–278.

Pigareva, Z. D. (1968). Oxidizing activity of the brain mitochondria in postnatal ontogenesis. In *Ontogenesis of the brain*, ed. L. Jilek and P. Trojan, pp. 245–254. Prague: Universitas Carolina Pragensis.

Pitts, F. N., Jr., and Quick, C. (1967). Brain succinate semialdehyde dehydrogenase—II. Changes in the developing rat brain. *J Neurochem* 14, 561–570.

Pollay, M., and Stevens, F. A. (1980). Starvation-induced changes in transport of ketone bodies across the blood-brain barrier. *J Neurosci Res* 5, 163–172.

Reichard, G. A., Haff, A. C., Skutches, C. L., Paul, P., Holroyde, C. P., and Owen, O. E. (1979). Plasma acetone metabolism in the fasting human. *J Clin Invest* 63, 619–626.

Reijnierse, G. L. A., Veldstra, H., and Van den Berg, C. J. (1975). Short-chain fatty acid synthases in brain. *Biochem J* 152, 477–484.

Robins, E., and Lowe, I. P. (1961). Quantitative histochemical studies of the morphogenesis of the cerebellum. I—Total lipid and four enzymes. *J Neurochem* 8, 81–95.

Robinson, A. M., and Williamson, D. H. (1980). Physiological roles of ketone bodies as substrates and signals in mammalian tissues. *Physiol Rev* 60, 143–187.

Robinson, B. H., Sherwood, W. G., Kahler, S., O'Flynn, M. E., and Nadler, H. (1981). Lipoamide dehydrogenase deficiency. *N Engl J Med* 304, 53–54.

Robinson, B. H., Taylor, J., and Sherwood, W. G. (1980). The genetic heterogeneity of lactic acidosis: Occurrence of recognizable inborn errors of metabolism in a pediatric population with lactic acidosis. *Pediatr Res* 14, 956–962.

Ruderman, N. B., Ross, P. S., Berger, M., and Goodman, M. N. (1974). Regulation of glucose and ketone-body metabolism in brain of anaesthetized rats. *Biochem J* 138, 1–10.

Sabster, J., Briones, P., Chabas, A., Pampols, T., Rodes, M., and Gonzalez-Sastre, F. (1979). Biochemical parameters in perinatal human brain. In *Neural growth and differentiation*, ed. E. Meisami and M. A. B. Brazier, pp. 433–459. New York: Raven Press.

Schwartz, R., and Kalhan, S. (1975). Energy sources for neonatal brain metabolism. In *Preventability of perinatal injury*, ed. K. Adamsons and H. A. Fox, pp. 187–200. New York: Alan Liss.

Settergren, G., Lindblad, B. S., and Persson, B. (1980). Cerebral blood flow and exchange of oxygen, glucose, ketone bodies, lactate, pyruvate and amino acids in anesthetized children. *Acta Paediatr Scand* 69, 457–465.

Sherman, T. G., and Wilson, J. E. (1978). Effects of prenatally-induced and postnatally-maintained ketosis on β-hydroxybutyrate dehydrogenase and hexokinase levels in the developing rat brain. *J Neurochem* 30, 639–641.

Sims, K. L., and Pitts, F. N., Jr. (1970). Brain glutamate decarboxylase: Changes in the developing rat brain. *J Neurochem* 17, 1607–1612.

Sims, K. L., Witztum, J., Quick, C., and Pitts, F. N., Jr. (1968). Brain 4-aminobutyrate: 2-oxoglutarate aminotransferase: Changes in the developing rat brain. *J Neurochem* 15, 667–672.

Sokoloff, L., Reivich, M., Kennedy, C., DesRosiers, M. H., Patlak, C. S., Pettigrew, K. D., Sakurada, O., and Shinohara, M. (1977). The [^{14}C]-deoxyglucose method for the measurement of local cerebral glucose utilization: Theory, procedure and normal values in the conscious and anesthetized albino rat. *J Neurochem* 28, 897–916.

Spatz, M., Micic, D., Mrsulja, B. B., Swink, M., and Micic, J. (1978). Changes in the capillary lactate and 2-deoxy-D-glucose uptake in developing brain. *Brain Res* 151, 619–622.

Stryer, L. (1981). *Biochemistry*. San Francisco: Freeman.

Stumpf, D. A., and Parks, J. K. (1978). Friedreich's ataxia: I. Normal pyruvate dehydrogenase complex activity in platelets. *Ann Neurol* 4, 366–368.

Swaiman, K. F. (1970). Energy and electrolyte changes during maturation. In *Developmental neurobiology*, ed. W. A. Himwich, pp. 318–330. Springfield, Ill.: Charles C. Thomas.

Swaiman, K. F., and Milstein, J. M. (1965). Oxidation of leucine, isoleucine and related ketoacids in developing rabbit brain. *J Neurochem* 12, 981–986.

Swiatek, K. R., Chao, K. L., Chao, H. L., and Dombrowski, G. J., Jr. (1979). The utilization of DL-[3-^{14}C]hydroxybutyrate by malnourished rat pups. *Neuroscience* 4, 1751–1760.

Thurston, J. H., Prensky, A. L., Warren, S. K., and Albone, K. R. (1971). The effects of undernutrition upon the energy reserve of the brain and upon other selected metabolic intermediates in brains and livers of infant rats. *J Neurochem* 18, 161–166.

Tildon, J. T., Ozand, P. T., and Cornblath, M. (1975). The effects of hyperketonemia on neonatal brain metabolism. In *Normal and pathological development of energy metabolism*, ed. F. A. Hommes and C. J. Van den Berg, pp. 143–154. New York: Academic Press.

Vannucci, R. C., Nardis, E. E., Vannucci, S. J., and Campbell, P. A. (1981). Cerebral carbohydrate and energy metabolism during hypoglycemia in newborn dogs. *Am J Physiol* 240, R192–R199.

Vannucci, R. C., and Vannucci, S. J. (1978). Cerebral carbohydrate metabolism during hypoglycemia and anoxia in newborn rats. *Ann Neurol* 4, 73–79.

Veech, R. L., Harris, R. L., Veloso, D., and Veech, E. H. (1973). Freeze-blowing: A new technique for the study of brain *in vivo*. *J Neurochem* 20, 183–188.

Wapnir, R. A., Tildon, J. T., and Cornblath, M. (1973). Metabolic differences in offspring of rats fed high-fat and control diets. *Am J Physiol* 224, 596–599.

Warshaw, J. B., and Terry, M. L. (1976). Cellular energy metabolism during fetal development. VI. Fatty acid oxidation by developing brain. *Dev Biol* 52, 161–166.

Weng, J. T., Nakamura, Y., and Spitzer, J. J. (1973). Arterial concentration and cerebral removal of metabolites in fasting puppies. *Am J Physiol* 225, 967–971.

Wilbur, D. O., and Patel, M. S. (1974). Development of mitochondrial pyruvate metabolism in rat brain. *J Neurochem* 22, 709–715.

Wilson, J. E. (1972). The relationship between glycolytic and mitochondrial enzymes in the developing rat brain. *J Neurochem* 19, 223–227.

Wu, C. (1964). Glutamine synthetase. III. Factors controlling its activity in the developing rat. *Arch Biochem Biophys* 106, 394–401.

Zubairu, S., Hothersall, J. S., McLean, P., and Greenbaum, A. L. (1982). Age-related changes in enzymes of rat brain. III. Hydrogen-transfer systems in relation to the disposition of acetyl groups in the brain. *Enzyme* 27, 130–136.

Ancillary Pathways of Energy Metabolism in Mammalian Brain: The Pentose Phosphate Pathway and Galactose Metabolism

C. J. Cummins, David J. Loreck, and David W. McCandless

The Pentose Phosphate Shunt

Introduction

The pentose phosphate shunt (also called the *hexose monophosphate pathway* or *phosphogluconate pathway*) is an alternate route for hexose metabolism in most tissues, including brain. The pathway is multifunctional, contributing reducing equivalents and sugar phosphates of three to seven carbons. The pentose phosphate pathway was discovered by Warburg, Christian, and Griese (1935) and Lippman (1936), reasoning that the Embden-Meyerhof-Parnas pathway was not adequate to account for the synthesis of substituted monosaccharides being formed within cells. Although they knew the Embden-Meyerhof-Parnas pathway was the route of the vast majority of hexose flux, the existence of pentoses, riboses, heptuloses, and their substituted derivatives clearly implied the function of another pathway.

The enzymes of the pathway were first documented by Warburg and coworkers (1935) and Lippman (1936) who showed the successive oxidation of glucose to ribulose-5-phosphate by glucose-6-phosphate dehydrogenase (*Zwischenferment*), and 6-phosphogluconate dehydrogenase. The subsequent steps (see below) were elucidated by a variety of investigators such as Racker, Dickens, Horecker, and others in the 1940s and 1950s.

Although the structure of the pentose phosphate pathway is well characterized in mammalian tissues including brain, the full extent of its function, metabolic role, relationship with other pathways, and regulation appears to be related to the specific tissue and is not completely clear for the mammalian central nervous system. Our knowledge of the pentose phosphate pathway is still in the formative stage. This chapter critically examines the major points of the experimental elucidation of the structure, function, regulation, and development of the pentose pathway in the mammalian brain. Because the development of the pentose phosphate pathway in the mammalian brain was reviewed by O'Neill

(1974), this chapter is largely concerned with the work appearing in the literature since that time.

Structure of the Pathway

The pentose pathway is commonly recognized as having an oxidative and a nonoxidative "arm," two separate routes for the synthesis of ribose sugars from glucose-6-phosphate (fig. 1). The so-called oxidative arm is the better characterized in terms of flux rates and contribution to the pool of ribose sugars. The first enzyme in this pathway is glucose-6-phosphate dehydrogenase (*d*-glucose-6-phosphate: NADP oxidoreductase, EC 1.1.1.49), which forms a lactone in the position (carbon 1) by transferring a pair of electrons to NADP. The triphosphopyridine nucleotide is the obligatory cofactor in this and the next oxidoreductase step. The lactone probably spontaneously hydrolyzes to form 6-phosphogluconate, although lactonases have been described for liver. The next step in the pathway is the second oxidative step, forming ribulose-5-phosphate and CO_2 from the carboxyl group in the 1 position, and that is catalyzed by 6-phosphogluconate dehydrogenase (6-phosphogluconate: $NAD[P]^+$ oxidoreductase, EC 1.1.1.43). Consideration of the energetics of the two oxidoreductase steps suggests that both reactions proceed to the right. Ribose-5-phosphate isomerase (*d*-ribose-5-phosphate ketol-

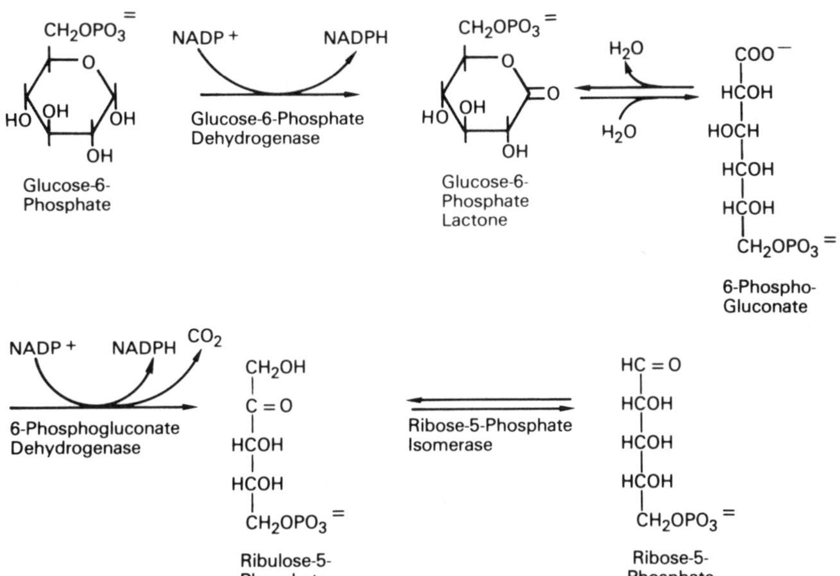

Figure 1. The intermediates and enzymes of the oxidative arm of the pentose phosphate pathway.

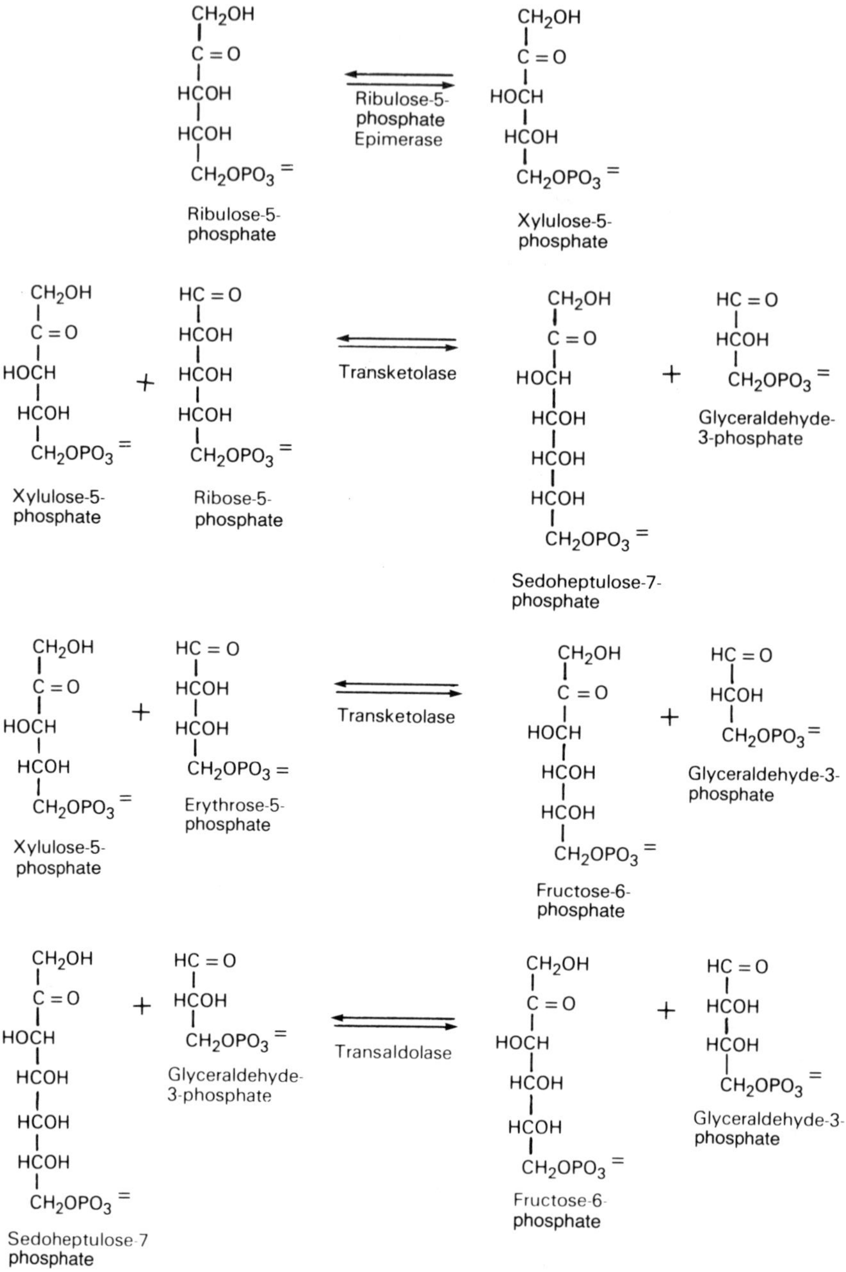

Figure 2. The intermediates and enzymes of the nonoxidative arm of the pentose phosphate pathway.

isomerase, EC 5.3.1.6) catalyzes the formation of ribose-5-phosphate from ribulose-5-phosphate.

The nonoxidative arm of the pentose phosphate pathway catalyzes several reactions, all of which appear readily reversible, suggesting that the flux through this part of the pentose pathway responds passively to glycolytic flux as well as ribose-5-phosphate levels (fig. 2). Xylulose-5-phosphate (X5P) and ribulose-5-phosphate are interconverted by epimerization about carbon 3 by ribulose phosphate epimerase (D-ribulose-5-phosphate 3-epimerase, EC 5.1.3.1). This forms an alternative pathway to remove five member sugar phosphates from the oxidative arm. Pools of X5P produced can then donate a glycoaldehyde group $\left(\begin{matrix}C-H_2-OH \\ | \\ C=O\end{matrix}\right)$ to the aldehyde groups of either ribose-5-phosphate or erythrose-4-phosphate to form glyceraldehyde-3-phosphate and sedoheptulose-7-phosphate or fructose-6-phosphate, respectively. The enzyme that catalyzes these reactions is transketolase (sedoheptulose-7-phosphate: *d*-glyceraldehyde-3-phosphate glycoaldehydetransferase, EC 2.2.1.1). Sedoheptulose-7-phosphate can then donate a dihydroxyacetone group to glyceraldehyde-3-phosphate, forming fructose-6-phosphate and erythrose phosphate. This enzyme, because its mechanism of action is considered analogous to fructose diphosphate aldolase (fructose-1, 6-diphosphate *d*-glyceraldehyde-3-phosphate lyase, EC 4.1.2.13), is named *transaldolase* (sedoheptulose-7-phosphate: D-glyceraldehyde-3-phosphate dihydroxyacetone transferase, EC 1.2.2.1.2).

Regulation of the Pentose Phosphate Pathway in the Adult Brain

The structure of the pentose phosphate pathway suggests that different features regulate the oxidative and nonoxidative arms; therefore, we will consider these separately.

Oxidative Arm. C-1 : C-6 RATIO METHOD. Flux through the oxidative portion of the pentose pathway can be measured by the administration of glucose labeled in the carbon 1 position (C-1). The action of 6-phosphogluconate dehydrogenase (6PGDH) liberates $^{14}CO_2$, which can be collected and compared with the amount of $^{14}CO_2$ liberated from glucose labeled in the carbon 6 position (C-6). Since $^{14}CO_2$ can be derived from C-1 in brain by either metabolism through the pentose phosphate pathway or by oxidation in the tricarboxylic acid (TCA) cycle but $^{14}CO_2$ from glucose at C-6 can be derived predominantly if not exclusively in brain from the TCA cycle, the ration of $^{14}CO_2$ obtained from glucose C-1 to $^{14}CO_2$ from glucose C-6 (C-1 : C-6) has been used as a crude estimate of the relative fluxes through the pentose phosphate pathway and through the TCA cycle. Although there is wide agreement that this ratio does not represent a true quantitative relationship (see Katz and Wood [1960]; Landau, Bartsch, Katz, and Wood [1964]; Landau and Katz [1964]; and Katz, Landau, and Bartsch [1966] for a critical discussion of the inherent assumptions of this method), the C-1 : C-6 ratio is widely used to qualitatively assess pentose phosphate pathway activity in brain. A C-1 : C-6 ratio approximating 1.0 ± 0.1 has been observed for the adult human brain in situ (Sacks, 1957, 1961), adult rabbit brain slices (Bloom, 1955), adult guinea pig brain cortex slices (O'Neill, Simon, and Shreeve, 1965), bovine brain cortex slices (Krass and Labella, 1967), and in adult rat brain cortex slices (DiPietro and Weinhouse, 1959; Kimura et al., 1974; McCandless, Curley, and Cassidy, 1976; Hothersall, Baquer, Greenbaum, and McLean, 1979; Brannan, Maker, and Karp, 1982; Hothersall, Baquer, and McLean, 1982).

In other nerve tissue, such as peripheral nerve, the C-1 : C-6 ratio approaches 2.0 (Hothersall, Baquer, and McLean, 1982). Synaptosomes appear to have a relatively rapid pentose phosphate pathway activity with C-1 : C-6 ratios of 2.0 or greater (Appel and Parrot, 1970; Bauer and Brand, 1982; Hothersall, Baquer, and McLean, 1982). As one may infer from ultrastructural and other biochemical evidence, the elements of the nervous system are not homogeneous with respect to the activity of the pentose phosphate pathway.

Katz, Landau, and Bartsch (1966) have suggested the following formulation for estimating flux rates quantitatively:

$$Flux = \frac{S}{3 - 2S},$$

where

$$S = \frac{\text{G-1 CO} - \text{G-6 CO}}{1 - (\text{G-6 CO}_2)},$$

where GCO_2 is the specific yield of CO_2. They define the specific yield as the amount of glucose consumed relative to the CO_2 produced. Using this formula, researchers have estimated glucose flux through the oxidative arm of the pentose phosphate pathway to be less than 5 percent of the total glucose flux in the adult mammalian brain (Krass and Labella, 1967; Hothersall, Baquer, and McLean, 1982).

The main regulatory step for the oxidative arm is likely to be the first and committed step, glucose-6-phosphate dehydrogenase (G6PDH). The liver enzyme is allosterically inhibited by 6-phosphogluconate and by NADPH (the reduced form of nicotinamide-adenine dinucleotide phosphate) (Eggleston and Krebs, 1974). The NADPH inhibition is partly relieved by oxidized glutathione, and partly by AMP (adenosine monophosphate) but not in the physiological range (Eggleston and Krebs, 1974). The purified brain enzyme is inhibited by Pi (phosphate), SO4 (sulfate), and ATP (adenosine triphosphate) (apparently competitive with G6P) and by NADPH (competitive with $NADP^+$, the oxidized form of NADP) (Passonneau, Schulz, and Lowry, 1966). The second dehydrogenase step may be regulatory as well since the levels of ribulose-5-phosphate and X5P changed in a direction opposite to 6-phosphogluconate after ischemic insult (Kauffman et al., 1967).

Flux through the oxidative arm may be largely controlled by the availability of the cofactor $NADP^+$. The ratios of cytosolic [$NADP^+$]: [NADPH] have been estimated to be in the range of 0.028–0.034 by Merrill and Guynn (1981). When NADP is added to brain homogenates, evolution of $^{14}CO_2$ from the glucose C-1 is increased, but $^{14}CO_2$ derived from C-6 is unaffected (DiPietro and Weinhouse, 1959; Guerra, Melgar, and Villavicencio, 1967). An artificial electron acceptor, such as phenazine methosulfate, and methylene blue increase the intracellular levels of $NADP^+$ and increase the fractional C-1 oxidized through the pentose phosphate pathway (O'Neill, Simon, and Shreeve, 1965; Hothersall et al., 1979; El-Hassan et al., 1981; Hothersall, Baquer, and McLean, 1982; Hothersall, Greenbaum, and McLean, 1982). Administration of pyrroline-5-carboxylate also increases the pentose phosphate flux in human brain, probably by conversion to proline by pyrroline-5-carboxylate reductase. This enzyme uses the cofactor NADPH, generating $NADP^+$ (Phang et al., 1982).

An interesting series of observations have recently been made by Brannan, Maker, and Karp (1982) and Hothersall, Greenbaum, and McLean (1982) that suggest that the pentose phosphate pathway in synaptosomes may be stimulated by catecholamines. The favored current theory suggests that the peroxides generated by catecholamine oxidation may be detoxified by glutathione peroxidase and glutathione reduction. Glutathione reductase uses NADPH as a cofactor, and metabolism of peroxide would increase the $NADP^+$ levels in synaptosomes, thereby stimulating the pentose phosphate pathway.

Electrical stimulation of brain slices also increases the flux through the pentose phosphate pathway (Kimura et al., 1974). Although the authors did not suggest a mechanism, catecholamine release, oxidation, release of peroxides, and elevation of $NADP^+$ levels could account for their observations.

C-2 RANDOMIZATION METHOD. Randomization of carbon 2 in glucose to C-1 and C-2 of glycogen has also been used to estimate pentose pathway flux (Hostetler and Landau, 1967). While not widely used for studies on brain since that time, results from using the C-2 randomization method agree substantially with results of the C-1:C-6 ratio method in demonstrating that flux through the oxidative arm may be less than 5 percent of the overall glycolytic rate (Hostetler et al., 1970).

Nonoxidative Arm. Flux through the nonoxidative arm of the pentose phosphate pathway is more difficult to determine. Since the enzymes of the nonoxidative arm catalyze readily reversible reactions, flux direction may be determined largely by the concentrations of the various intermediates (see "Galactose Metabolism" below). Kauffman et al. (1967) have calculated that the maximum flux could approach 0.2 mmol of triose/min/kg wet wt, or 20–30 percent of the normal rate of triose consumption. Under normal circumstances the brain is not synthesizing ribonucleotides at a de novo rate that would support a flux of this magnitude. Their estimate must therefore be appreciated as the maximal catalytic capacity of the nonoxidative arm and not a measure of the true instantaneous flux.

Enzyme Activities and Intermediates in the Adult Brain

Although the flux rates of the pentose pathways are not substantially understood, the activities of the enzymes of the pathway, their kinetic constants, and the levels of the intermediates are known at least in part for several adult systems. The best characterized is the adult mouse brain (Kauffman et al., 1967; Kauffman, 1972). The levels of intermediates are listed in table 1, and the enzymatic activities are shown in table 2, taken from Kauffman et al. (1967) and Kauffman (1972). The activity measurements are roughly comparable to the results of El-Hassan et al. (1981) and Hothersall, Baquer, and McLean (1982), considering the different techniques of analysis and different species.

Kauffman (1972) noted that the maximum catalytic capacity of the pentose phosphate pathway could approach 2–4 mmol/kg wet wt/h. The V_{max} of all enzyme activities is in excess of the predicted flux, and thus the flux is not limited by the maximal catalytic capacity at any step. The steady state levels for all substrates, except for G6P, are below the K_m values for the individual enzymes. If one assumes Michaelis-

Table 1. *Levels of Pentose Phosphate Pathway Intermediates in Adult Mouse Brain*

Intermediates	*Level (mmol/kg wet wt)*
Glucose-6-phosphate	60.4
6-Phosphogluconate	9.7
Ribulose-5-phosphate	7.2
Xylulose-5-phosphate	14.2
Sedoheptulose-7-phosphate + ribose-5-phosphate	57.1
Erythrose-4-phosphate	2.0
Fructose-6-phosphate	9.9
Glyceraldehyde-phosphate	4.7
Dihydroxyacetone-phosphate	38.0

Source: Data from Kauffman et al. (1967).

Table 2. *Enzyme Kinetic Constants for Hexose Monophosphate Pathway in Adult Mouse Brain Homogenates*

Enzyme	*Substrates*	K_m *(mM)*	V_{max} *(mmol/kg dry wt/h)*
Glucose-6-phosphate dehydrogenase	G6P	0.019	112
6-Phosphogluconate dehydrogenase	6-PG	0.015	75
Ribose-5-phosphate isomerase	Ribose-5-P	4.85	349
Ribulose-5-phosphate epimerase	Ribulose-5-P	0.143	389
Transketolase	Xylulose-5-P	0.018	48
Transaldolase	Erythrose-4-P	0.004	46
	Fructose-6-P	0.055	

Source: Data from Kauffman (1972).
Note: G6P, glucose-6-phosphate; 6-PG, 6-phosphogluconate; P, phosphate.

Menten kinetics and takes into account the relative concentrations of the intermediates, ribose-5-phosphate isomerase and ribulose-5-phosphate epimerase have the lowest activities.

Pentose Phosphate Pathway as a Cycle

Though the pentose pathway can be represented as a linear pathway (oxidative arm) and a branched pathway (nonoxidative arm), consideration of the summed reactants and products suggests that under some circumstances the pentose phosphate pathway may function as a cycle, oxidizing G6P to CO_2 and reducing NADP+ (Hollman as cited in O'Neill, 1974):

1. $6G6P + 12NADP^+ + 6H_2 \rightarrow 6Ribulose\text{-}5 + 6CO_2 + 12NADPH + 12H$
2. 2Ribulose-5-phosphate → 2Ribose-5-phosphate
3. 4Ribulose-5-phosphate → 4Xylulose-5-phosphate
4. 2Xylose-5-phosphate + 2Ribose-5-phosphate → 2F-6-phosphate + 2Erythose-4-phosphate
5. 2Xylose-5-phosphate + 2Erythose-4-phosphate → 2-F6-phosphate + 2Glyceraldehyde-3-phosphate
6. 4F-6-phosphate → 4G6P
7. 2Glyceraldehyde-3-phosphate → G6P + Pi

If one cancels out common terms,

$$1G6P + 12NADP^{+} \rightarrow 6CO_2 + 12NADPH^{+} + 12H^{+} + Pi.$$

Such a cycle, while theoretically possible, has been very difficult to prove experimentally.

Functions of the Pentose Phosphate Pathway

Several functions have been ascribed to this pathway based upon considerations of the products. Pentose sugars are produced by this pathway, and thus one would expect the pentose phosphate pathway to be "driven" under conditions of high RNA (ribonucleic acid) synthesis. Although brain has a high rate of RNA turnover, little de novo ribonucleotide synthesis probably occurs.

Lipid synthesis requires NADPH, and as we have seen, the oxidative arm of the pathway is regulated by $NADP^{+}$ levels. During critical periods of active lipid synthesis, the pentose phosphate pathway would be expected to increase, thereby maintaining the ratios of [$NADP^{+}$]:[NADPH]. Kauffman (1972) has pointed out that the enzymes of the pentose phosphate pathway are remarkably constant from brain region to brain region, despite a substantial diversity (fourfold) of lipid. The only enzyme correlating with lipid content is G6PDH (Buell et al., 1958; McDougal et al., 1961; Kauffman, 1972). Thus, a simple functional correlation between the pentose phosphate pathway and lipid synthesis in the adult brain is probably not wholly correct.

The pentose phosphate pathway also responds to the alteration of [$NADP^{+}$]:[NADPH] arising from the detoxification of peroxides by the glutathione system. Given the rapid turnover of catecholamines in the adult CNS, this may constitute a major function of the pentose phosphate pathway.

RNA and lipid synthesis occurs in the adult nervous system at a constant rate, which can be accommodated by a pentose phosphate pathway flux of 1–5 percent of the total glucose flux. However, the developing nervous system has critical periods of RNA and lipid synthesis, and thus brain development may be reflected in a patterned development of the pentose phosphate pathway.

Developmental Aspects of the Pentose Phosphate Pathway

Most of the reports describing the development of brain pentose phosphate pathway enzymes are from studies using mammals, and most deal with only the postpartum period. Only Guerra, Melgar, and Villavicencio (1967) have investigated the pentose phosphate pathway in mammalian fetal brain (rat). G6PDH, 6PGDH, ribose-5-phosphate, and heptulose utilization were measured on fetal rat brain homogenates. Both enzymes had a lower activity in fetal brain, but ribose-5-phosphate and heptulose utilization were significantly greater in fetal tissue than in adult tissue. The C-1 : C-6 ratio for fetal brain minces was 2.1, suggesting elevated flux in relation to the rate of TCA cycle activity. Added NADP stimulated the appearance of $^{14}CO_2$ from C-1 three-fold in adult brain homogenate, but over 14-fold in the fetal tissues, suggesting that (1) the pentose phosphate pathway of fetal brain tissues is also limited by $NADP^+$ and (2) other regulatory factors are involved, since the magnitude of the stimulation is different in the fetal brain homogenate. Since the cellular levels of other known cofactors were not determined and the allosteric properties of the fetal enzyme are not known, it is impossible to assess the contributions by Pi, ATP, and NADPH.

Villee and Loring (1961) determined a C-1 : C-6 close to 1.0 in fetal human brain cortex slices in vitro eight to 25 weeks old; they interpreted these data to suggest that the pentose phosphate pathway was not significantly different in either fetal or adult human brain. However, prolonged ischemia seems to deleteriously affect the operation of the pentose phosphate pathway (Hothersall et al., 1979) and may have affected the preparations used by Villee and Loring (1961).

The early developmental pattern of a selected pentose phosphate pathway enzyme was examined in the chick by Burt and Wenger (1961) and Burt (1965). They measured the change in activity in G6PDH in the developing chick brain and brachial spinal cord at Hamburger-Hamilton stages 20–46 (four to eight days after fertilization). Aldolase was selected as a model glycolytic enzyme and did not change throughout development. The activity of brain G6PDH at the earliest stages examined was 3.95 nmol/h/mg protein and decreased rapidly to 40 percent of this level at hatching (Burt and Wenger, 1961). Burt (1965) found essentially the same pattern in the brachial spinal cord. The developmental time course in both these tissues followed the time of the greatest proliferative capacity of these structures but preceded the time course of maximal RNA synthesis. Burt and Wenger (1961) and Burt (1965) suggested that the pentose phosphate pathway played a major role in making ribose equivalents available for de novo ribonucleotide synthesis. If one assumes that the regulation of the chick brain pentose phosphate pathway is the same as in mammals and further assumes that G6PDH development is typical of the developmental events of the pathway, their hypothesis has some credibility.

During postnatal development in the rat brain, little change is seen in the activities of four of the six pentose phosphate pathway enzymes; 6PGDH, ribose-5-phosphate isomerase, ribulose-5-phosphate epimerase, and transketolase have attained their adult levels by day 1 postpartum (Baquer, McLean, and Greenbaum, 1973; Hothersall et al., 1979; and El-Hassan et al., 1981). G6PDH increases 26–40 percent after birth to attain adult levels (Kuhlman and Lowry, 1956; El-Hassan et al., 1981). Transaldolase declines by day 5 postpartum to attain adult levels (Baquer, McLean, and Greenbaum, 1973). Despite the relative constancy of the enzymatic machinery, Hothersall et al. (1979) have shown that $^{14}CO_2$ evolved from glucose—C-1 and C-6 both increase after birth in the rat, with the C-6 oxidative rate increasing faster than the C-1 oxidative rate (fig. 3). The consequence of this is that the C-1 : C-6 ratio of newborn rat brain slices approaches 2.0 and declines to adult levels in the first ten days of life. The calculated pentose phosphate pathway flux for day 1 rat brain cortex slices is 6 percent of the total glucose flux.

That the control of the pentose phosphate pathway may differ in the developing rat brain from that in adult rat brain is suggested by phenazine methosulfate (PMS) stimulation experiments on cortical slices from rats of different ages (Hothersall et al., 1979). The oxidation of C-6 is unaffected by PMS at any developmental stage, but 0.1 mM of PMS stimulates C-1 production to a maximum of 20 mM/g/h irrespective of

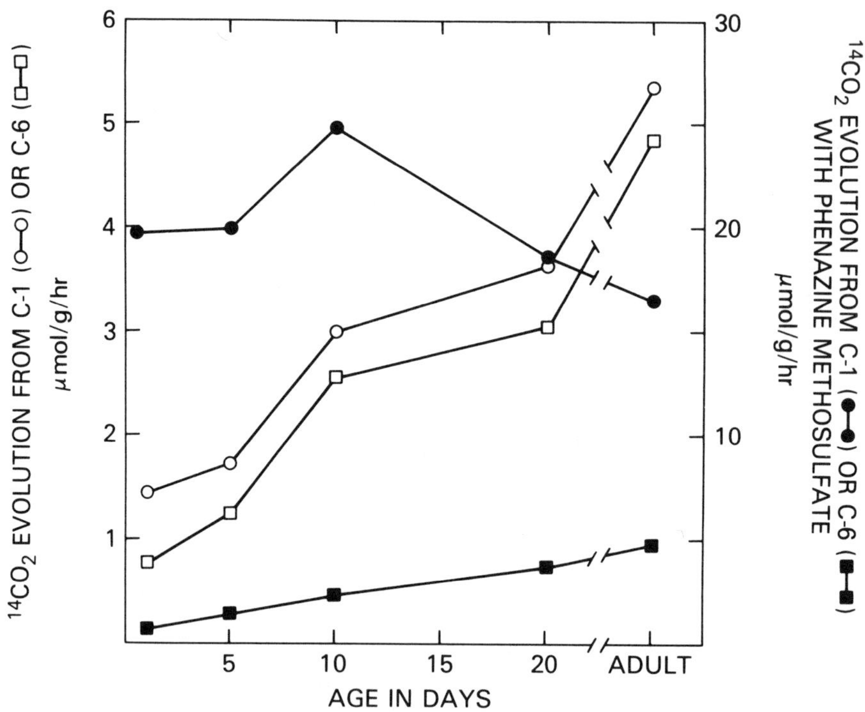

Figure 3. The relative oxidation (*open symbols*) of glucose ($1-{}^{14}C$) and glucose ($6-{}^{14}C$) by rat brain slices in vitro. The oxidation rates for the two labeled glucoses are essentially parallel. The effects of phenazine methosulfate (*closed symbols*) are largely specific for C-1 oxidation. Note the difference in the left and right vertical scales. (Based on data from Hothersall et al., 1979.)

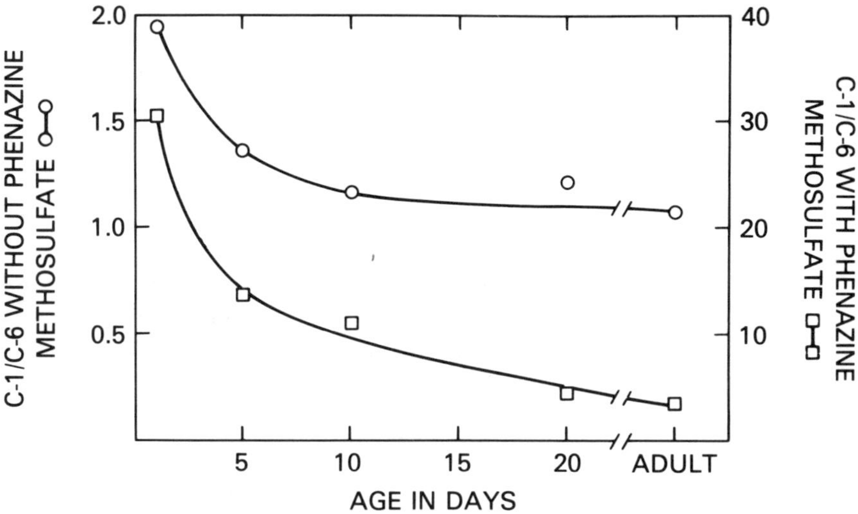

Figure 4. C-1 : C-6 ratios obtained in cerebral cortical slices from developing rats. Note the differences in scale. (Data from Hothersall et al., 1979, and fig. 1.)

the developmental stage (fig. 4). PMS stimulates C-1 oxidation by 30-fold in the one-day-old animals and by three-fold in the 20-day-old and adult animals. The involvement of the pentose phosphate pathway in lipogenesis is also suggested by the observation that incorporation of glucose carbon into lipids also falls within a comparable developmental time course (Hothersall et al., 1979).

Hakim, Moss, and Scuderi (1980) have demonstrated that the levels of 6-phosphogluconate increase shortly after birth in the rat cortex, reaching a maximum at day 3 postpartum. Thereafter, the level of this intermediate declines to adult levels by day 20. The turnover rates for this intermediate are also maximal at day 3, despite a constancy in the level of the 6PGDH.

The relationship of the pentose phosphate pathway to myelination has been elegantly demonstrated in newborn mouse cerebellum cultures by Hoskin and Allerand (1968) who showed that the C-1 : C-6 ratios increased during the premyelination phase of the explant cultures and then declined to near adult levels as myelination was completed.

The developing retina has been the subject of a few studies of the pentose phosphate pathway. Graymore (1970) has suggested that the activity of the pentose phosphate pathway is low in the adult rat retina and is slightly higher in the developing rat retina. C-1 : C-6 ratios of 2–5 have been determined in developing rabbit retinas; adult ratios are 1.2 (Cohen and Noell, 1960). Masterson, Israel, and Chader (1978) have shown that in the retinas of the developing chick, pentose phosphate pathway activity is high in both the neural retina and pigmented epithelium at Hamburger-Hamilton stages 29–30 (C-1 : C-6 of 4.0). At this time, differentiation is incomplete, and the cells of both layers are undergoing rapid proliferation. The pentose phosphate activity decreases until hatching (C-1–C-6 of 1.0).

Conclusions

The pentose phosphate pathway of developing and adult neural tissue appears to have three primary functions. In development, pentose phosphates are formed as part of the synthetic events supporting two central aspects of biochemical differentiation: de novo formation of ribose sugars for nucleic acids and production of reducing equivalents for lipogenesis. In the adult nervous system these functions are probably minimal, and the flux rates of the pathway reflect this fact. The elevated activity of this pathway in synaptic structures probably reflects a slightly different function—maintenance of reduced pyridine nucleotide pools to facilitate peroxide detoxification.

The regulation of this pathway seems to involve, through all stages of development, the pool sizes of triphosphopyridine nucleotides. Maintenance of this pool in a relatively reduced state may be said to be a general function of the pentose phosphate pathway. If this is true, then in the absence of significant demand for ribose sugar, the oxidative arm of the pathway may be viewed as the synthetic side (major products being NADPH and ribose-5-phosphate) and the nonoxidative arm the degradative side, returning ribose-5-phosphate to glycolysis to minimize the ATP cost. While the basic structure of the pentose phosphate pathway is well defined, the complete picture of its development, regulation, and metabolic function in brain is still obscure.

Galactose Metabolism

Introduction

The major dietary source of galactose is milk through the hydrolysis of its lactose. The resultant monosaccharides, glucose and galactose, differ only in the structure of the hydroxyl groups at C-4. The major pathway for galactose involves the conversion to glucose through epimerization of the C-4 hydroxyl group. The mechanism whereby this is achieved is depicted in figure 5. Galactose also participates in other pathways, for example, in galactocerebroside synthesis. The rate of synthesis of galactocerebroside closely parallels myelin synthesis and therefore is highest during myelin deposition in newborn animals (Cohn and Segal, 1973). Other pathways for galactose metabolism also exist and may be impor-

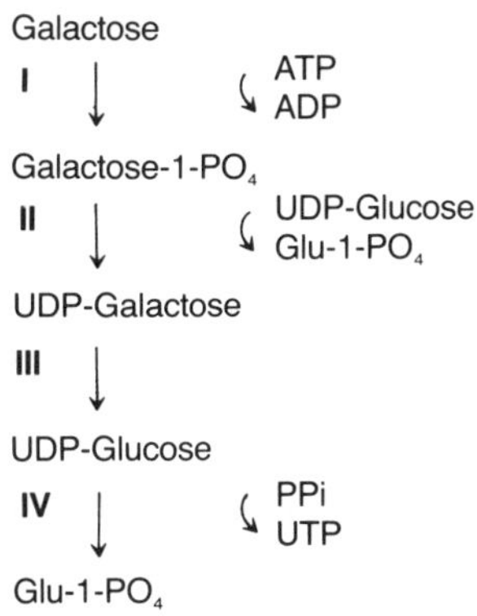

I Galactokinase

II Galactose-1-PO_4 uridyltransferase

III UDP Galactose-4-epimerase

IV UDP Glucose pyrophosphorylase

Figure 5. Schematic representation of the intermediates and enzymes of galactose metabolism as it relates to glucose (*UDP*, uridine diphosphate; *PPi*, inorganic phosphate; *UTP*, uridine triphosphate).

tant in situations of high galactose ingestion. These pathways include the reduction of galactose to galactitol (Hers, 1960) and the oxidation of galactose to galactonate (Cuatrecasas and Segal, 1966). The normal human adult is capable of handling high levels of ingested galactose, and, for example, galactose tolerance tests may be used to evaluate hepatic function (Colcher, Patek, and Kendall, 1946).

Galactosemia

Galactosemia is an inherited disorder of galactose metabolism, characterized by failure to thrive. Signs such as vomiting, diarrhea, and jaundice usually present during the first week of life. Hemolysis may occur in galactosemic patients, which worsens the jaundice. Both cataracts and mental retardation are frequent features of the untreated galactosemic patient. The mode of treatment includes providing the patient with a galactose-free diet. Although absolute elimination of galactose usually does not occur, reasonably successful results can be obtained using soybean milk and Nutramigen.

In 1956, Isselbacher and coworkers showed that the red blood cells of galactosemic patients lacked the enzyme galactose-1-phosphate uridyl transferase. Blockage of galactose metabolism of this site leads to an accumulation of large amounts of galactose-1-phosphate. The assay of red blood cell galactose-1-phosphate uridyl transferase has become the diagnostic criterion for galactosemia. Deficiency of this enzyme seems to occur in nearly all body tissues. More recently, patients have been described suffering galactosemia in which the defective enzyme is galactokinase (Gitzelmann, 1965). As is the case with the transferase variety of galactosemia, the galactokinase deficiency disorder is transmitted as an autosomal recessive trait. Parents of afflicted children may exhibit about one half the typical red blood cell enzyme activity.

In untreated galactosemia, the metabolite preceding the enzymatic deficiency is elevated in most studies. It is natural, therefore, that many studies have been directed toward investigating the toxicity of this metabolite, galactose-1-phosphate. A major problem with studying a disorder such as galactosemia is that it is currently not possible to duplicate the precise metabolic derangement in laboratory animals. Therefore, most investigators have employed a dietary method for simulating galactosemia in which animals are weaned to a diet containing a high concentration (usually 40 percent w/w) of D-galactose. It is important to understand that such models represent only an approximation of the human disorder at best; therefore, one must interpret data carefully when trying to determine human applicability.

Toxicity

Although there have been a variety of studies on the mechanism of toxicity in experimental models of galactosemia, the precise way in which brain function is compromised is unknown. In one recent study, weanling rats were placed on a 40 percent galactose diet, and sacrificed two to nine weeks later (Berry, Yandrasitz, and Segal, 1981). Brain synaptosomes were used in evaluating the acetylcholine-stimulated myo-[2-^{3}H]inositol labeling of phosphatidylinositol. Results showed a highly significant 40 percent decrease in the acetylcholine-stimulated phosphatidylinositol labeling by myoinositol. These data may indicate a possible important role for deranged synaptic transmission in the pathogenesis of galactosemia.

In another series of studies the effects of galactose neurotoxicity on the cerebral energy metabolism in newborn chicks was assessed (Kozak and Wells, 1969; 1971; Granett et al., 1972; Knull, Wells, and Kozak, 1972). These studies employed the 40 percent galactose diet method for producing toxicity, and in newborn chicks, this regimen produced convulsive activity and death in 48 hours. Results from these studies showed that whereas galactose and galactitol levels were quite high, galactose-1-phosphate levels were elevated at ten hours after initiation of diet but gradually fell and were almost undetectable in some chicks. This finding detracts somewhat from the model in that in the human disorder, galactose-1-phosphate levels are quite high.

When the brains of chicks that were fed a 40 percent galactose diet were analyzed for high-energy compounds, it was found that both ATP and phosphocreatine were significantly decreased. It was further shown in this series of studies that the ATP levels were decreased as early as nine hours after initiation of the diet. Turnover studies using the Lowry "closed system" technique were consistent with the concept that the glycolytic rate was diminished; furthermore, ATP turnover in the galactose-intoxicated chick was faster than that in controls.

These results and others have led to the advancement of the so-called futile ATPase cycle to account for the toxicity of galactose. In this cycle, galactose is rapidly phosphorylated to galactose-1-phosphate at the expense of ATP. The galactose-1-phosphate is subsequently dephosphorylated by a phosphatase to galactose and organic phosphate. In this way (futile cycle) the galactose is continually replenished at the expense of ATP, which is presumably then lost for other purposes. It is thought that this cycle is probably trivial in normal brain because of the low galactose concentration; however, in a situation in which galactose levels are elevated (galactosemia), such a pathway may assume importance. It is also likely that galactose-1-phosphate, which is elevated in galactosemia, could enter this cycle. It is noteworthy that cerebral levels of glucose were also diminished.

The toxicity of the galactose diet can be reversed by the intraperitoneal injection of glucose. This may act to replenish diminished cerebral levels of glucose, thereby permitting a replenishment of high-energy phosphates. When the symptoms were eliminated by glucose injection, the levels of depressed metabolites returned to normal. One reason for decreased cerebral glucose levels may be an interference by galactose of glucose transport across the blood-brain barrier. It should also be pointed out that hyperosmolality may be an important feature of galactose intoxication; however, it may be a nonspecific finding (Malone, Wells, and Segal, 1971).

Summary

The metabolic encephalopathy galactosemia is interesting from the standpoint that while pathways have been reasonably well defined, the precise nature of the toxicity is still somewhat uncertain. The fact that galactosemic patients can apparently develop a normal brain with appropriate dietary treatment means that early and accurate diagnosis is important. The chick model of the disorder is not an ideal one; the development of a mammalian model for the disorder would be highly beneficial. Most biochemical studies have involved whole brain, and therefore important data would be derived from doing regional studies.

Such experiments could show even more pronounced changes in subregions of the brain in experimental galactosemia.

References

Appel, S., and Parrot, B. (1970). Hexose monophosphate pathway in synapses. *J Neurochem* 17, 1619–1626.

Baquer, N., McLean, P., and Greenbaum, A. (1973). Enzymic differentiation in pathways of carbohydrate metabolism in developing brain. *Biophys Biochem Res Commun* 53, 1282–1288.

Bauer, U., and Brand, K. (1982). Carbon balance studies of glucose metabolism in rat cerebral cortical synaptosomes. *J Neurochem* 39, 239–243.

Berry, G., Yandrasitz, J. R., and Segal, S. (1981). Experimental galactose toxicity: Effect on synaptosomal phosphatidylinositol metabolism. *J Neurochem* 37, 888.

Bloom, B. (1955). Catabolism of glucose by mammalian tissues. *Proc Soc Exp Biol Med* 88, 317–318.

Brannan, I., Maker, H., and Karp, E. (1982). Norepinephrine stimulation of pentose cycle in rat brain prisms. *Res Commun Chem Pathol Pharmacol* 38, 509–512.

Buell, M. V., Lowry, O. H., Roberts, N. R., Chang, M. W., and Kapphahn, J. I. (1958). The quantitative histochemistry of the brain: V. Enzymes of glucose metabolism. *J Biol Chem* 232, 979–993.

Burt, A. (1965). Glucose metabolism and chick neurogenesis: I. Glucose-6-phosphate dehydrogenase activity in the embryonic brachial spinal cord. *Dev Biol* 12, 213–232.

Burt, A., and Wenger, B. (1961). Glucose-6-phosphate dehydrogenase activity in the brain of the developing chick. *Dev Biol* 3, 84–95.

Cohen, L., and Noell, W. (1960). Glucose catabolism of rabbit retina before and after development of visual function. *J Neurochem* 5, 253–276.

Cohn, R., and Segal, S. (1973). Galactose metabolism and its regulation. *Metabolism* 22, 627.

Colcher, H., Patek, A. J., and Kendall, F. E. (1946). Galactose disappearance from the blood stream: Calculation of a galactose removal constant and its application as a test of liver function. *J Clin Invest* 25, 768.

Cuatrecasas, P., and Segal, S. (1966). Galactose conversion to D-xylulose: An alternate route of galactose metabolism. *Science* 153, 549.

DiPietro, D., and Weinhouse, S. (1959). Glucose oxidation in rat brain slices and homogenates. *Arch Biochem Biophys* 80, 268–282.

Eggleston, L., and Krebs, H. A. (1974). Regulating the pentose phosphate cycle. *Biochem J* 138, 425–435.

El-Hassan, A., Zubairu, S., Hothersall, J., and Greenbaum, A. (1981). Age related charges in enzymes of rat brain. I. Enzymes of glycolysis, the pentose phosphate pathway and lypogenesis. *Enzyme* 26, 107–112.

Gitzelmann, R. (1965). Deficiency of erythrocyte galactokinase in a patient with galactose diabetes. *Lancet* 2, 670.

Granett, S. E., Kozak, L. P., McIntyre, J. P., and Wells, W. W. (1972). Studies on cerebral energy metabolism during the course of galactose neurotoxicity in chicks. *J Neurochem* 19, 1659.

Graymore, C. (1970). Biochemistry of the retina. In *Biochemistry of the eye*, ed. C. Graymore, pp. 645–735. New York: Academic Press.

Guerra, R., Melgar, E., and Villavicencio, M. (1967). Alternative pathways of glucose metabolism in fetal rat brain. *Biochim Biophys Acta* 148, 356–361.

Hakim, A., Moss, G., and Scuderi, D. (1980). The pentose phosphate pathway in brain during development. *Biol Neonate* 37, 15–21.

Hers, H. G. (1960). L'aldose-reductase. *Biochim Biophys Acta* 37, 120.

Hoskin, F., and Allerand, C. D. (1968). Metabolism of specifically labelled glucose by explants of newborn mouse cerebellum. *J Neurochem* 15, 427–432.

Hostetler, K., and Landau, B. (1967). Estimation of the pentose cycle contribution to glucose metabolism in tissue *in vivo*. *Biochemistry* 6, 2961–2964.

Hostetler, K., Landau, B., White, R., Albin, M., and Yashon, D. (1970). Contribution of the pentose cycle to the metabolism of glucose in the isolated perfused brain of the monkey. *J Neurochem* 17, 33–39.

Hothersall, J., Baquer, N., Greenbaum, A., and McLean, P. (1979). Alternative pathways of glucose utilization in brain: Changes in the pattern of glucose utilization in brain during development and the effect of phenazine methosulfate on the integration of metabolic routes. *Arch Biochem Biophys* 198, 478–492.

Hothersall, J., Baquer, N., and McLean, P. (1982). Pathways of carbohydrate metabolisms in peripheral nervous tissue. I. The contribution of alternative routes of glucose utilization in peripheral nerve and brain. *Enzyme* 27, 259–267.

Hothersall, J., Greenbaum, A., and McLean, P. (1982). The functional significance of the pentose phosphate pathway in synaptosomes: Protection against peroxidative damage by catecholamines and oxidants. *J Neurochem* 39, 1325–1332.

Isselbacher, K. J., Anderson, E. P., Kurahaski, K., and Kalckar, H. M. (1956). Congenital galactosemia, a single enzymatic block in galactose metabolism. *Science* 123, 635.

Katz, J., Landau, B., and Bartsch, G. (1966). The pentose phosphate cycle, triose phosphate isomerization and lipogenesis in rat adipose tissue. *J Biol Chem* 241, 727–740.

Katz, J., and Wood, H. (1960). The use of glucose-C14 for the evaluation of the pathways of glucose metabolism. *J Biol Chem* 235, 2165–2177.

Kauffman, F. (1972). The quantitative histochemistry of enzymes of the pentose phosphate pathway in the central nervous system of the rat. *J Neurochem* 19, 1–9.

Kauffman, F., Brown, J., Passonneau, J. V., and Lowry, O. H. (1967). Effects of changes in brain metabolism on levels of pentose phosphate pathway intermediates. *J Biol Chem* 214, 3647–3653.

Kimura, H., Naito, K., Nakagawa, K., and Kuriyama, K. (1974). Activation of hexose monophosphate pathway in brain by electrical stimulation in vitro. *J Neurochem* 23, 79–84.

Knull, H. R., Wells, W. W., and Kozak, L. P. (1972). Galactose toxicity in the chick: Hyperosmolality or depressed brain energy reserves? *Science* 176, 815.

Kozak, L. P., and Wells, W. W. (1969). Effect of galactose on energy and phospholipid metabolism in the chick brain. *Arch Biochem Biophys* 135, 371.

Kozak, L. P., and Wells, W. W. (1971). Studies on the metabolic determinants of D-galactose induced neurotoxicity in the chick. *J Neurochem* 18, 2217.

Krass, M., and Labella, F. (1967). Hexosemonophosphate shunt in endocrine tissues: Quantitative estimation of the pathway in bovine pineal body, anterior pituitary, posterior pituitary and brain. *Biochim Biophys Acta* 148, 384–391.

Kuhlman, R., and Lowry, O. H. (1956). Quantitative histochemical changes during the development of the rat cerebral cortex. *J Neurochem* 1, 173–180.

Landau, B., Bartsch, G., Katz, J., and Wood, H. (1964). Estimation of pathway contribution of glucose metabolism and the rate of isomerization of hexose-6-phosphate. *J Biol Chem* 239, 686–696.

Landau, B., and Katz, J. (1964). A quantitative estimation of the pathways of glucose metabolism in rat adipose tissue *in vitro*. *J Biol Chem* 239, 697–704.

Lippman, F. (1936). Fermentation of phosphogluconic acid. *Nature* 138, 588–589.

McCandless, D., Curley, A., and Cassidy, C. (1976). Thiamin deficiency and pentose phosphate cycle in rats: Intracerebral mechanisms. *J Nutr* 106, 1144–1151.

McDougal, D., Schulz, D., Passonneau, J. V., Clark, J., Reynolds, M., and Lowry, O. H. (1961). Quantitative studies of white matter: 1. Enzymes involved in glucose-6-phosphate metabolism. *J Gen Physiol* 44, 487–498.

Malone, J. I., Wells, J. H., and Segal, S. (1971). Galactose toxicity in the chick: Hyperosmolality. *Science* 174, 953.

Masterson, E., Israel, P., and Chader, G. (1978). Pentose shunt activity in developing chick retina and pigment epithelium: A switch in biochemical expression in cultured pigment epithelial cells. *Exp Eye Res* 27, 409–416.

Merrill, D., and Guynn, R. (1981). The calculation of the cytoplasmic free [$NADP^+$]/[NADPH] ratio in brain: Effects of electroconvulsive seizure. *Brain Res* 221, 307–318.

O'Neill, J. (1974). Carbohydrate metabolism in the developing nervous system. In *Biochemistry of the developing brain*, ed. W. Himwich, pp. 69–119. New York: Marcel Dekker.

O'Neill, J., Simon, S., and Shreeve, W. (1965). Alternate glycolytic pathways in brain: A comparison between the actions of artificial electron acceptors and electrical stimulation. *J Neurochem* 12, 797–802.

Passonneau, J. V., Schulz, D., and Lowry, O. H. (1966). The kinetics of glucose-6-phosphate dehydrogenase. *Fed Proc* 25, 219.

Phang, J., Downing, S., Yeh, G., Smith, R., Williams, J., and Hagedorn, C. (1982). Stimulation of the hexosemonophosphate-pentose pathway by pyrroline-5-carboxylate in cultured cells. *J Cell Physiol* 110, 255–261.

Sacks, W. (1957). Glucose in normal human subjects. *J Appl Physiol* 10, 37–44.

Sacks, W. (1961). Cerebral metabolism of glucose-3-C^{14}, pyruvate-1-C^{14} and lactate-1-C^{14} in mental disease. *J Appl Physiol* 16, 175–180.

Villee, C., and Loring, J. (1961). Alternative pathways of carbohydrate metabolism in fetal and adult tissue. *Biochem J* 81, 488–494.

Warburg, O., Christian, W., and Griese, A. (1935). Wasserstottubertragendes Co Ferment, Seine Zusammensetzung und Wirkung Weise. *Biochemische Zeitschrift* 282, 157–205.

Coordination of Brain Metabolism with Function during Development

R. Wayne Albers

This chapter contains (1) a survey of studies of energy production in developing and mature brains and (2) an evaluation of the major energy demands at various stages of development. Primary attention will be directed toward building a concept of overall energy balance, to the extent that it can be assessed, at a few generalized stages of brain development.

Previous chapters have established that the mature brain is rather stringently committed to aerobic glycolysis. Although the reasons for this are not so clear, its corollary is that the brain is highly conservative of its carbon supply. Most of the glucose extracted from the bloodstream by the brain is oxidized completely to CO_2. The amount that is not completely oxidized is mostly returned to the blood as lactate. The apparent simplicity of this balance is misleading because concurrently neurotransmitters are being synthesized and degraded, membrane vesicles are synthesized and transported, and a fairly high rate of protein turnover is maintained. This balance does not exist in a rapidly growing tissue, and thus it is to be expected that brain energy metabolism undergoes profound alterations during development and maturation.

Since we will be concerned for the most part with estimates of energy production and demand, it is convenient to employ units that permit ready comparison of both sides of this account using data obtained by a variety of techniques. We will use the common currency of bioenergetics, ~P. Rates of ~P production will be expressed as mM min^{-1}, which is equivalent to $\mu mol\ g^{-1}\ min^{-1}$, based upon wet weight of tissue. Because of the diversity of techniques and the approximations involved, the derived values will usually be rounded to only one or two significant figures.

Metabolic Demands of Cell Proliferation

Mitogenesis is experimentally inducible by a variety of agents (lectins, growth factors, antibodies) whose common factor is an interaction with plasma membrane receptors. The earliest consequence reported is an increased membrane permeability to Na^+ (Rosengurt and Mendoza, 1980; Segal, 1981). Mitogenesis does not appear to ensue from increased cytoplasmic Na^+ or from membrane depolarization, because in many cases these effects, when brought about by the sodium pump inhibitor ouabain, prevent cell division (Quastel, Wright, and Kaplan, 1972). It may be that the observed Na^+ influx reflects activation of Na-coupled transport of cell nutrients, such as amino acids or nucleosides (Heinz, 1981). The mitogen-induced Na^+ influx can be inhibited by the diuretic amiloride. This agent is known to inhibit the Na^+/H^+ exchange pump, and thus it has been suggested that the initial Na^+ influx brings about a critical increment in intracellular pH that triggers further events in mitosis (Moolenaar et al., 1981). Although sodium pump capacity in embryonic tissues is relatively low, there is substantial evidence that regulation of pump activity is important in regulating cell proliferation and differentiation. Na^+ pump inhibitors block stimulation by mitogens in lymphocytes (Quastel, Wright, and Kaplan, 1972). Among the earliest discernible effects of nerve growth factor (Skaper and Varon, 1980) and other growth factors (Segal, 1981; Rosengurt and Mendoza, 1980) is a stimulation of the Na^+ pump.

Warner (1981) has shown that early evidence of neuronal differentiation is the development of a high membrane potential. Exposure of embryos to cardiac glycosides can inhibit this process. Intercellular communication by gap junctions occurs in a circumscribed region corresponding to future brain before overt differentiation of the neural plate. Warner has proposed that the combination of intercellular communication and local differentiation of Na^+ pumps can produce electrophoretic currents, causing movement of growth factors and other charged molecules.

The repression of mitosis that is characteristic of neuronal maturation may also be regulated by intracellular Na^+, according to the hypothesis of Cone (1980). His experimental support for this theory comes from observations of the behavior of neurons in primary cultures from tryptic dispersions of chick spinal cord. Few cells showed ^3H-thymidine incorporation in control cultures; however, 30 percent of the neurons incubated for six hours in the presence of ouabain were labeled. After 24 hours, 20 percent appeared binucleate.

Several other cell lines were found to behave similarly. However, as noted above, some cells, notably lymphocytes, which can be stimulated to divide by various mitogens, are prevented from doing so by ouabain (Segal, 1981; Quastel, Wright, and Kaplan, 1972). Clearly, further delineation of the factors regulating neuronal cell division is of major importance.

Energy Demands for Growth and Maturation

During periods of intense biosynthetic activity, the cells must provide not only ~P but also raw materials for synthetic activity. The extent to which these materials arrive preformed at the cell membrane will be a major factor in the overall cellular energy equation: transport processes by and large require less energy than does de novo synthesis of the required molecules. Very little is known about the extent to which preformed amino acids, fatty acids, and other small molecules may be provided by maternal circulation in early brain development. Postnatal brain has acquired the ability to synthesize many of the major building block molecules, but incorporation of exogenous amino acids and fatty acids has also been demonstrated.

Protein Synthesis

Rapidly growing cells devote a major part of their energy to protein synthesis: in the case of *Escherichia coli* in logarithmic phase, 95 percent of energy metabolism is attributed to biosynthesis. Rapidly growing animal cells typically contain 5 percent protein, which is equivalent to about 0.4 M of peptide bonds. Thus for each doubling, about 1.6 M of ~P is required for this amount of protein synthesis from preformed amino acids.

Some data are available concerning overall oxygen consumption and ~P production in newborn and fetal brains (Lowry et al., 1964). From these one may calculate the minimum generation time for a cell as it is constrained by available ~P. Thus for 3 mM/min ~P production, the requirement of 1.6 M ~P per generation of cells for protein limits the cell cycle to about nine hours. In fact, generation times somewhat less than this have been calculated for chick neural epithelia (Fujita, 1962), although in most cases periods of nearly 24 hours are reported.

Myelination

The period of rapid myelination in rats extends from the first to the third postnatal week. Norton and Poduslo (1973) have provided some quantitative data from which one may estimate the energy requirement of this process. Between days 15 and 20, total brain lipids increase by about 3.3 g/kg brain/d. In the case of lipid synthesis from 2-carbon precursors, this corresponds to about 700 mM ~P per day. If the energy production of a 15-day-old rat is not less than that of a ten-day-old mouse (Lowry et al., 1964), this peak rate of myelination requires less than 4 percent of total brain energy production.

Osmotic Work and Ion Transport

Na^+ extrusion is considered the major energy-consuming process in mature brain. If one assumes that embryonic cells maintain comparably high ionic gradients, the minimal requirement for Na^+clearance will arise from the doubling of intracellular volume in each generation: this would require the transformation of a volume of extracellular composition, say 115 mM Na^+, to one of intracellular composition, say 10 mM Na^+. From the 3 Na^+/ATP (adenosine triphosphate) stoichiometry of the Na^+ pump, this would require only 35 mM ~P, equivalent to ten or 15 minutes' worth of cell metabolism per cell generation.

The clear implication of these estimates is that under conditions of rapid cell division, the osmotic work of volume expansion per se is insignificant relative to that required for biosynthesis. However, a variety of other metabolite fluxes may utilize significant amounts of energy. For example, amino acid and other nutrient uptake processes are important for rapidly growing cells.

Most evidence suggests that in rapidly growing cells, generally, active ion transport is at a low level. Consistent with this, intracellular Na^+ is relatively high in embryonic and fetal tissues including brain (Cameron et al., 1980). As discussed above, the ionic milieu (or the consequent membrane potential) is hypothetically a factor in controlling cell growth as well as mitotic rate. Significant Na^+ and K^+ gradients evidently exist in early embryonic cells since membrane potentials of −30 to −40 mV are recorded in amphibian neural plate cells and this potential is abolished by ouabain (Warner, 1981).

The roles of Na^+, K^+, and Ca^{++} in early development are undoubtedly important but are not so well defined as they are for the functioning of mature brain. Ionic gradients are postulated to be important factors in the regulation of the rate of protein synthesis (Lubin, 1967).

The electrolyte composition of rat brain was determined over the postnatal development period (Vernadakis and Woodbury, 1962). Brain chloride space declines from about 47 percent on day 1 to near 24 percent on day 21. The Na^+ space closely parallels the chloride space. These observations suggest that a low internal Na^+ concentration is maintained throughout this period, and thus the Na^+ regulation by active transport is fully competent from postnatal day 1.

There have been numerous studies of the activity of Na,K-ATPase as a function of development. Although there is general agreement that activity in rat brain is difficult to measure at day 1 and increases rapidly over the first 30 postnatal days, there is remarkable diversity in the descriptions of the exact time course. The period of most rapidly increasing Na^+ pump activity has been found to correlate with the maturation of electroencephalographic activity in the chick (Bignami, Palladini, and Venturini, 1966), the rat (Abdel-Latif, Brody, and Ramahi, 1967), and the kitten (Huttenlocker and Rawson, 1968). The specific activity of Na,K-ATPase in mature rat brain is usually reported to be in the range of 4–7 μmol mg^{-1} h^{-1}. This corresponds to 6.5–11.5 mM min^{-1}, or 25–50 percent of total ~P production.

Energy Production by Rapidly Growing Brain

Quantitative Aspects

The earliest relevant in vivo data derive from measurements on whole embryos maintained under organ culture conditions (Shepard, Tanimura, and Robkin, 1970; see table 1). These were calculated from measurements of utilization of U-^{14}-C-glucose. Less than 5 percent of the radioactivity was incorporated into the tissue, and less than 5 percent appeared as CO_2. More than 90 percent was recovered as lactate.

In considering these early stages of development, one expects that

Table 1. *High-energy Phosphate Production by Developing Brain: Summary of Data Recalculated to Standard Units*

Subject	*Phosphate Production (mM/min)*	*Source*
Rat embryo (whole)		Shepard, Tanimura, and Robkin (1970)
10.3 days	2.0	
Chick embryo brain		Gonya-Magee and Vannucci (1982)
9 days	2.8	
14 days	3.2	
19 days	5.8	
22 days	9.0	
Mouse brain		Lowry et al. (1964)
10 days	13	
Adult	25	
Human		Kennedy and Sokoloff (1957)
6 years	15	
Adult	10	

Note: Calculation for rat embryos from data of Shepard, Tanimura, and Robkin (1970): Embryos were cultured with radioactive glucose. Mean glucose consumption at this age was 12 μmol/g protein/min, of which 90 percent appeared as lactate, 6 percent was incorporated into macromolecules, and 4 percent was metabolized to CO_2, largely through the pentose shunt. Two ~P are generated from glucose by glycolysis, and, with the assumption that protein is 10 percent of wet weight (probably too high in the case of embryos), the calculation is $.9 \times 12$ mM/kg protein/min $\times$ .1 kg protein/kg tissue $\times$ 2 ~P/glucose = 2 mM ~P/min.

The data for chick embryo and mouse brains were calculated in the referenced papers from measurements of the rate of disappearance of ~P immediately after decapitation.

The data for human brains were calculated based on the measured rates of glucose consumption and the fact that the brain respiratory quotient is about .95.

Table 2. *High-energy Phosphate Requirements by Developing Brain: Summary of Data Recalculated to Standard Units*

Subject	*Phosphate Requirement (mM/min)*	*Source*
Protein synthesis		
Rat embryo (whole)		Shepard, Tanimura, and Robkin (1970)
10.3 Days	5.7	
11.0 Days	1.2	
Rat brain		
Newborn	0.43	Lajtha et al. (1979)
15–20 Days	0.11	Norton and Poduslo (1973)
20–30 Days	0.05	Norton and Poduslo (1973)
Adult	0.09	Lajtha et al. (1979)
Lipid synthesis		
Rat brain		Norton and Poduslo (1973)
15–20 Days	0.5	
20–30 Days	0.15	

Note: The calculations for protein synthesis assume 4 ~P/amino acid residue and that preformed amino acids are used. Lipid content was given in units of weight and the factor estimated for average lipid was 216 mmol ~P/g.

most of the energy requirement is for biosynthesis. The ten-day-old rat embryo increases its protein content fourfold in 24 hours (Shepard, Tanimura, and Robkin, 1970; see table 2). If one assumes 5 percent protein, the energy requirement to sustain this rate of peptide bond synthesis is 2 mM min^{-1}. This is equivalent to the total ~P production of these embryos estimated from their apparent rate of glucose consumption. Because membrane synthesis requires approximately an equal weight of phospholipid, which is even more costly than peptide in ~P, the energy production must have been somewhat underestimated in this study.

Energy production in chick embryo brain was measured from nine days until hatching (Gonya-Magee and Vannucci, 1982; see table 1). This was determined by the method of Gatfield et al. (1966), a method based on measuring the rate of depletion of energy reserves subsequent to decapitation. An increase of nearly threefold in cerebral metabolic rate was found to occur over this period. Lowry and coworkers (1964) determined that the cerebral metabolic rate (CMR) in the ten-day-old postnatal mouse is about 50 percent that of the adult mouse.

In humans CMR is well established for the adult by application of the nitrous oxide technique and its variants. The rate in six-year-old children is about 50 percent greater than the mean adult value (Kennedy and Sokoloff, 1957).

Energy Sources in Developing Brain

As noted above, in rat embryos ten to twelve days old about 90 percent of the radioactivity in metabolites of uniformly labeled glucose was found in lactate. This is consistent with the general view that early stages of metabolism are characterized by a reliance on anaerobic glycolysis. This has long been considered the principal source of ~P for the immature brain, on indirect evidence (Fazekas, Alexander, and Himwich, 1941; Himwich, 1970). However, the rate of lactate production is insufficient to account for the CMR in chick embryo brains: lactate production was only 30 percent of ~P consumption in nine-day-old embryos, increasing to a maximum of 78 percent at 16 days and declining to 56 percent at hatching (Gonya-Magee and Vannucci, 1982).

Animals of many species are known to be more tolerant of anoxia as infants than as adults (Dawes et al., 1963), and this is often attributed to a greater capacity for anaerobic glycolysis. As pointed out by Thurston and McDougal (1969), the energy reserves in newborn mouse brain are nearly equal to those of adult brain. A major factor in the resistance to anoxia is evidently just the lower rate of consumption of energy reserves by immature brain.

Much of the energy requirement of de novo biosynthesis is for reducing equivalents in the form of $NADPH^+$. The primary $NADPH^+$ generating system is the pentose phosphate pathway. Tanimura and Shepard (1970) found a high proportion of glucose metabolized by the pentose path in rat embryos 10.3 days old, whereas somewhere between 1 percent and 10 percent enters the pentose shunt in adult brain (see p. 166 in Siesjo [1978]). There are also numerous studies of various aspects of the development of mitochondria and mitochondrial metabolism in brain.

In the course of developing their technique for in vitro culture of rat embryos, New, Coppola, and Cockroft (1976) found that embryos 9.5 days old require culture for 24 hours in low (5 percent) O_2 to avoid development of misshapen or unclosed neural tubes. Moriss and New (1979) found that morphologically normal embryos cultured under these conditions have "anaerobic" mitochondria with smooth profiles and few cristae. Defective embryos raised under 20–40 percent oxygen possessed "aerobic" mitochondria with many cristae. Samson, Balfour, and Jacobs (1960) estimated the number, dry weight, and protein content of mitochondria in rat brain over the early postnatal period. Over the first 21 postnatal days, mitochondrial protein content was found to increase from 15 mg/g wet brain to 26 mg/g wet brain.

Wilbur and Patel (1974) investigated the metabolism of pyruvate in brain homogenates and mitochondria as a function of postnatal age. Pyruvate decarboxylation rate increased from about .25 mM min^{-1} at day

1 to .9 mM min^{-1} at day 2. This corresponds to a potential increase in the rate of ~P generation by oxidation of glucose of about threefold, from 5 mM min^{-1} to 16 mM min^{-1}. Land et al. (1977) found that pyruvate dehydrogenase (PDH) is delayed in its development relative to other tricarboxylic acid (TCA) enzymes in rat brain. In rats PDH was almost absent in mitochondria from day 1 and attained only 50–60 percent of adult activity by day 20.

Presumably, the extent of de novo biosynthetic activity is coordinated with the availability of reducing equivalents (e.g., through the pentose phosphate pathway) and of carbon sources (ketone body metabolism, TCA cycle). The withdrawal of acetyl CoA from the TCA cycle for biosynthetic purposes thus has a double effect on the energy balance: some 12 ~P/acetyl CoA are lost from the production side, and several ~P and reducing equivalents are added to the energy requirement for biosynthesis.

Energy Requirements for Functions of Mature Brain

Calculation of the Energy Required for Na^+ Transport

Sodium channel density is apparently quite variable in different excitable membranes; however, the conductance of an open channel is relatively invariant. At the usual resting membrane potential, the peak current would be 4,000–5,000 Na^+/ms (Hille, 1981). The sodium pump is much slower: probably less than 0.5 Na^+/ms when fully activated. The total number of voltage-dependent Na^+ channels can be estimated from the number of binding sites for tetrodotoxin or saxitoxin (Catterall, Morrow, and Hartshorne, 1979). Similarly, the number of sodium pump sites are measured by ouabain binding or Na^+-dependent ^{32}P incorporation (Choi and Akera, 1978). From these data the ratio of pumps to channels is about ten to one in mammalian brain. By equating these capacities one may conclude that brain activity, which maintains an average of more than 0.1 percent of the channels in an open state, could not be sustained for more than a few seconds. Squid axon studies that estimate the actual Na^+ influx per action potential indicate either that only a small fraction of channels open or that each channel is open for only a small segment of the spike width (only about 100 ions flow per channel during one impulse). If this holds for spikes in mammalian brain, the maximal activity that could be sustained would be equivalent to about 25 impulses per second. This should be an overestimate because it assumes that all active sodium transport is available to generate the membrane potential of neurons.

Synthesis and Release of Neurotransmitters and Peptides

Much of the energy budget for neurotransmitter function is subsumed in that for ion transport, that is, to the extent that the neurotransmitter functions are involved in regulating ion channels. Additional energy is required for neurotransmitter biosynthesis, intracellular transport, and secretion. Since these are molecules that act as regulators, the effects of their actions involve large amplification factors: one receptor-ligand complex generally modulates the flow of thousands of ions, or it activates an enzyme molecule that in turn alters the flow of a stream of metabolites. For this reason, one does not expect the production of regulator molecules to be a major energy-consuming factor.

The case is not so clear for some of the candidate amino acid neurotransmitters. Glutamate, aspartate, and GABA (γ-aminobutyric acid) are all present in millimolar concentrations in brain, and they all turn over rapidly. The situation is further complicated by the existence of multiple kinetically defined "compartments" of these amino acids, the anatomical basis of which is not clearly established (Balazs, Machiyama, and Patel, 1972). However, some considerable fraction of the pyruvate that enters the TCA cycle proceeds through glutamate, GABA, and succinic semialdehyde, constituting the "GABA shunt" pathway (Balazs, Machiyama, and Patel, 1972). It seems likely that, to the extent that these compounds participate as neurotransmitters, a relatively small fraction of their total metabolic flux is involved. However, a contrary view is implicit in a variety of applications of enzyme histochemistry to studies designed to inform about neurotransmitter functions. For example, the distribution of glutamic decarboxylase is commonly accepted as a marker for "GABAergic" neurons. Recently it has been proposed that aspartate:oxoglutarate aminotransferase and glutaminase may be used to inform about the neurotransmitter functions of aspartate, glutamate, or both (Altschuler et al., 1982). Thus the possible quantitative significance of amino acid neurotransmitter metabolism is an unresolved problem.

Integration of Energy Production and Demand

We have seen that in the earliest developmental stages most of the energy production is used for biosynthesis. As development proceeds, energy production increases and biosynthetic requirements decrease (per unit volume). Even at the stage of rapid myelination, biosynthesis is a minor component of energy use. The major increasing demand is for active transport: the restorative processes that must balance the ionic currents mediating electrical activity as well as the electrically silent fluxes of a variety of secondary transport processes.

Little is known of the mechanisms that regulate energy balance. At very early stages, genetic programming of the appearance of enzymes in the various catabolic pathways is evidently a major factor. As the brain becomes functional, the essence of its functioning is responsiveness to external information. The new regulatory mechanisms that operate in mature brain are dramatically documented by studies of regional blood flow and metabolism (Sokoloff, 1981). Significant control probably resides at several levels: blood flow, metabolite flux at cell membranes, as well as regulation of intracellular metabolic rates. Brain arteries receive extensive sympathetic innervation, and regional differences in the responses of cerebral blood flow to β-adrenergic stimulation have been reported (Kogure et al., 1979). Cerebral microvasculature may be regulated by catecholamines and other neurotransmitters (Dora and Kovach, 1983).

There is little evidence for humoral control of glucose transport across cell membranes in brain. In fact, transport of glucose into brain seems to be mediated largely by a facilitative mechanism; that is, the rate of entry is determined by the local transmembrane concentration gradient unless the capacity of the transport system is exceeded. The major postulated metabolic control mechanisms, through phosphofructokinase and through the supply of ADP to mitochondria, are in turn functions of energy utilization.

Thus to a large extent CMRs may respond passively to neural activity at the cell level. It may be that neural mechanisms can also "anticipate" functional central nervous system requirements, perhaps in a fashion analogous to the anticipatory actions of epinephrine on peripheral metabolism. There is ample demonstration that various experimentally imposed stresses (hypoxia, for example) produce changes in cyclic nucleotides and other factors that may link central nervous system metabolism to extrinsic neural control (Magistratti et al., 1981).

Considerations such as these may help to formulate useful experimental approaches to the understanding of the unique characteristics of cerebral metabolism. Careful delineation of the ontogeny of these characteristics should continue to be an important aspect of neurobiological research.

References

Abdel-Latif, A. A., Brody, J., and Ramahi, H. (1967). Studies on Na,K-ATPase of the nerve endings and appearance of electrical activity in developing rat brain. *J Neurochem* 14, 1133–1141.

Altschuler, R. A., Mosinger, J. L., Harmison, G. G., Parakkal, M. H., and Wenthold, R. J. (1982). Aspartate aminotransferase–like immunoreactivity as a marker for aspartate/glutamate in guinea pig photoreceptors. *Nature* 298, 657–659.

Balazs, R., Machiyama, Y., and Patel, A. J. (1972). Compartmentation and the metabolism of γ-aminobutyrate. In *Metabolic compartmentation in the brain,* ed. R. Balazs and J. E. Cremer, pp. 57–70. London: Macmillan.

Bignami, A., Palladini, G., and Venturini, G. (1966). Na,K-ATPase in the developing chick brain. *Brain Res* 3, 207–209.

Cameron, I. L., Smith, N. K. R., Pool, T. B., and Sparks, R. L. (1980). Intracellular concentration of sodium and other elements as related to mitogenesis and oncogenesis *in vivo*. *Cancer Res* 40, 1493–1500.

Catterall, W. A., Morrow, C. S. and Hartshorne, R. P. (1979). Neurotoxin binding to receptor sites associated with voltage-sensitive sodium channels in intact, lysed and detergent-solubilized brain membranes. *J Biol Chem* 254, 11379–11387.

Choi, Y. R., and Akera, T. (1978). Membrane (Na^+ + K^+)-ATPase of canine brain, heart, and kidney. *Biochim Biophys Acta* 508, 313–327.

Cone, C. D. (1980). Ionically mediated induction of mitogenesis in CNS neurons. *Ann NY Acad Sci* 339, 115–138.

Dawes, G. S., Mott, J. C., Shelley, H. J., and Stafford, A. (1963). The prolongation of survival time in asphyxiated immature foetal lambs. *J Physiol* 168, 43–64.

Dora, E., and Kovach, A. G. B. (1983). Effect of topically administered epinephrine, norepinephrine, and acetyl choline on cerebrocortical circulation and the NAD/NADH redox state. *J Cereb Blood Flow Metab* 3, 161–169.

Fazekas, J. F., Alexander, F. A. D., and Himwich, H. E. (1941). Tolerance of newborn anoxia. *Am J Physiol* 134, 281–287.

Fujita, S. (1962). Kinetics of cellular proliferation. *Exp Cell Res* 28, 52–60.

Gatfield, P. O., Lowry, O. H., Schulz, D. W., and Passonneau, J. V. (1966). Regional energy reserves in mouse brain and changes with ischemia on anesthesia. *J Neurochem* 13, 185–195.

Gonya-Magee, T. and Vannucci, R. C. (1982). Ontogeny of cerebral oxidative metabolism in the chick embryo. *J Neurochem* 38, 1387–1392.

Heinz, E. (1981). Electrical potentials in biological membrane transport. *Mol Biol Biochem Biophys* 33, 1–85.

Hille, B. (1981). Excitability and ionic channels. In *Basic neurochemistry*. 3rd ed., ed. G. J. Siegel, R. W. Albers, B. W. Agranoff, and R. Katzman, pp. 95–106. Boston: Little, Brown.

Himwich, H. E. (1970). Historical review. In *Developmental neurobiology*, ed. W. A. Himwich. Springfield, Ill.: Charles C. Thomas.

Huttenlocker, P. R., and Rawson, M. D. (1968). Neuronal activity and ATPase in immature cerebral cortex. *Exp Neurol* 22, 18–29.

Kennedy, C., and Sokoloff, L. (1957). An adaptation of the nitrous oxide method to the study of cerebral circulation in children: Normal values for cerebral blood flow and cerebral metabolic rate in childhood. *J Clin Invest* 36, 1130–1137.

Kogure, K., Scheinberg, P., Kishikawa, H., Utsunomiya, Y., and Busto, R. (1979). Adrenergic control of CBF and energy metabolism in the rat. *Stroke* 10, 179–184.

Lajtha, A., Dunlop, D., Patlak, C., and Toth, J. (1979). Compartments of protein metabolism in the developing brain. *Biochim Biophys Acta* 561, 491–501.

Land, J. M., Booth, R. F. G., Berger, R., and Clark, J. B. (1977). Development of mitochondrial energy metabolism in rat brain. *Biochem J* 164, 339–348.

Lowry, O. H., Passonneau, J. V., Hasselberger, F. X., and Schulz, D. W. (1964). Effect of ischemia on known substrates and cofactors of the glycolytic pathway in brain. *J Biol Chem* 239, 18–30.

Lubin, M. (1967). Intracellular K^+ and macromolecular synthesis in mammalian cells. *Nature* 213, 451–453.

Magistratti, P. J., Morrison, J. H., Shoemaker, W. J., Sapin, V., and Bloom, F. E. (1981). VIP induces glycogenolysis in mouse cortical slices: A possible regulatory mechanism for local control of energy metabolism. *Proc Natl Acad Sci USA* 78, 6535–6539.

Moolenaar, W. H., Mummery, C. L., van der Saag, P. T., and de Last, S. W. (1981). Rapid ionic events and the initiation of growth in serum-stimulated neuroblastoma cells. *Cell* 23, 789–798.

Moriss, G. M., and New, D. A. T. (1979). Effect of oxygen concentration on morphogenesis of cranial neural folds and neural crests in cultured rat embryo. *J Embryol Exp Morph* 54, 17–35.

New, D. A. T., Coppola, P. T., and Cockroft, D. L. (1976). Improved development of head-fold rat embryos in culture resulting from low oxygen and modifications of the culture serum. *J Reprod Fertil* 48, 219–222.

Norton, W. T., and Poduslo, S. E. (1973). Myelination in rat brain: Changes in myelin composition during brain maturation. *J Neurochem* 21, 759–773.

Quastel, M. R., Wright, P., and Kaplan, J. G. (1972). K^+ uptake and lymphocyte activation: Generality of the effect of ouabain and a model of the events at the lymphocyte surface induced by phytohemaglutins. In *Proceedings of the Sixth Leukocyte Culture Conference*, pp. 185–214. New York: Academic Press.

Rosengurt, E., and Mendoza, S. (1980). Monovalent ion fluxes and the control of cell proliferation in cultured fibroblasts. *Ann NY Acad Sci* 339, 175–190.

Samson, F. E., Balfour, W. M., and Jacobs, R. J. (1960). Mitochondrial changes in developing brain. *Am J Physiol* 199, 693–696.

Segal, G. B. (1981). Membrane alterations in lymphocyte proliferation. *American Journal of Pediatric Hematology/Oncology* 3, 433–438.

Shepard, T. H., Tanimura, T., and Robkin, M. A. (1970). Energy metabolism in early embryos. *Dev Biol* [*Suppl*] 4, 42–58.

Siesjo, B. K. (1978). *Brain energy metabolism*. New York: John Wiley & Sons.

Skaper, S. D., and Varon, S. (1980). Properties of the sodium extrusion

mechanism controlled by nerve growth factor in chick embryo dorsal root ganglionic cells. *J Neurochem* 34, 1654–1660.
Sokoloff, L. (1981). Localization of functional activity in the CNS by measurement of glucose utilization with radioactive deoxyglucose. *J Cereb Blood Flow Metab* 1, 7–36.
Tanimura, T., and Shepard, T. H. (1970). Glucose metabolism by rat embryos *in vitro. Proc Soc Exp Biol Med* 135, 51–54.
Thurston, J. H., and McDougal, D. B. (1969). Effect of ischemia on metabolism of the brain of the newborn mouse. *Am J Physiol* 216, 348–352.
Vernadakis, A., and Woodbury, D. M. (1962). Electrolyte and amino acid changes in rat brain during maturation. *Am J Physiol* 203, 748–752.
Warner, A. E. (1981). The early development of the nervous system. In *Development in the nervous system*, ed. D. R. Garrod, and J. D. Feldman, pp. 109–127. Cambridge: Cambridge University Press.
Wilbur, D. O., and Patel, M. S. (1974). Development of mitochondrial pyruvate metabolism in rat brain. *J Neurochem* 22, 709–715.

Neurotransmitters

Michael V. Johnston

Neurons communicate with each other using chemical neurotransmitters released at synapses. A neurotransmitter's chemical structure and its quantity and duration of release present a message that is decoded by receptor mechanisms on adjacent neurons. Chemical machinery for neurotransmission, localized primarily in nerve terminals, includes transmitter synthetic pathways, storage pools, depolarization-release mechanisms, and methods for transmitter inactivation in the synaptic cleft (fig. 1). This presynaptic transmitter-related activity consumes a major share of neuronal energy supplies, and proper regulation of steps in this process is important for normal brain function. Although the details of neurotransmitter machinery (e.g., identity of synthetic enzymes) differ for each of the many transmitter-specific neuronal types, certain common patterns are followed throughout the nervous system. This chapter describes developmental patterns for presynaptic transmitter chemistry and the following one deals with receptor mechanisms.

Much of the work on mapping neurotransmitter systems and determining their developmental timetables is based on methods for visualizing and measuring the key elements of synaptic chemical machinery (Cooper, Bloom, and Roth, 1982). These biochemical phenotypic expressions of neuronal genes are often localized to one neuronal group and characteristic of it. Highly sensitive methods for measuring these markers (e.g., endogenous terminal transmitter stores) have made it possible to determine dates when neurotransmitter characteristics are first expressed in fetal or postnatal life (Black, 1982).

Because our aim is to emphasize developmental features of neurotransmitter systems, most of the experimental evidence supporting a transmitter role for these chemicals has been omitted. However, to clarify the multimodality approach used to study neurotransmitter development, key features of synaptic chemistry will be reviewed with emphasis on catecholamine and acetylcholine systems that have been studied extensively.

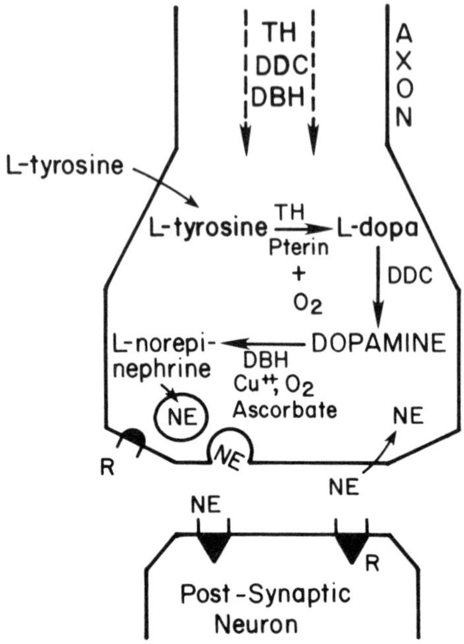

Figure 1. Diagram of a noradrenergic synapse with biosynthetic machinery. Biosynthetic enzymes are synthesized in the perikaryon and transported down the axon to the nerve terminal (*TH*, tyrosine hydroxylase; *DDC*, L-dopa decarboxylase; *DBH*, dopamine-β-hydroxylase; *NE*, norepinephrine; *R*, receptors for norepinephrine on presynaptic and postsynaptic sites).

Catecholamine Neurotransmission

Transmitter Metabolism in Catecholamine Neurons

Catecholamine systems are among the earliest neuronal groups to replicate and differentiate. In adults they are important regulators of movement, attention, emotion, and vegetative functions. Widespread norepinephrine axonal arbors are projected upward from the pons and medulla into diverse regions including cerebral cortex, hypothalamus, cerebellum, and caudally to the spinal cord (Moore, 1982). Peripheral noradrenergic neurons in autonomic sympathetic ganglia terminate on viscera and blood vessels. Dopamine perikarya in the midbrain project axons into the forebrain in two major groups—a nigrostriatal projection to the caudate and putamen and a mesolimbic projection to limbic cortical regions. A few brain stem neurons as well as peripheral neurally derived adrenal chromaffin cells use epinephrine. All three groups share common synaptic chemical features as shown in figure 1 for a noradrenergic nerve terminal.

Synthetic Enzymes. Transmitter synthetic enzymes are produced in the body and are transported by axoplasmic flow to nerve endings. Tyrosine hydroxylase (TH), the rate-limiting enzyme in catecholamine neurons, converts tyrosine and molecular oxygen to L-dopa (L-dihydroxyphenylalanine) using a biopterin cofactor. TH is confined to catecholamine neurons and is localized in the cytosol. The enzyme's activity is regulated by several factors including biopterin and tyrosine levels, end-product inhibition by L-dopa, and nerve impulse traffic. End-product inhibition occurs through competition for enzyme binding of pteridine cofactor. The enzyme pteridine reductase provides the active, reduced form of biopterin needed for TH activity. Defects in this enzyme in humans reduce TH activity and catecholamine synthesis. Transient enzyme activation follows nerve depolarization, probably related to a reversible enzyme phosphorylation. During ontogeny, the specific activity of TH in target areas rises considerably in concert with development of catecholamine nerves (see below), and the enzyme becomes progressively compartmentalized into nerve endings (Coyle, 1977).

The next synthetic step, decarboxylation of L-dopa to dopamine, is catalyzed by L-dopa decarboxylase (DDC), an enzyme that is ubiquitous in the nervous system. Although important, it is not rate limiting and is not specific for catecholamine neurons. Norepinephrine neurons contain TH and DDC but also contain dopamine-β-hydroxylase (DBH), which converts dopamine to norepinephrine. The enzyme contains copper, and chelators of this metal or diseases affecting it (e.g., Menkes' disease) drastically lower DBH activity (Grover et al., 1982). DBH is enclosed in synaptic vesicles along with endogenous stores of norepinephrine. Like increases in TH, the ontogenetic increases in DBH activity accompanying differentiation of the noradrenergic arbor are related to addition of enzyme rather than changes in cofactor requirements or kinetic characteristics (Coyle and Axelrod, 1972). However, levels of endogenous brain inhibitors released during homogenization do affect the in vitro assay of this enzyme in developing tissues. DBH is confined to noradrenergic neurons, and antibodies directed against this enzyme allow perikarya and terminals to be localized with immunohistochemical methods (Grzanna and Molliver, 1980). Anti-TH antisera may be used to visualize both noradrenergic and dopaminergic neurons, but staining does not distinguish between the two (Specht et al., 1978; Pearson, Brandeis, and Goldstein, 1980). Immunochemical techniques can also be used to stain selectively epinephrine-containing neurons or chromaffin cells with antibody against phenylethanolamine *N*-methyltransferase, the final enzyme converting norepinephrine to epinephrine. Measurement of synthetic enzyme activity in regional tissue samples may also provide an estimate of innervation density by specific neuronal types.

Transmitter Storage, Release, and Inactivation. Neurotransmitters commonly exist in three pools within an axon terminal: in vesicles close to the synaptic membrane where they are available for immediate release by depolarization; in vesicles farther from the membrane, acting as a reservoir for the active compartment; and in an unbound state that is constantly being inactivated by degradative enzyme. Endogenous stores of catecholamines can be visualized using formaldehyde-induced fluorescent techniques, and concentrations can be measured using fluorometric and radiometric assays. Currently, many laboratories also utilize high-pressure liquid chromatography with electrochemical detection (Anderson et al., 1980; Nielson and Johnston, 1982).

With the arrival of a nerve impulse at nerve terminals, and in the presence of calcium, vesicles travel to the membrane and release their contents (exocytosis) into the synaptic cleft. At noradrenergic terminals, DBH and ATP (adenosine triphosphate) are released with transmitter from synaptic vesicles. Released transmitter diffuses across the synaptic cleft where contact with postsynaptic receptors produces an alteration in membrane potential (depolarization or hyperpolarization) or other action (e.g., alteration in adenyl cyclase activity). Without an efficient means for rapidly lowering the concentration of transmitter within the cleft, effective information transfer would be impossible. In catecholamine neurons and many others, energy-dependent high-affinity reuptake systems are located on presynaptic nerve endings and can rapidly recapture the released transmitter. This method also has an obvious advantage of conserving transmitter material.

Dopamine, norepinephrine, and epinephrine are pumped inward at specific reuptake sites on axon terminals, and uptake of tritiated catecholamines into synaptosomes (pinched off nerve endings) can be assayed relatively easily. Selective catecholamine transport into synaptosomes can be distinguished from nonspecific uptake using blockers such as desmethylimipramine for norepinephrine sites and benztropine for dopamine uptake sites. The maximum velocity of tritiated catecholamine uptake into a tissue is generally proportional to the quantity of catecholamine axonal nerve endings, although a variety of other factors (e.g., neuron activity) may alter this process (Coyle and Molliver, 1977).

Degradative enzymes are also used to end transmitter action, although the time course is slower than for the uptake systems. Enzyme-generated neurotransmitter catabolites appear in brain extracellular fluid and eventually pass into the cerebrospinal fluid (CSF). Catecholamine synapses use monoamine oxidase (MAO) and catechol-o-methyltransferase (COMT) to degrade dopamine and norepinephrine to their respective acid metabolites, homovanillic acid (HVA) and 3-methoxy-4-hydroxyphenylethylene glycol (MHPG), which can be measured in the CSF and brain tissue. In rat brain, 3, 4-dihydroxyphenylacetic acid (DOPAC) is also a major dopamine metabolite. MAO is localized intracellularly in mitochondria while COMT is distributed widely on cell surfaces. The intracellular location provides a means for removing transmitter that leaks out of storage granules. Measurement of these metabolites in CSF provides an index of neurotransmitter turnover if rates of egress from the spinal fluid through transport systems remain steady.

Ontogeny of Cortical Catecholamine Neurons

Over the last decade, investigation of the spectrum of synaptic neurochemical activities has provided detailed information about maturation of central noradrenergic projections. Dating of neurogenetic "birthdays" can be accomplished using tritiated thymidine autoradiography. In rats, neurons originating this projection in the locus coeruleus of the pons divide after 12–14 days of gestation (total gestation, 21–22 days), and catecholamine histofluorescence appears shortly thereafter in perikarya (Lauder and Bloom, 1974). TH and DBH activities are also first detected in the rat fetus by day 15 of gestation. Tritiated thymidine autoradiographic timing of the locus coeruleus cell division in the monkey shows that it occurs between gestational days 27 and 36 in the monkey (total gestation, 165 days) (Levitt and Rakic, 1982). In the human brain, catecholamine histofluorescence appears by the seventh week of gestation in the brain stem (Olson, Boreus, and Seiger, 1973).

At the time of birth in rodents, a widespread axonal arbor has extended to telencephalic target areas (Coyle and Molliver, 1977). In cerebral cortex, approximately 80 percent of the biochemical maturation of this system occurs postnatally (Rakic and Goldman-Rakic, 1982). All studies indicate that immature axonal terminals are extended well in advance of maturation of norepinephrine synthetic and storage processes. Capacity for tritiated norepinephrine uptake into axonal terminals matures in advance of TH and DBH activity as well as accumulation of endogenous norepinephrine stores (fig. 2) (Coyle and Axelrod, 1971). Although the fetal brain exhibits nonsaturable norepinephrine uptake at early stages, the first indication of specific, high-affinity saturable uptake, inhibitable by desmethylimipramine, occurs at day 18 of gestation. The maximum velocity of high-affinity tritiated norepinephrine uptake into synaptosomes increases dramatically during late fetal and early postnatal life, but the affinity of the pump remains relatively constant ($Km = 3 \times 10^{-7}M$) throughout development.

Major organizational changes accompany the quantitative changes in biochemical markers that occur during differentiation of the noradrenergic arbor. The distribution of noradrenergic axons in cortex has been studied using histofluorescence (Lidov, Molliver, and Zecevic, 1978). Although this technique does not distinguish between dopamine and nor-

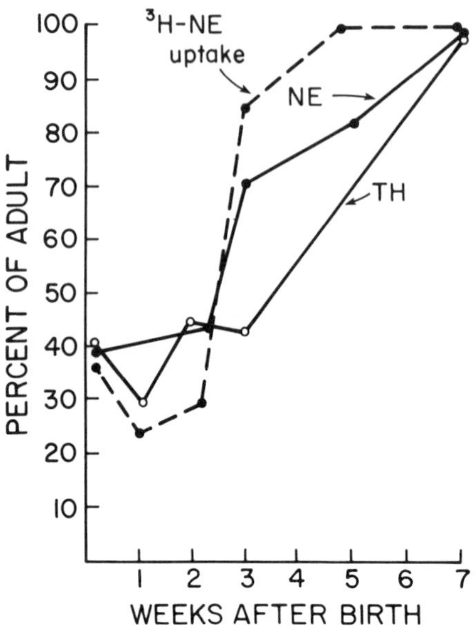

Figure 2. Postnatal ontogeny of neurotransmitter biochemical markers for noradrenergic axons in rat neocortex. Points are means of values from five or more animals from two or more litters (*^{3}H-NE,* tritiated norepinephrine; *TH,* tyrosine hydroxylase). (Data from Johnston and Coyle, 1980.)

epinephrine, the vast majority of catecholamine terminals in lateral cortex are noradrenergic. Despite the fact that neonatal rat cortex contains only 20 percent of the total catecholamine innervation of the adult cortex, the relative density of aminergic synaptic profiles is markedly higher than at later times. In certain regions of the immature cortex, 70 percent of synaptic profiles contain dense small (50 nM) granular vesicles (SGVs), which are storage sites for amines seen on electron microscopy. Histofluorescence microscopy indicates that a general pattern similar to that in the adult is established after one week of age in rat. During development, the apparent high density of catecholamine synapses is diluted by rapid addition of other types (Levitt and Moore, 1979). In human neocortex, a definite catecholamine fiber pattern can be detected after 23 weeks of gestation.

Reuptake and storage capacity for catecholamines in the neonatal cortex greatly exceeds the small pool of endogenously synthesized transmitter. Administration of 5-hydroxydopamine, an electron opaque "false" transmitter taken up into catecholamine nerve endings, results in a marked increment in SGVs over levels in controls of the same age (Coyle and Molliver, 1977). Assay of catecholamine levels after loading infant rats with 5-OH-dopamine or L-dopa also demonstrates a large latent storage capacity that is five to ten times higher than endogenous stores. Approximately 50 percent of this capacity is not localized to vesicles because it is not displaceable with reserpine, a specific inhibitor of catecholamine vesicular storage. A relatively large nonvesicular storage pool is also found in rat pups until they are well past two weeks of age. Increasing confinement of nerve terminal transmitter to storage vesicles and rising endogenous transmitter synthesis are features of later postnatal development.

In subhuman primates, the cortical maturation of noradrenergic innervation as reflected by transmitter levels and synthetic rates unfolds over a two- to three-year period of postnatal growth (Goldman-Rakic and Brown, 1982). As in rodents, about 80 percent of the ontogenetic changes take place postnatally. However, the primate exhibits considerably more regional variation than lower animals, major shifts occurring in different cortical regions over postnatal life. The most striking increases in catecholamines occur in association areas including prefrontal and parietal areas. In the primary visual cortex of the monkey, norepinephrine levels increase more than twofold within the first two months after birth. This is particularly interesting because norepinephrine may play a critical role in regulating early plastic synaptic changes in visual cortex related to development of binocular vision. Norepinephrine and dopamine levels in prefrontal cortex increase more gradually, corresponding to the more protracted development of prefrontal function. These coordinated time relationships as well as the early appearance of catecholamine systems in the brain suggest that they may play an early trophic role in neuronal development. Receptors for catecholamines may depend in part on development of presynaptic catecholamine terminals since neonatal lesions of these axons retard the accumulation of α-adrenergic and dopamine receptors (Deskin et al., 1981).

Nigrostriatal Dopamine Innervation

Dopaminergic axons projected from neurons in the substantia nigra in the midbrain form an extremely dense innervation into the caudate and putamen (corpus striatum in the rat) (Golden, 1972). Tritiated thymidine autoradiography indicates that the dopaminergic neurons are "born" on gestational days 13–15 in the rat. In monkeys, the neurons divide on fetal days 36–43. As is true for noradrenergic axons, the dopamine reuptake system, synthetic activity of TH, and endogenous transmitter stores undergo about 90 percent of their maturation in the postnatal period (Coyle and Campochiaro, 1976). Although the ontogenetic increase in dopamine markers in the striatum is relatively linear, TH activity in the mouse midbrain overshoots the adult levels by 25 percent at 15 days of age (Baker, Joh, and Reis, 1982). TH activity in the midbrain is correlated to the number of perikarya originating nigrostriatal projections, and the overshooting may be related to a readjustment of the number of these neurons. This readjustment might occur through programmed cell death or conversion of some dopamine neurons to other neurotransmitter types.

Development of the intensity of the histofluorescence pattern of striatal dopamine terminals in postnatal animals generally parallels the increase in other biochemical markers. However, the histofluorescent technique reveals an early pattern of "islands" of dopamine containing terminals in the postnatal animal that is invisible in the confluent fluorescence in the adult brain (Tennyson et al., 1972; Olson, Seiger, and Fuxe, 1972). Accumulation of fluorescent dopamine correlates with the differentiation of axonal growth cones into mature nerve terminals. The islands persist as special groupings of axons in the adult brain that can be visualized if dopamine synthesis is inhibited. This insular topographic arrangement of dopamine terminals is also a feature of the organization of other transmitter systems in the striatum (see below).

Dopamine Turnover

Measurement of brain levels of HVA and DOPAC in immature rodents indicates that dopamine turnover is relatively high even in the early postnatal period. Whole brain HVA is already at adult levels at birth, rising to higher levels at 12 days of age with a subsequent slow fall back to adult levels by 60 days of age. CSF HVA in rat pups is also 64 percent higher at 12 days than in adult rats. The ratio of striatal HVA to endogenous dopamine, which provides an estimate of dopamine turnover related to neuronal activity, is twice as high at four days of age as it is in adulthood (Keller, Bartholini, and Pletscher, 1973). This suggests that impulse flow is established in immature neurons soon after the nigrostriatal pathways are established. On the other hand, feedback inhibition of turnover by the direct dopamine agonist apomorphine is minimal at four days of age and slowly increases to adult levels by 30 days of age. Feedback stimulation of dopamine turnover by receptor blockade with haloperidol is first apparent at four days of age but also slowly increases with age (Coyle and Campochiaro, 1976).

Regulation of Development of Central Catecholamine Neurons

Experimental work to ascertain the factors that determine the qualitative and topographic differentiation of catecholamine projections is complex. For this reason, much of the information about these factors has come from work on simpler peripheral autonomic systems (see below). Nevertheless, several experimental approaches using animals suggest that the quantitative features of axonal arbor development are relatively independent of their terminal fields. On the other hand, the distribution of the immature axons appears to be heavily determined by the field. Experiments using two kinds of lesioning techniques support this statement.

One paradigm uses the toxin 6-hydroxydopamine, which destroys catecholamine axon terminals. When this toxin is injected into the forebrain of immature animals, telencephalic catecholamine neuronal markers are drastically reduced. However, the brain stem and cerebellum, which receive collateral innervation from the same neurons innervating the forebrain, then develop hyperplastic noradrenergic innervations with greatly increased axonal density (Sacks and Jonsson, 1975). This result suggests that "pruning" of the forebrain axonal arbor results in a compensatory increase in other fields unaffected by the toxin.

A second lesioning experiment utilized fetal treatment with antimitotic agents such as methylazoxymethanol acetate (MAM). Since the neurons originating noradrenergic projections are "born" at least several days before most neurons in neocortical target fields, injections of the antimitotic agent at an appropriate time in fetal life produced a markedly reduced target field with sparing of developing noradrenergic nuclei (Johnston and Coyle, 1981). Evaluation in adulthood of rats exposed to MAM on days 14–16 of gestation demonstrated that the total axonal arbor was preserved, but its density was considerably increased. Therefore, the total number of terminals in developing noradrenergic axonal arbors appears to be insensitive to loss of neurons in the target field. However, the distribution of the terminals is markedly disrupted and presumably determined by the terminal field (fig. 3).

The dependence of the distribution of noradrenergic axons on the terminal field is also demonstrated by experiments in which irises implanted in the brain develop a pattern of innervation from central noradrenergic neurons typical for the iris and not for brain tissue. Furthermore, in a cell culture system, dopaminergic mesencephalic neurons from fetal mouse brains form distinct patterns appropriate for either frontal cortex or striatum when presented with these targets (Hemmendinger et al., 1981). Therefore, data from a variety of preparations indicate that the central catecholamine neurons are programmed early in development for the total number of their axons to be formed though their distribution is target sensitive.

Cholinergic Neurotransmission

Cholinergic neurons are extremely important throughout the body and include spinal motoneurons, preganglionic neurons innervating autonomic ganglia, postganglionic parasympathetic neurons, and central pathways in the brain. In many ways, cholinergic systems provide a counterpoint to catecholamine systems. Several simple neuronal circuits involve catecholamine-cholinergic "balance" or at least interaction. For ex-

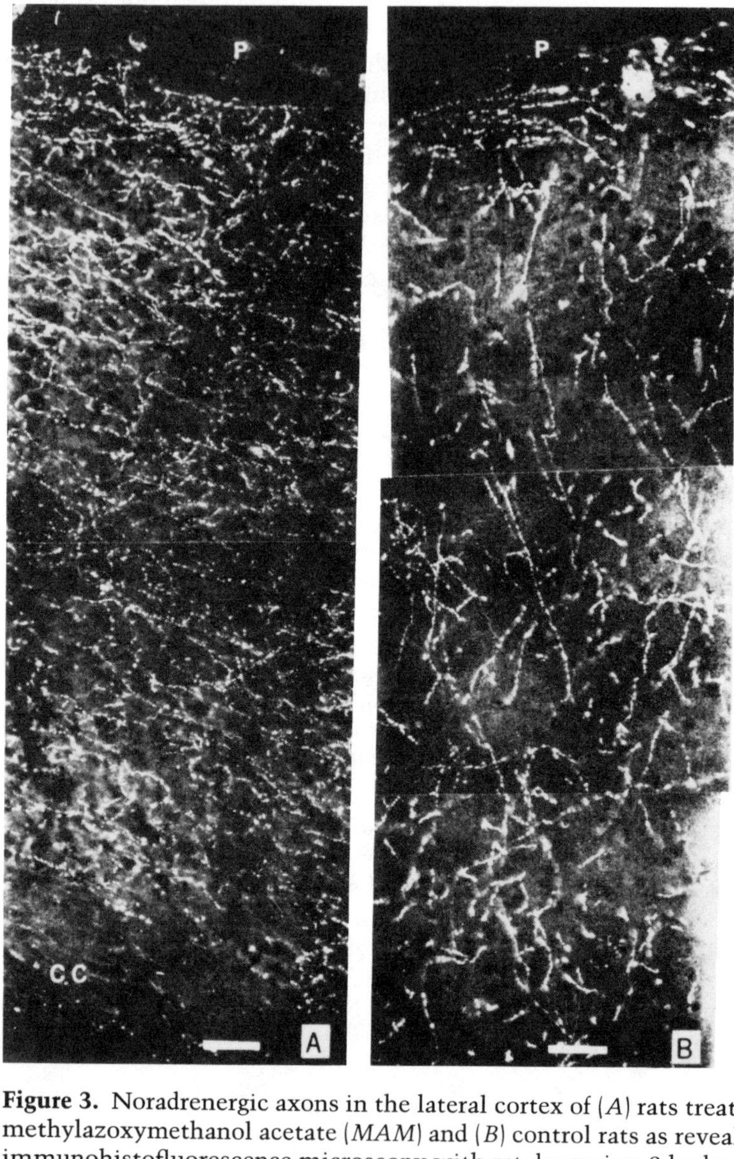

Figure 3. Noradrenergic axons in the lateral cortex of (*A*) rats treated with methylazoxymethanol acetate (*MAM*) and (*B*) control rats as revealed by immunohistofluorescence microscopy with rat dopamine-β-hydroxylase (*P*, pial surface; *CC*, corpus callosum; bar = 50μm). In (*A*) the plexus of noradrenergic axons is denser than in control cortex. Control cortex has fibers that run parallel to the surface in superficial layers and radially oriented fibers in deeper layers. The MAM-induced fetal lesion disrupted the distribution of noradrenergic fibers coming from the locus coeruleus. (From Johnston, Grzanna, and Coyle, 1982.)

ample, in most sympathetic ganglia the preganglionic cholinergic axons synapse on second-order noradrenergic neurons. Similarly, cholinergic neurons synapse on catecholamine-secreting chromaffin cells in the adrenal medulla. In the brain, neurochemical interactions between dopaminergic projections and cholinergic interneurons in the striatum are thought to be functionally relevant to normal movement and pathological conditions (Coyle and Campochiaro, 1976). These central interactions resemble the well-known balances between cholinergic and noradrenergic innervation in the heart or iris. During development, the central cholinergic neurons innervating sympathetic ganglia also have potent influences on development and maintenance of ganglionic neurons. In the brain, cholinergic pathways generally lag behind catecholamine systems with respect to developmental timetables (Johnston and Coyle, 1981).

Transmitter Metabolism in Cholinergic Neurons

All cholinergic neurons appear to share common systems for synthesizing, storing, releasing, and catabolizing acetylcholine (ACh). Cholinergic nerve endings resemble aminergic ones in having a specific cytosolic synthetic enzyme, multiple storage pools, and Ca^{++}-mediated release mechanisms (fig. 4). However, unlike that of catecholamines, the synaptic action of ACh is rapidly terminated by hydrolytic activity of acetylcholinesterase (AChE). The choline liberated by this cleavage is then pumped back into the terminal by a sodium-dependent high-affinity system in a fashion similar to the way intact catecholamine transmitters are taken up into aminergic terminals. The ACh synthetic enzyme, choline acetyltransferase (CAT), endogenous ACh, and the saturable Na^{+}-dependent choline uptake pump are highly localized to cholinergic nerve terminals and are considered specific biochemical markers for them. AChE is also synthesized by cholinergic neurons, and histochemical staining for AChE has been used to map presumptive cholinergic networks (Krnjevic and Silver 1966; Butcher and Hodge, 1976). In some systems, notably in dopamine neurons in the brain stem, noncholinergic neurons may contain impressive quantities of this enzyme. Recently, specific antibodies for CAT have been used successfully to map them in a manner similar to immunocytochemical mapping of catecholamine neurons (Kimura et al., 1981; Levey et al., 1983).

CAT and AChE are synthesized in cholinergic perikarya and transported down the axon. CAT is concentrated in the cytosol of the nerve terminal in a free state, though some may be attached to other surfaces of synaptic vesicles. CAT acetylates choline using acetyl-coenzyme A (AcCoA) as a substrate with K_m values in the order of 1 mM for choline and 10 μM for AcCoA. Endogenous choline is thought to be supplied by a pool of plasma choline since it cannot be synthesized by the brain. In the brain, a small extracellular pool of free choline and catabolism of

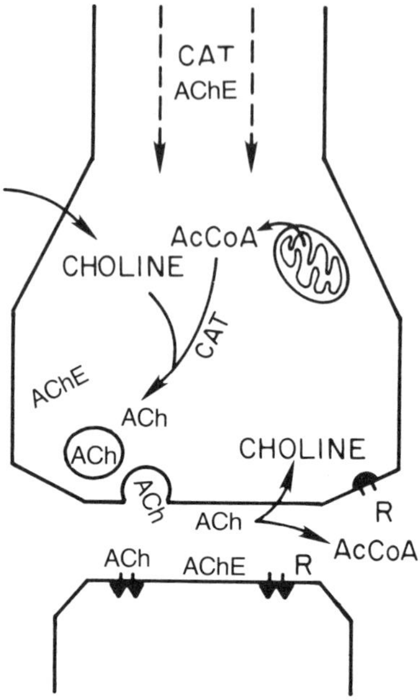

Figure 4. Diagram of cholinergic synapse. Choline is partially derived from the periphery and is used along with acetylcoenzyme A (*AcCoA*) to synthesize acetylcholine (*ACh*) by choline acetyltransferase (*CAT*). Choline is returned to the nerve terminal by a reuptake pump. Acetylcholinesterase (*AChE*) is present in the nerve terminal and in the synaptic cleft to hydrolyze ACh (*R*, presynaptic and postsynaptic receptors for ACh).

phospholipids provide a substrate level that is suboptimal for maximal enzyme activity. More than half the choline formed by hydrolysis of released ACh is recaptured by the high-affinity choline uptake system ($K_m = 1\ \mu M$), and uptake is accelerated by heightened neuronal impulse flow. There is evidence that choline supply may be rate limiting for ACh synthesis, especially at high neuronal activity, and raising plasma levels to increase transit into CSF may raise ACh synthesis (Blusztajn and Wurtman, 1983). AcCoA is synthesized in mitochondria from pyruvate generated from glucose. Defects in enzymatic processing of pyruvate or reduced supplies of pyruvate (hypoglycemia, hypoxia) lead to reduced ACh synthesis (Gibson and Duffy, 1981).

ACh appears to be stored in both free and vesicular pools in the same terminal, and the vesicles appear to be separated into a readily releasable group near the synaptic surface and a more stable pool farther back. Newly synthesized ACh is most easily released. Neurophysiological data from peripheral cholinergic nerves suggest that ACh is released in discrete quanta in response to influx of Ca^{++}. It has been assumed that these quanta correspond to exocytosis of the contents of discrete numbers of synaptic vesicles. Evidence from freeze-fracture electron micrographs of the neuromuscular junction supports this hypothesis, but the alternative—gated release of free transmitter—has not been excluded.

Less is known about the intimate details of the ontogeny of cholinergic biochemical machinery than is known about catecholamine terminals. In general, the V_{max} values of CAT and choline uptake processes increase steadily with age in concert with neuronal differentiation while the substrate K_m values remain the same (Coyle and Yamamura, 1976; Marchi et al., 1980). Developmental curves for CAT and AChE display similar characteristics in brain and spinal cord, having low activities in the first postnatal week and accelerating activities thereafter.

Cholinergic Development of the Neuromuscular Junction

The cholinergic innervation of striated muscle has been studied extensively because developing nerve-target interactions can be examined in relative isolation from other systems (Thoenen and Edgar, 1982; Dennis, 1981). A good deal of evidence indicates that the developing cholinergic terminals exert important initial influences on developing muscle and that temporal patterns of neuronal impulse flow determine several aspects of muscular gene expression. On the other hand, once contacts are established, the muscle field size influences the quantitative aspects of motoneuron innervation.

Axons innervating skeletal muscles arrive in the periphery before muscle cells appear, and myoblast differentiation occurs from the beginning in the presence of motor terminals. Generally, motor nerves mature in advance of muscle, and there is little evidence to support the possibility that growing axons are guided by target muscles. In fact, neuromuscular junctions will form when receptors are blocked with drugs. In rats, innervation of muscle fibers occurs over the last week of gestation. End plate potentials related to release of ACh appear quite soon after initiation of innervation on day 14 of gestation in rats. These synaptic potentials, related to low rates of spontaneous transmitter release, occur at a time when nerve terminals are small, immature, and contain few vesicles. The first morphologically identifiable contacts are seen in rats at day 16 of gestation. More mature junctions with a relatively thickened postsynaptic density appear by day 18 of gestation, and recognizable synaptic folds appear by the time of birth (day 21). Histochemically detectable AChE is present at day 17.

Nicotinic postsynaptic receptors are unevenly distributed on immature muscle prior to innervation, and their aggregation into characteristic high-density areas postsynaptic to nerve terminals occurs subsequent to innervation (Fischbach and Cohen, 1973). These receptors can be studied by electrophysiological methods and with autoradiography using radioactive ligands such as α-bungarotoxin. Study of neurons in culture indicates that spinal cord neurons appear to release a factor that induces clustering of the postsynaptic receptors.

Nerve impulse activity appears to regulate the development of AChE activity in the synaptic cleft as well as the number of postsynaptic receptors. Chronic blockade of neuromuscular transmission by curare inhibits the development of AChE. Enhanced neuronal activity is associated with a reduction in receptors but blockade of muscle activity by tetrodotoxin enhances their synthesis. Innervation patterns also influence the expression of glycolytic and oxidative metabolic enzymes as well as functional characteristics (fast versus slow) of developing muscle.

After establishment of neuromuscular connections, the muscle field appears to send signals that regulate the quantity of innervation. Data from chicks and other animals indicate that programmed cell death among lumbar motoneurons results in the loss of half the neuronal population shortly after their neuromuscular connections are established by all axons. The natural degeneration of motoneurons is closely associated with competition for a limited number of muscle cells as targets. One explanation for the phenomenon is that axons in excess of the number of developing muscle fibers are elaborated, and elimination of multiply innervated fibers ensues. This is supported by experiments showing that limb bud removal results in accentuation of cell death, but transplantation of supernumerary hind limbs markedly decreases it. What is interesting is that blockade of neuromuscular activity by pharmacological agents completely prevents the cell death, and increased muscular activity induced by chronic electrical stimulation increases it. These experiments suggest that once contacts are established, axons are dependent on critical levels of a trophic substance released by the muscles.

Central Cholinergic Neurons

Development of central cholinergic neurons is studied by assessing the regional ontogeny of specific cholinergic neuronal markers in brain. Biochemical studies and histochemical staining for AChE provide a reasonably complete picture, although precise analysis of neuronal interactions is more difficult for central than for peripheral neurons.

Cholinergic neurons include brain stem pathways, ascending telencephalic projections to cortex from nuclei in the basal forebrain, and intrinsic cholinergic neurons in the corpus striatum. In adults, ACh levels are highest in striatum and lowest in cerebellum. The basal forebrain nuclei include medial septal nuclei projecting into the hippocampus and a group of magnocellular nuclei extending from the nucleus of the diagonal band backward to the basal nucleus of Meynert. In rats, some cholinergic perikarya may reside in cortex but in higher animals and primates almost all the innervation arises from the basal nuclei (Johnston, McKinney, and Coyle, 1981).

As is the case for catecholamine projections, levels of markers such as CAT mature in rodent brain stem regions in advance of the telencephalon. In contrast to that of catecholamine systems, however, the ontogenetic curve for ACh synthetic enzyme elevation in rodents and chick is relatively flat during the first postnatal week and then rises tenfold in the next three weeks (Coyle and Yamamura, 1976; Nadler et al., 1974). Some elevation continues into adulthood. The relatively delayed time course, in contrast to earlier catecholamine patterns, appears to be followed in development of human cortex also (Brooksbank et al., 1978). The elevation in CAT activity appears to parallel regional formation of synaptic connections and cholinergic terminals.

In catecholamine terminals, the synthetic enzyme and reuptake capacity for choline develop in advance of or coincident with endogenous transmitter, but in cholinergic terminal fields transmitter levels are already relatively high early in development (fig. 5). At birth, endogenous ACh in rodent brain is at 25–85 percent of adult levels in various regions, but CAT activity is less than 10 percent of adult activity. Rela-

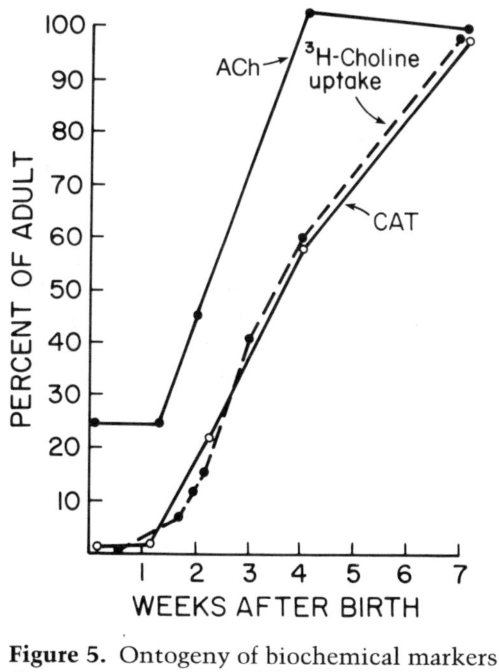

Figure 5. Ontogeny of biochemical markers for cholinergic neurons in the corpus striatum of the rat (*ACh,* acetylcholine; *CAT,* choline acetyltransferase; ^{3}*H-Choline,* tritiated choline).

tively low activities of AChE and low ACh turnover rates related to low neuronal firing may contribute to the high ACh levels in early life. The development of high-affinity sodium-dependent ^{3}H-choline uptake displays a developmental curve that is concurrent with or slightly behind that of CAT (Shelton, Nadler, and Cotman, 1979). In part, the ontogeny of the uptake system may reflect development of neuronal impulse flow that heavily influences this process in adult animals.

Compared with what is known about the neuromuscular junction, relatively little is known about early growth of central cholinergic axons and interaction with their target fields. In the rat hippocampus, which receives its entire cholinergic innervation from the septum, staining for AChE is first visible on day four after birth and moves progressively peripherally at later times until the entire area is stained at two weeks. CAT activity in the hippocampus is low at 11 days of age but attains adult levels one week later. Choline uptake increases threefold from six days of age until adulthood with a period of activity higher than an adult, at 30 days of age. An earlier invasion of immature cholinergic fibers into the hippocampus is suggested by experiments using retrograde transport of the fluorescent marker bisbenzimide (Crutcher, 1982). Intrahippocampal injection on the day of birth is followed by fluorescence evaluation in perikarya in the medial septum, the site of origin of the cholinergic projection. This suggests that immature cholinergic terminals invade their field in advance of expression of other definitive neurotransmitter characteristics.

AChE histochemical staining appears to be a relatively specific marker for cholinergic neurons in the septohippocampal pathway and the cortical projections from the nucleus basalis of Meynert and other forebrain basal nuclei (Mesulam et al., 1983; Levey et al., 1983). In cats, the AChE positive basal forebrain projections begin to spread outward at about day 40 of gestation and invade deep layers of cortex by the time of birth (total gestation, 63 days) (Krnjevic and Silver, 1966). In rats, only a few cortical somata (Cajal-Retzius cells) stain for AChE in the first postnatal week. By the end of the first week, AChE-rich bands are established that correspond to cortical lamina with high synaptic density (Kristt, 1979). This time course corresponds well to the ontogenetic pattern of CAT and ^{3}H-choline uptake in sensorimotor cortex.

Lesion experiments suggest that cholinergic cortical afferents are similar to catecholamine ones in their response to reduction of the terminal field. Relative cholinergic hyperinnervation of lateral cortex, hippocampus, and cerebellum results from early toxic or radiation lesions that kill dividing cells (Johnston and Coyle, 1980).

Striatal ACh Neurons. ACh in the striatum is primarily localized to relatively large interneurons. In rodents, neurogenesis in the striatum continues in the last week in gestation and into the neonatal period, but cholinergic neurons are generated well before birth. The time when neurogenesis of intrinsic neurons ceases in neonates corresponds to the phase of elevation in CAT and choline uptake activity in the second postnatal week.

AChE histochemical staining in the developing striatum suggests a complex organizational pattern of cholinergic innervation and considerable rearrangement during ontogeny. In the adult striatum, the AChE pattern is diffuse, although an uneven trabecular pattern of AChE staining can be observed in some preparations. A definite insular pattern is more clear-cut in immature rodents. In fetal cat, relatively homogeneous dark staining fields progress to a pattern of dark patches on a light background. AChE staining is also concentrated in densely stained patches in the human fetus and neonate (fig. 6) (Graybiel and Ragsdale, 1980). The AChE islands in primates may correspond to islands of densely packed striatal neurons observed in Nissl-stained sections of monkey striatum (Goldman-Rakic, 1981).

The distinct islands are surrounded by less cellular anuli, which appear to be the terminals of corticostriatal afferent axons. In adult animals, AChE-poor areas are enriched in opiate receptors, and other receptors are also localized in specific trabecular patterns. The precise relationship of the AChE-containing areas to dopamine-containing islands in the striatum remains to be worked out. However, these studies suggest that reorganization of cholinergic and other innervation is a prominent feature in the developing striatum.

It should also be emphasized that the AChE activity may also be localized to noncholinergic neurons in early life. Major shifts in AChE activity in cortex and thalamus of developing brain indicate that this marker may either be transiently localized to noncholinergic neurons or possibly to cholinergic neurons that later switch their transmitter identity (Kristt and Kasper, 1983).

Ontogeny of Dopaminergic-Cholinergic Interactions. In the mature brain, dopaminergic projections to the striatum generally have effects that counter those of later differentiating cholinergic interneurons in the striatum. Dopamine release inhibits cholinergic neurons and decreases ACh turnover. On the other hand, diminished dopaminergic activity results in increased ACh turnover; and the behavioral and motor effects of decreased dopamine action can be blocked by muscarinic antagonists (Tarsy, 1977).

Development of dopamine-cholinergic interactions can be studied with biochemical and behavioral methods. One method for studying the biochemical influence of dopamine pathways on ACh neurons is to measure drug effects on endogenous ACh levels (Coyle and Campochiaro, 1976). Increased turnover of ACh related to cholinergic activity results in lowered endogenous ACh levels, but reduced activity results in ele-

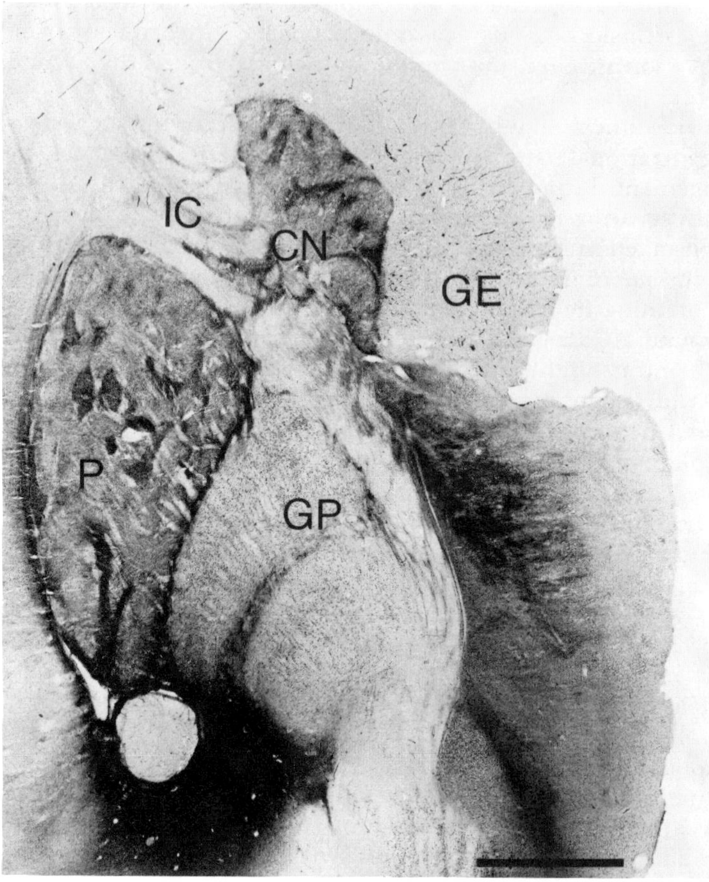

Figure 6. AChE histochemical stain of the caudate nucleus (*CN*) and putamen (*P*) of the 20-week-old human fetus (*IC*, internal capsule; *GE*, ganglionic eminence; *GP*, globus pallidum). (From Graybiel and Ragsdale, 1980.)

vated endogenous concentrations. Blockade of dopamine receptors with the neuroleptic drug fluphenazine reduces ACh levels in adult rats but not in animals less than eight days of age. The timing of influence of dopamine on the cholinergic neurons corresponds to the time when CAT activity begins a rapid ascent in the second week of life.

Behavioral studies using drugs also suggest that cholinergic links in the dopaminergic-cholinergic circuits develop later than the dopamine neurons (Burt et al., 1982). Several behaviors mediated by dopamine pathways, such as hyperactivity, stereotypic behavior, and catalepsy can be induced soon after birth in rats. Behaviors mediated by dopamine excess can be antagonized by cholinergic agonists in adults, but the effect is not detectable in immature rats. Pilocarpine, a cholinergic agonist, antagonizes excitatory effects of amphetamine, a dopamine releaser, at 20 days after birth but not at earlier times. Scopolamine, a muscarinic blocker, potentiates amphetamine-induced stereotypic or hyperactive behavior but not until two weeks of age. Catalepsy, a behavior produced by dopamine blockade in adults, can be antagonized by anticholinergic agents only after the rat is 15 days of age. Cholinergic agonists also counteract dopamine effects and produce catalepsy in rodents more than two weeks old. These results suggest that establishment of a functional dopaminergic-cholinergic link lags behind development of the dopamine innervation. This is one example of what appears to be a general theme in transmitter circuit development—that segments of multineuronal circuits have distinct developmental timetables. The biochemical characteristics and functional activity of these circuits, therefore, remain different from those of adult circuits until mature connections are established. The shifts in relative influence of various neurotransmitter groups are relevant to understanding behavioral development and response of the developing brain to neuroactive drugs.

Autonomic Neurotransmitter Development

The complexity of the central nervous system makes it difficult to determine the molecular factors that regulate formation of specific neurotransmitter circuits. The foregoing account of central neurotransmitter ontogeny is largely descriptive, but more precise information about expression of neurotransmitter phenotypes and interneuronal relationships is available from experiments with simpler peripheral autonomic neurons.

Norepinephrine and ACh are two of several neurotransmitters used by autonomic ganglia. In the adult superior cervical sympathetic ganglia, cholinergic axons emanating from the spinal cord synapse on noradrenergic cell bodies in the ganglia. It is noteworthy that the peripheral system, although much simpler than the brain, contains two basic neuronal types that are also interconnected in the central nervous system. Neuronal transmitter markers (synthetic enzymes, uptake processes, transmitters) are the same as those for central catecholamine and cholinergic neurons (Black, 1982).

Factors that govern the choice of neurotransmitter characteristics during development of autonomic neurons have been studied in animals and in culture. Final neurotransmitter expression is influenced by interactions among genetic programs, local environmental signals, and neuronal activity. Experiments performed with avian embryos demonstrate the influence of local environments. Neural crest, which produces cells destined for autonomic neurons, ordinarily gives rise to adrenomedullary chromaffin tissue if located in the embryonic trunk region. In contrast, rostral crest in the so-called vagal region gives rise to cholinergic enteric ganglia. Transplantation experiments indicate that embryonic crest from the vagal region displays catecholamine features when moved to the trunk, while trunk-to-vagal transplants differentiate along cholinergic lines (Le Douarin et al., 1975).

In vitro studies with dissociated sympathetic ganglia from newborn rats also demonstrate a profound influence of environment. Neurons grown in the absence of ganglionic nonneuronal cells synthesize catecholamine but little ACh. Addition of nonneuronal ganglion cells stimulates ACh synthesis (Patterson and Chun, 1974). Neuronal excitation, whether from electrical stimulation or depolarization with potassium or veratridine, has an effect opposite from nonneuronal cells in culture and favors noradrenergic development (Walicke, Campenot, and Patterson, 1977).

The final choice of neurotransmitter identity sometimes shifts early in development. Noradrenergic traits (TH activity, DBH, endogenous norepinephrine, tritiated norepinephrine uptake) that are expressed in autonomic primordial cells destined to become sympathetic neurons or chromaffin cells are also seen in early cells destined to lose these markers. Embryonic gut cells in the rat contain noradrenergic markers at 11.5 days but lose them later. In rat eccrine sweat glands, early noradrenergic markers are lost, and cholinergic attributes are acquired (Black, 1982).

Central and autonomic noradrenergic systems share certain similarities in biochemical maturation because expression of early appearing uptake systems occurs independently of TH and DBH. Also, the autonomic system resembles central transmitter systems in that the bulk of the maturational increase in synthetic enzymes and other markers occurs after migration and cell division have ceased. However, this temporal relationship may be fortuitous to some extent rather than implying a strictly coordinated set of events. The earliest noradrenergic traits expressed by many autonomic neurons appear to develop largely independently of other major events such as neuronal migration, neuronal elongation, completion of afferent or efferent neuronal connections, and completion of DNA synthesis.

Some of the factors regulating the quantitative aspects of autonomic transmitter expression during development also help maintain a steady state in adults. Decentralization of adult rat superior sympathetic ganglia or separation from peripheral targets lowers ganglion TH activity, indicating dependence of this trait on transsynaptic central influences. Stimulation of presynaptic sympathetic neurons induces TH and DBH through stimulation of postsynaptic nicotinic receptors in a manner similar to induction of these enzymes in developing ganglia. Loss of presynaptic stimulation has a potent but opposite effect on the ganglionic level of the peptide substance P, which rises dramatically after decentralization in newborn and adult animals. Levels are reduced by depolarizing agents such as veratridine. Nerve growth factor, which stimulates large increments in noradrenergic neuronal markers in developing ganglia, may also help maintain them in adults (Kessler and Black, 1980; Kessler et al., 1980). Furthermore, adrenocortical steroids promote increases in the epinephrine synthetic enzyme phenylethanolamine *N*-methyltransferase in the adrenal medulla of both immature and adult animals.

Although certain factors such as transsynaptic impulse flow appear to influence both neurotransmitter choice and quantity of transmitter characteristics, others act on only one set of variables. Thus the influence of nerve growth factor or corticosteroids on sympathetic neurons promotes accumulation of transmitter markers, but neither has any known influence on early determination of transmitter choice. There are undoubtedly many factors yet to be defined. It will be particularly important to define factors that direct some autonomic neurons to change transmitters if influenced by an appropriate target. Even certain adult neurons are mutable since adult chromaffin cells will extend neuronal processes in culture or if transplanted to brain.

The relevance of this information derived from the autonomic system to the central nervous system remains to be defined. Certain differences are now apparent, such as the fact that nerve growth factor has little effect on central noradrenergic neurons; nevertheless, it seems probable that major themes may be followed throughout the nervous system.

Other Neuronal Systems

Serotonin

Serotonin (5-hydroxytryptamine) (5-HT), like catecholamines and ACh, has actions throughout the body in the cardiovascular, respiratory, and gastrointestinal systems. In the brain it is involved in diverse functions such as pain, sleep, control of hypothalamopituitary functions, appetite, and other behavioral and vegetative functions. Serotonergic projections in the brain originate from raphe nuclei in the brain stem, diffuse cortical projections emanating from the midbrain and projections to the spinal cord and cerebellum coming from more caudal nuclei. Other terminal fields include the hippocampus, striatum, and hypothalamus.

Although less is known about 5-HT nerve endings, their basic machinery resembles that for catecholamine neurons (Hedner and Lundborg, 1980; Tissari, 1973). The amino acid tryptophan, largely derived from the diet, is taken into the neuron and hydroxylated to 5-hydroxytryptophan (5-HTP) by tryptophan hydroxylase. This enzyme, like TH, requires both elemental oxygen and reduced pteridine cofactor tetrahydrobiopterin. The enzyme K_m value (5×10^{-5} M) is above ordinary concentrations of tryptophan so that dietary augmentation can increase 5-HT synthesis. The transmitter is formed by decarboxylation of 5-HTP by an enzyme very similar to if not the same as DDC. The major catabolic pathway is by MAO to 5-hydroxyindolacetic acid (5-HIAA). The 5-HT is taken up by specific processes and stored in granules. Drugs that inhibit tryptophan hydroxylase (*p*-chlorophenylalamine), deplete 5-HT granules (reserpine), or inhibit uptake (tricyclic antidepressants) are available. In animals, brain 5-HT is altered by physiological changes in plasma amino acids, and similar effects may be responsible for dietary effects on sleep patterns in human infants (Zeisel et al., 1981; Yogman and Zeisel, 1983).

Raphe 5-HT neurons undergo early cell division from day 11 to day 15 of gestation in the rat, and histofluorescence is detectable on day 12. Whole-brain tryptophan hydroxylase activity, measured by in vitro assays or accumulation of 5-HTP levels after decarboxylase inhibition, reaches 50 percent of adult levels at birth. Adult activity is reached by four weeks of age. Maturation follows a caudal-to-rostral gradient with brain stem activity at two-thirds of adult level at birth and hemisphere activity at about 20 percent. Endogenous concentrations of 5-HTP and activity of the 5-HT uptake process follow a similar developmental curve. Brain 5-HT turnover estimated by a fall in endogenous transmitter after inhibition of the synthetic enzyme is similar in newborn animals to turnover in 14- or 28-day-olds, suggesting that the neuronal impulse flow is near adult levels even in the neonatal period. Levels of 5-HIAA in brain tissue increase progressively over the first four weeks in all brain regions except the brain stem where levels decrease slightly with age. The pattern of CSF 5-HIAA in the rat shows less of an increase over time, possibly related to an increase in capacity to clear the metabolite from the CSF. The ontogenetic changes in 5-HT markers appear to reflect changes in synthetic capacity in maturing nerve terminals rather than substrate availability. Blood and brain serotonin are essentially at adult levels at birth but rise to levels above those of the adult at one week of age (Hedner and Lundborg, 1980).

Amino Acids

GABA (γ-aminobutyric acid) is an extremely important inhibitory transmitter, which is documented as acting throughout the brain. Cerebellar Purkinje cells use GABA as a neurotransmitter in dense projections to the brain stem and cerebellar nuclei. The striatum contains GABAergic interneurons, as well as projections to the substantia nigra with an important role in controlling movement. GABAergic aspinous stellate local circuit neurons in the cerebral cortex have an important inhibitory role, and lesions of these neurons are implicated in epilepsy. GABAergic neurons interact with the other neurotransmitter systems, especially with dopaminergic and cholinergic neurons in the striatal-nigral-striatal circuit. Comparison of the ontogenetic patterns of GABAergic neuronal markers with cholinergic and catecholaminergic ones reveals that they generally fall midway between the other two, catecholamine markers maturing earlier and cholinergic ones lagging behind.

GABA. GABA, for neurotransmitter use, may be compartmentalized in the synaptic terminal separated from other pools that serve as intermediates in energy metabolism. Both pools are formed primarily from glucose by the GABA shunt. The synaptic machinery in GABAergic neurons includes glutamate decarboxylase (GAD), the rate-limiting, pyridoxine-requiring synthetic enzyme; endogenous GABA; and a high-affinity sodium-dependent reuptake mechanism. Glia also have a GABA uptake system, but the two pumps are blocked by different drugs. GABA is quickly removed by the uptake mechanism and is metabolized intracellularly by GABA transaminase (GABA-T) to succinic semialdehyde and then by a succinic semialdehyde dehydrogenase to succinic acid. GABA-T is localized in mitochondria and occurs in both GABAergic neurons and glia. GAD is the most specific marker for GABAergic neurons, and antibodies directed against it make possible the histochemical study of these neurons.

Like other neurotransmitter markers, the GABAergic markers mature in a caudal-to-rostral gradient. In rat, the synthetic enzyme GAD, the uptake process, and endogenous GABA have somewhat different maturational time courses (fig. 7). GAD is detectable in cortex in the third gestational week and is present at 12 percent of adult activity at birth (Coyle and Enna, 1976). In contrast, cortical GABA levels are at 19 percent of the adult level at day 15 of gestation but at 60 percent of adult concentration at birth. Both of these markers reach near adult levels in whole brain and cortex by 30 days of age. In contrast to the steady rise in markers for GAD and GABA, ^{3}H-GABA uptake into synaptosomes from whole brain rises abruptly in the neonatal period to more than twice the adult level by one week of age and remains high for several weeks. There is regional variation, elevations in cortex being a bit lower. Kinetic

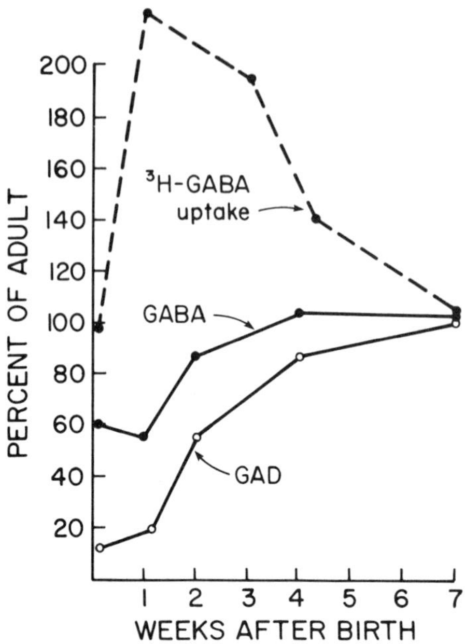

Figure 7. Postnatal ontogeny of biochemical markers for GABA-containing neurons in the whole brain of the rat (*GAD*, glutamate decarboxylase; *GABA*, γ-aminobutyric acid; *³H-GABA*, tritiated GABA).

analysis indicates that the high activity is related to V_{max} rather than an affinity change, and pharmacological studies show a response to uptake blockers appropriate for a synaptosomal GABA uptake rather than a glial process. The explanation for the "overshoot," which is more dramatic than the overshoot for noradrenergic neurons and unlike the choline uptake pattern for cholinergic ontogeny, is uncertain. One possible explanation is that excess GABAergic nerve endings are generated and then reduced. It is noteworthy that in human cortex, a developmental overshoot is seen in GAD activity during the first year of life (Diebler, Farkas-Bargeton, and Wherle, 1979). The pattern of overproduction of neurons and programmed cell death, discussed above, is well established in several other systems.

Glutamate, Aspartate, and Glycine. Compared with other transmitters discussed in this chapter, relatively little is known about the ontogeny of glutamate, aspartate, and glycine neurotransmitter systems, although they appear to serve major roles in brain function and are subjects of major research efforts (Greenamyre, Young, and Penney, 1983). Glycine appears to be the transmitter in inhibitory interneurons in the spinal cord. Glutamate and aspartate are strongly excitatory amino acids, and glutamate has been connected with a variety of pathways including corticostriatal and corticospinal pathways. Major problems with defining the ontogeny of these systems are related to the lack of specific synthetic enzyme markers. Furthermore, all play a role in intermediary metabolism so that endogenous transmitter is not specifically localized. Measurement of ^{3}H-amino acid uptake into synaptosomes may be used as biochemical markers, although the uptake process is found to some extent in glia. Nevertheless, lesion studies in adults and studies of developing systems indicate that the uptake process is a reasonable marker for assessing terminals for excitatory amino acid neurons. For example, studies of developmental rises in endogenous glutamate and ^{3}H-glutamate uptake correlate well to morphological evidence of development of corticostriatal pathways (Campochiaro and Coyle, 1978; Hitzemann and Loh, 1978).

Brain Peptides. Neurons that release peptides have been known for years, but recently study of central peptidergic systems has become one of the most rapidly expanding areas of neuroscience (Guillemin, 1978; Snyder and Innis, 1979; Krieger and Martin, 1981).

The first such systems to be identified were hypothalamic neurons that release regulatory peptides, for example, thyrotropin releasing hormone (TRH), into the portal circulation of the anterior pituitary and vasopressin- and oxytocin-containing neurons projecting axons to the posterior pituitary. The list has now expanded considerably to include enkephalins, cholecystokinin, vasoactive intestinal polypeptide, substance P, β-lipotropin, and many others. Immunocytochemical techniques have been used to localize these neurons. They have widespread distributions and projections in the brain, suggesting that they serve alongside "conventional" transmitter systems as communication circuits. There is considerable evidence that some peptides, such as substance P and cholecystokinin, may coexist in the same neuron with other transmitters (Hokfelt et al., 1980). Whether all peptides do serve a definite transmitter role in the central nervous system remains uncertain. Some may act in a relatively slow fashion, producing a regulatory hormonelike response in target neurons. Peptidergic systems may play important roles in the developing brain just as peripheral peptide hormones (e.g., insulin) exert potent effects on peripheral development. However, very little is known about this area. It is clear, however, that the best studied systems—the hypothalamic neurons influencing the pituitary—have important neuroendocrine roles in parturition, growth, and development (Gluckman, Grumbach, and Kaplan, 1981; Cooper and Martin, 1980). Synthetic, secretion, and inactivation mechanisms for peptidergic systems differ in certain ways from other systems (i.e., aminergic systems) (Krieger and Martin, 1981). Synthetic mechanisms for neuropeptides appear to share major features with synthesis of other proteins in the body. Proteins such as insulin are synthesized by enzymatic cleavage of a much larger molecule called a *prohormone*. Similarly, vasopressin is cleaved from a larger molecule, which also contains neurophysin. The precursor molecule is packaged into vesicles in the neuronal nucleus soon after ribosomal synthesis on the endoplasmic reticulum. Intravesicular synthesis occurs as the vesicles travel down the axon toward the nerve ending. A similar method may be used by neurons containing adrenocorticotropic hormone, α-melanocyte– stimulating hormone, β-lipotropin, luteinizing hormone-releasing factor (LRF), β-endorphin, and metenkephalin, all of which are contained in the molecule β-lipotropin. These molecules are probably derived from one or more large molecules called *proopiocorticotrophins*. Delineating the ontogeny of propeptide processing and determining the relative concentrations of synthetic products versus promolecules in the developing brain are important issues for investigation (Alessi et al., 1983; Khachaturian et al., 1983). Study of rats indicates that there are significant changes in posttranslational processing in opiate peptides during development. In newborn rats, approximately 25 percent of pituitary β-endorphin is present in the acetylated form, but in adults, very little is acetylated.

Release of vesiculated neuropeptides in response to neuronal depolarization resembles ordinary neurotransmitter secretion and is mediated by calcium influx. Unlike amino acid and catecholamine transmitters, reuptake has not been demonstrated as a significant inactivation mechanism for peptides. Relatively specific extraneuronal peptidases appear to degrade them, although these are not specific markers for separate groups of neurons. These facts imply that peptidergic neurosecretion can be studied using conventional methods including depolarization-mediated secretion in brain slices. On the other hand, the absence of high-affinity reuptake as well as uncertainty about synthetic enzymes prevents detection and study of developing nerve terminals in the same detail possible for catecholamine systems.

Presently, endogenous peptide content is the primary biochemical marker useful for studying these neurons. Recent work has also identified specific messenger RNAs and DNA nucleotide sequences specific for certain neurons, and these may be accepted as markers in the future. Endogenous peptides have been studied using radioimmunoassay and immunocytochemistry to provide some details for several systems in the brain. For example, immunoreactive arginine vasopressin is found in the hypothalamic supraoptic and paraventricular nuclei by fetal day 16 in the rat and by 11 weeks of pregnancy in the human. Oxytocin-neurophysin–staining neurons appear in week 14 in humans (Swaab and TerBorg, 1981). Various opiate precursor peptides are also present in the rat brain by day 15 of gestation as is substance P (Pickel, Sumal, and Miller, 1982). LRF- and TRH-staining cells and fibers are present in the fetal human brain before nine weeks of gestation. The early appearance of these hypothalamic and brain stem peptide systems is consistent with their roles in regulating basic vegetative function (e.g., water balance by vasopressin) since these functions must be well developed by the time of birth.

Peptides localized to later maturing cortical regions also follow a later ontogenic pattern. Vasoactive intestinal polypeptide containing intrinsic neurons are found in the cerebral neocortex in adult animals. Immunoreactive vasoactive intestinal polypeptide is not found until the postnatal period in rats, first in the early maturing deep cortical layers and then in superficial neurons. Cholecystokinin peptides, also found in cerebral cortex, increase rapidly only in the postnatal period.

This information suggests a large number of questions about the development of these systems. Do any of the peptide systems influence neuronal plasticity during development in a fashion similar to ACh neurons? What factors influence the "choice" of peptide in a given neuronal group during differentiation? How rigid is the choice in the mature brain once it has been made? Are any diseases or developmental anomalies related to abnormal ontogeny of peptide brain systems?

Information already available suggests that the variety of peptidergic systems may be extremely complex with seemingly limitless numbers of peptide transmitters possible. The next few decades could even see peptides overshadow the "conventional" transmitters with respect to their importance in brain development.

References

Alessi, N., Khachaturian, H., Watson, S., and Akil, H. (1983). Postnatal ontogeny of acetylated and unacetylated beta-endorphin in rat pituitary. *Life Sci* [*Suppl*] 33, 1, 57–60.

Anderson, G., Butter, D., Young, G., Shaywitz, B., and Cohen, D. (1980). Simplified liquid chromatographic-electrochemical determination of norepinephrine and dopamine in rat. *J Chromatogr* 181, 453–455.

Baker, H., Joh, T. H., and Reis, D. J. (1982). Time of appearance during development of differences in nigrostriatal tyrosine hydroxylase activity in two inbred mouse strains. *Dev Brain Res* 4, 157–166.

Black, I. (1982). Stages of neurotransmitter development in autonomic neurons. *Science* 215, 1198–1204.

Blusztajn, J. K., and Wurtman, R. J. (1983). Choline and cholinergic neurons. *Science* 221, 614–620.

Brooksbank, B. W. L., Atkinson, D. J., and Balazs, R. (1981). Biochemical development of the human brain: II. Some parameters of the GABAergic system. *Dev Neurosci* 4, 188–200.

Brooksbank, B. W. L., Martinez, M., Atkinson, D. J., and Balazs, R. (1978).

Biochemical development of the human brain: I. Some parameters of the cholinergic system. *Dev Neurosci* 1, 267–284.

Burt, D. K., Hungerford, S. M., Crowner, M. L., and Baem, L. A. (1982). Postnatal development of a cholinergic influence on neuroleptic-induced catalepsy. *Pharmacol Biochem Behav* 16, 533–540.

Butcher, L., and Hodge, G. (1976). Postnatal development of acetylcholinesterase in the caudate-putamen nucleus and substantia nigra of rats. *Brain Res* 106, 223–240.

Campochiaro, P., and Coyle, J. (1978). Ontogenic development of kainate neurotoxicity: Correlates with glutamatergic innervation. *Proc Natl Acad Sci USA* 75, 2025–2029.

Cooper, J. R., Bloom, F. E., and Roth, R. H. (1982). *The biochemical basis of neuropharmacology*. New York: Oxford University Press.

Cooper, P., and Martin, J. B. (1980). Neuroendocrinology and brain peptides. *Ann Neurol* 8, 551–557.

Coyle, J. (1977). Biochemical aspects of neurotransmission in the developing brain. *Int Rev Neurobiol* 20, 65–102.

Coyle, J., and Axelrod, J. (1971). Development of the uptake and storage of L-(3H)norepinephrine in the rat brain. *J Neurochem* 18, 2061–2075.

Coyle, J., and Axelrod, J. (1972). Dopamine-beta-hydroxylase in the rat brain: Developmental characteristics. *J Neurochem* 19, 449–459.

Coyle, J. T., and Campochiaro, P. (1976). Ontogenesis of dopaminergic-cholinergic interactions in the rat striatum: A neurochemical study. *J Neurochem* 27, 673–678.

Coyle, J., and Enna, S. (1976). Neurochemical aspects of the ontogenesis of GABAergic neurons in the rat brain. *Brain Res* 111, 119–133.

Coyle, J., and Molliver, M. (1977). Major innervation of newborn rat cortex by monoaminergic neurons. *Science* 196, 444–447.

Coyle, J., and Yamamura, H. (1976). Neurochemical aspects of the ontogenesis of cholinergic neurons in the rat brain. *Brain Res* 118, 429–440.

Crutcher, K. A. (1982). Development of the rat septohippocampal projection: A retrograde fluorescent tracer study. *Dev Brain Res* 3, 145–150.

Dennis, M. (1981). Development of the neuromuscular junction: Inductive interactions between cells. *Ann Rev Neurosci* 4, 43–68.

Deskin, R., Seidler, F. J., Whitmore, W. L., and Slotkin, T. A. (1981). Development of alpha-noradrenergic and dopaminergic receptor systems depends on maturation of their presynaptic nerve terminals in the rat brain. *J Neurochem* 36, 1683–1690.

Diebler, M. F., Farkas-Bargeton, E., and Wherle, R. (1979). Developmental changes of enzymes associated with energy metabolism and the synthesis of some neurotransmitters in discrete areas of human neocortex. *J Neurochem* 32, 429–435.

Emson, P. C., Gilbert, R. F. T., Loren, I., Fahrenkrug, J., Sundler, F., and

Schaffalitzky de Muckadell, O. B. (1979). Development of vasoactive intestinal polypeptide (VIP) containing neurons in the rat brain. *Brain Res* 177, 437–444.

Fischbach, G., and Cohen, S. (1973). The distribution of acetylocholine sensitivity over uninnervated and innervated muscle fibers grown in cell culture. *Dev Biol* 31, 147–162.

Gibson, G. E., and Duffy, T. E. (1981). Impaired synthesis of acetylcholine by mild hypoxic hypoxia or nitrous oxide. *J Neurochem* 36, 28–33.

Gilad, G., and Reis, D. (1979). Collateral sprouting in central mesolimbic dopamine neurons. *Brain Res* 160, 17–36.

Gluckman, P. D., Grumbach, M. M., and Kaplan, S. L. (1981). The neuroendocrine regulation and function of growth hormone and prolactin in the mammalian fetus. *Endocr Rev* 2, 363–395.

Golden, G. (1972). Embryologic demonstration of a nigrostriatal projection in the mouse. *Brain Res* 44, 278–282.

Goldman-Rakic, P. (1981). Prenatal formation of the cortical input and development of cytoarchitectonic compartments in the neostriatum of the rhesus monkey. *J Neurosci* 1, 721–735.

Goldman-Rakic, P., and Brown, R. (1982). Postnatal development of monoamine content and synthesis in the cerebral cortex of rhesus monkeys. *Brain Res* 4, 339–349.

Graybiel, A., and Ragsdale, C. (1980). Clumping of acetylcholinesterase activity in the developing striatum of the human fetus and young infant. *Proc Natl Acad Sci USA* 77, 1214–1218.

Greenamyre, J. T., Young, A. B., and Penney, J. B. (1983). Quantitative autoradiography of L-3H-glutamate binding to rat brain. *Neurosci Lett* 37, 155–160.

Grover, W. D., Henkin, R. E., Schwartz, M., Brodsky, N., Hobdell, E., and Stolk, J. M. (1982). A defect in catecholamine metabolism in kinky-hair disease. *Ann Neurol* 12, 263–266.

Grzanna, R., and Molliver, M. E. (1980). The locus coeruleus in the rat: An immunohistochemical delineation. *Neurosci* 5, 21–40.

Guillemin, R. (1978). Peptides in the brain: The new endocrinology of neuron. *Science* 202, 390–402.

Hedner, T., and Lundborg, P. (1980). Serotonergic development in the postnatal rat brain. *J Neural Transm* 49, 257–279.

Hemmendinger, L. M., Garber, B. B., Hoffman, P. C., and Heller, A. (1981). Target neuron-specific process formation by embryonic mesencephalic dopamine neurons in vitro. *Proc Natl Acad Sci USA* 78, 1264–1268.

Herkenham, M., and Pert, C. (1981). Mosaic distribution of opiate receptors, parafascicular projections and acetylocholinesterase in rat striatum. *Nature* 291, 415–418.

Hitzemann, R. J., and Loh, H. (1978). High affinity GABA and glutamate transport in developing nerve ending particles. *Brain Res* 159, 29–40.

Hokfelt, T., Lundberg, J., Schultzberg, M., Johansson, O., Ljungdahl, A., and Rehfeld, J. (1980). Coexistence of peptides and putative transmitters in neurons. In *Neural peptides and neuronal communication*, ed. E. Costa, and M. Trabucchi. New York: Haven Press.

Johnston, M., and Coyle, J. (1980). Ontogeny of neurochemical markers for noradrenergic, GABAergic and cholinergic neurons in neocortex lesioned with methylazoxymethanol acetate. *J Neurochem* 34, 1429–1441.

Johnston, M., and Coyle, J. (1981). Development of central neurotransmitter systems. In *The fetus and independent life*. Ciba Foundation Symposium 86, pp. 251–270. London: Pitman.

Johnston, M., Grzanna, R., and Coyle, J. (1982). Methylazoxymethanol treatment of fetal rats results in abnormally dense noradrenergic innervation of neocortex. *Science* 203, 369–371.

Johnston, M., McKinney, M., and Coyle, J. (1981). Neocortical cholinergic innervation: A description of extrinsic components in the rat. *Exp Brain Res* 43, 159–172.

Keller, H. H., Bartholini, G., and Pletscher, A. (1973). Spontaneous and drug-induced changes of cerebral dopamine turnover during postnatal development of rats. *Brain Res* 64, 371–378.

Kessler, J., Adler, J., Bohn, M., and Black, I. (1980). Substance P in principal sympathetic neurons: Regulation by impulse activity. *Science* 214, 335–336.

Kessler, J., and Black, I. (1980). Nerve growth factor stimulates the development of substance P in sensory ganglia. *Proc Natl Acad Sci USA* 77, 649–652.

Khachaturian, H., Alessi, N., Munfahh, N., and Watson, S. (1983). Ontogeny of opiate and related peptides in rat CNS and pituitary. *Life Sci* 33, 61–64.

Kimura, H., McGeer, P. L., Peng, J. H., and McGeer, E. G. (1981). The central cholinergic system studied by choline acetyltransferase immunohistochemistry in the cat. *J Comp Neurol* 200, 151–171.

Krieger, D., and Martin, J. (1981). Brain peptides. *N Engl J Med* 304, 876–885.

Kristt, D. (1979). Development of neocortical circuitry: Histochemical localization of acetylcholinesterase in relation to the cell layers of rat somatosensory cortex. *J Comp Neurol* 186, 1–16.

Kristt, D. A., and Kasper, E. (1983). High density of cholinergic muscarinic receptors accompanies high intensity acetylcholinesterase-staining in layer IV of infant rat somatosensory cortex. *Brain Res* 8, 373–376.

Krnjevic, K., and Silver, A. (1966). Acetylcholinesterase in the developing forebrain. *J Anat* 100, 63–89.

Lauder, J. M., and Bloom, F. E. (1974). Ontogeny of monoamine neurons in the locus coeruleus, raphe nuclei and substantia nigra of the rat. *J Comp Neurol* 155, 469–482.

Le Douarin, N. M., Renaud, D., Teilett, M. A., and Le Douarin, G. H. (1975). Cholinergic differentiation of presumptive adrenergic neuroblasts in interspecific chimeras after heterotopic transplantations. *Proc Natl Acad Sci USA* 72, 728–732.

Levey, A. I., Wainer, B. H., Mufson, E. J., and Mesulam, M. M. (1983). Colocalization of acetylcholinesterase and choline acetyltransferase in the rat cerebrum. *Neuroscience* 9, 9–22.

Levitt, P., and Moore, R. Y. (1979). Development of the noradrenergic innervation of neocortex. *Brain Res* 162, 243–259.

Levitt, P., and Rakic, P. (1982). The time of genesis, embryonic origin and differentiation of the brainstem monoamine neurons in the rhesus monkey. *Brain Res* 4, 35–57.

Lidov, H. G. W., Molliver, M. E., and Zecevic, N. R. (1978). Characterization of the monoaminergic innervation of immature neocortex: A histofluorescence study. *J Comp Neurol* 81, 663–680.

Marchi, M., Hoffman, D., Giacobini, E., and Fredrickson, T. (1980). Development and aging of cholinergic synapses: IV. Acetylcholinesterase and choline acetyltransferase activities in autonomic ganglia and iris of the chick. *Dev Neurosci* 3, 235–247.

Mesulam, M., Mufson, E., Levey, A., and Wainer, B. (1983). Cholinergic innervation of the cortex by the basal forebrain: Cytochemistry and cortical connections of the septal area, diagonal band nuclei, nucleus basalis (substantia innominata) and hypothalamus in the rhesus monkey. *J Comp Neurol* 214, 170–197.

Moore, R. (1982). Catecholamine neuron systems in brain. *Ann Neurol* 12, 321–327.

Nadler, J., Matthews, D., Cotman, C., and Lynch, G. (1974). Development of the cholinergic innervation in hippocampal formation of the rat. *Dev Biol* 36, 142–154.

Nielson, J. A., and Johnston, C. A. (1982). Rapid analysis of dopamine, 5-hydroxytryptamine, other precursors and metabolites using HPLC-EC. *Life Sci* 31, 2847–2856.

Olson, L., Boreus, L. O., and Seiger, A. (1973). Histochemical demonstration and mapping of 5-hydroxytryptamine- and catecholamine-containing neuron systems in the human fetal brain. *Zeitschrift für Anatomie and Entwicklungsgeschichte* 139, 259–282.

Olson, L., Seiger, A., and Fuxe, K. (1972). Heterogeneity of striatal and limbic dopamine fluorescent islands in developing and adult rats. *Brain Res* 44, 283–288.

Patterson, P., and Chun, L. (1974). The influence of non-neural cells on catecholamine and acetylcholine synthesis and accumulation in cultures of dissociated sympathetic neurons. *Proc Natl Acad Sci USA* 71, 3607–3610.

Patterson, P., and Chun, L. (1977). The induction of acetylcholine synthesis in primary cultures of dissociated rat sympathetic neurons. *Dev Biol* 56, 263–280.

Pearson, J., Brandeis, L., and Goldstein, M. (1980). Appearance of tyrosine hydroxylase immunoreactivity in the human embryo. *Dev Neurosci* 3, 140–150.

Pickel, V., Sumal, K., and Miller, R. (1982). Early prenatal development of substance P and enkephalin-containing neurons in the rat. *J Comp Neurol* 210, 411–422.

Rakic, P., and Goldman-Rakic, P. (1982). Development and modifiability of the cerebral cortex. Neurosciences Research Program Bulletin, vol. 20. Cambridge: MIT Press.

Sacks, C., and Jonsson, G. (1975). Effects of 6-hydroxydopamine on central noradrenaline neurons during ontogeny. *Brain Res* 99, 277–291.

Shelton, D. L., Nadler, J. V., and Cotman, C. W. (1979). Development of high affinity choline uptake and associated acetylcholine synthesis in the rat fascia dentata. *Brain Res* 163, 263–275.

Snyder, S. H., and Innis, R. B. (1979). Peptide neurotransmitters. *Annu Rev Biochem* 48, 755–775.

Specht, L. A., Pickel, V. M., Joh, T. H., and Reis, D. J. (1978). Immunocytochemical localization of tyrosine hydroxylase in processes within the ventricular zone of the prenatal rat brain. *Brain Res* 156, 315–321.

Swaab, D. F., and TerBorg, J. P. (1981). Development of peptidergic systems in the rat brain. In *The fetus and independent life.* Ciba Foundation Symposium 86, pp. 271–294. London: Pitman.

Tarsy, D. (1977). Dopamine-acetylcholine interactions in the basal ganglia. In *Neurotransmitter function,* ed. W. S. Fields, pp. 213–246. New York: Stratton.

Tennyson, V., Barret, R., Cohen, G., Cote, L., Heikkila, R., and Mytilineow, C. (1972). The developing neostriatum of the rabbit: Correlation of fluorescence histochemistry, electron microscopy, endogenous dopamine levels and (3H) dopamine uptake. *Brain Res* 46, 251–285.

Thoenen, H., and Edgar, D. (1982). The regulation of neuronal gene expression. *Trends in Neuroscience* 5, 311–313.

Tissari, A. H. (1973). Serotonergic mechanisms in ontogenesis. In *Fetal pharmacology,* ed. L. Boreus, pp. 237–257. New York: Raven Press.

Walicke, P., Campenot, R., and Patterson, P. (1977). Determination of transmitter function by neuronal activity. *Proc Natl Acad Sci USA* 74, 5767–5771.

Yogman, M. W., and Zeisel, S. (1983). Diet and sleep patterns in newborn infants. *N Engl J Med* 309, 1147–1149.

Zeisel, S., Mauron, C., Watkins, C. J., and Wurtman, R. J. (1981). Developmental changes in brain indoles, serum tryptophan and other serum neutral amino acids in the rat. *Brain Res* 1, 551–564.

Ontogenetic Development of Central Nervous System Neurotransmitter Receptors

Shaila D. Telang and S. J. Enna

Introduction

One of the more fundamental questions in developmental neurobiology concerns the relationship between presynaptic and postsynaptic elements. Attempts to address this issue have been frustrated by difficulties associated with monitoring developmental changes in neurotransmitter receptor properties and function. Although a few elegant electrophysiological studies have been performed to examine receptor development vis-à-vis innervation (Woodward et al., 1971), by necessity these projects were limited in scope. Although biochemical methods capable of assessing nerve terminal development (enzyme activity, neurotransmitter content, and accumulation) have been available for a number of years, it has only been in the past decade that ligand-binding assays could be used to study the biochemical properties of receptor recognition sites. This technical advance has therefore made it possible to examine directly recognition site development, yielding new insights into the factors that influence these membrane constituents.

The aim of the present communication is to summarize briefly the more recent studies conducted on the developmental characteristics of central nervous system neurotransmitter receptors using ligand-binding assays. For the purposes of this chapter a receptor is defined as a site on the external surface of the neuron that mediates a change in membrane polarity because of the attachment of a neurotransmitter. Several criteria must be met before a ligand-binding assay can be considered appropriate for a neurotransmitter receptor (Yamamura, Enna, and Kuhar, 1978). Thus, the binding site should be saturable at ligand concentrations that can reasonably be expected to occur in vivo; the organ, cellular, and subcellular distribution of the binding site should parallel the known distribution of the physiologically relevant receptor; the relative potencies of a series of recognition site agonists and antagonists as inhibitors of ligand attachment should be similar to their relative biologic potencies. It must be borne in mind, however, that ligand-binding assays reveal nothing about the functional properties of a recognition site. That is, recognition site activation sets into motion a sequence of events that result in a change in ion flux (Enna and Strada, 1983). This may be accomplished by the opening of a channel that is directly coupled to the recognition site or by means of a second messenger, such as a cyclic nucleotide. With ligand-binding assays only the number and kinetic properties of recognition sites are examined. Since it is conceivable that the recognition site–coupled event could be modified independently, a modification in receptor ligand binding does not necessarily prove that there has been a significant change in receptor function. Physiological studies must be performed to answer that question.

General Properties and Electrophysiological Characteristics of the Developing Synapse

Neurotransmitter receptors are located in the active zone on the external surface of the postsynaptic membrane. Clusters containing several hundred receptors are distributed in patches within the zone. The recognition sites themselves are embedded deep in the lipid matrix of the membrane. In some cases these sites are directly coupled to an ion channel, such as the nicotinic-cholinergic receptor, whereas others are affiliated with a second messenger, such as the adrenergic receptor.

As detailed in other chapters of this text, dendritic spines have the highest concentration of postsynaptic active zones in the cerebral cortex. It has been estimated that a single cortical neuron has up to 50,000 such spines. Transmitter-sensitive regions of muscle are known to be assembled from small, prefabricated arrays of receptor membrane, although it is not known whether a similar phenomenon accounts for development in the central nervous system. While the postsynaptic active zone is generally more stable than other parts of the membrane, the cortical spines are known to change shape under certain conditions, and this alteration modifies the degree of activity. Thus the active zone is a dynamic entity.

The nervous system develops from two different cell types, neuroblasts and spongioblasts. After multiplication in the germinal layer, the cells differentiate and migrate to appropriate regions. The primitive architecture of the nervous system is laid down in the form of small, densely packed neurons with relatively few glia. During development the cell body increases in size, followed by the outgrowth of axons and dendrites. These primitive axons (growth cones) are the first part of the neuron to associate with target cells, suggesting that these components have the capability of recognizing and interacting with an appropriate target to form a synaptic complex. If an axon fails to make the necessary contact, the parent neuron degenerates. Similarly, a target neuron deprived of its complementary axon fails to survive (Cowan, 1978). It is unclear, therefore, whether the developing axon seeks out the target cell or vice versa. Since it appears that dendritic arborization usually follows the establishment of a synaptic connection, it would appear that dendrite formation may be regulated in this way (Jones, 1981). Overall, however, the factors determining the type and distribution of synapses on a given target cell are not understood. One possibility is that the synaptic association between two cells is genetically predetermined and therefore independent of trophic factors. In this case the rate of development of the presynaptic and postsynaptic elements may be independent of one another. On the other hand, it is also possible that the developing axon is intrinsically undirected and that it is guided to the target, or maintains a particular synaptic contact, by some signal from the postsynaptic cell.

Numerous attempts have been made to study this question at histological, electrophysiological, and biochemical levels. One of the more important contributions was made by Bloom and his associates (Woodward et al., 1971) working with cerebellar Purkinje cells. The aims of this study were to define the stage of neuronal differentiation when synaptic transmission commences and to characterize the sensitivity of the postsynaptic membrane during the different periods of development. The results of these electrophysiological tests revealed that Purkinje cells are responsive to neurotransmitters before the development of synaptic contacts. This conclusion lent support to the proposal that the postsynaptic membrane matures prior to the presynaptic contact and therefore may play a role in directing the oncoming axon to the target cell (Woodward et al., 1971). Alternatively, it is also possible that the more precocious postsynaptic membrane has little to do with the establishment of a specific synaptic contact but rather may simply be a genetic expression of cellular sensitivity. However, it has been reported that Purkinje cells that fail to receive their normal complement of parallel fiber synapses have an abnormal number of spines (Jacobson, 1978), and it has also been observed that there is a decline in spine number on cerebral cortical pyramidal cells deprived of synaptic contact (Jones, 1981). Taken together these data indicate that the early appearance of receptor sensitivity is not merely a genetic expression of the cell but rather reflects a more complex phenomenon having functional significance with regard to the formation of synapses. While these observations indicate that the development of postsynaptic components, such as receptors or receptor-mediated events, precedes the development of presynaptic elements, in muscle it has been demonstrated that receptors are added to the postsynaptic cell after synapse formation (Benjamins and McKhann, 1981). This makes it appear likely that synapse formation can serve as a signal for the continued development of receptors on the postsynaptic membrane.

Biochemical Analysis of Receptor Development

Monoamine Receptors

Norepinephrine. Based on pharmacological and physiological data, adrenergic receptors may be divided into two major classes, α and β. More recent studies have indicated that within each of these groups there are pharmacologically, physiologically, and anatomically distinct subsets of receptors. These subsets have been designated as α_1, α_2 and β_1, β_2. Evidence has been obtained to indicate the presence of all four types of receptors in the central nervous system. Using ligand-binding assays, scientists have detected both β_1- and β_2-receptor subtypes in numerous brain areas, the regional distribution of the former varying more markedly than that of the latter (Pittman, Minneman, and Molinoff, 1980). In general, β_1 receptors appear to predominate in the cerebral cortex, whereas the number of β_2 sites is greater in the cerebellum. Similarly, the ratio of α_1 to α_2 receptors varies throughout the central nervous system (Snyder and Goodman, 1980).

Studies aimed at examining the development of adrenergic receptors have concentrated mainly on rat brain cerebral cortex and cerebellum. The rate of development of the total β-receptor population (β_1 and β_2) in cortex is maximal during the second week of postnatal life (Harden, Wolfe, Sporn, Perkins, and Molinoff, 1977; Harden, Wolfe, Sporn, Poulos, and Molinoff, 1977; Pittman, Minneman, and Molinoff, 1980). By 21 days of age the concentration of β receptors in this brain region is maximal, but it begins to decline after 40 days of age. A similar pattern of development was found when analyzing β_1 and β_2 receptors separately, and it was also discovered that the β_1:β_2 ratio is constant at all ages (Pittman, Minneman, and Molinoff, 1980). Equilibrium binding data revealed that the change in recognition site binding during development is due solely to a change in the number of receptors, with no change in the affinity of the recognition site for the radioligand. In contrast to β receptors, norepinephrine accumulation, a presynaptic marker for adrenergic neurons, does not reach adult levels in the cerebral cortex until the eighth week of life (Harden, Wolfe, Sporn, Perkins, and Molinoff, 1977). It would appear, therefore, that in this system the postsynaptic marker matures more rapidly than does the presynaptic.

The rate of β-adrenergic receptor development in the cerebellum differs from that found in the cerebral cortex. There the β receptors are primarily of the β_2 type. In this brain region the concentration of β receptors begins to increase during the first week of postnatal life but is not maximal until the sixth week. This value remains constant for up to six months (Pittman, Minneman, and Molinoff, 1980; Smith, Patel, Kingsbury, Hunt, and Balazs, 1980). Moreover, unlike the cerebral cortex, the β_1:β_2 ratio in the cerebellum changes markedly during development. During the second week of life almost 20 percent of the receptors are of the β_1 type, but by six months of age only 2 percent are β_1 (Pittman, Minneman, and Molinoff, 1980). These data suggest that the pharmacological specificity of β receptors varies with age, at least in some brain regions. A subsequent study revealed that the cerebellar β-receptor populations are differentially affected by X irradiation during development (Minneman et al., 1981). It was found that the β_1 receptors were much more resistant to the effects of irradiation than the β_2. These authors speculated that the differences were based on the anatomical location of the different receptor populations. Thus, irradiation is known to be more lethal to cerebellar interneurons than to Purkinje cells. Inasmuch as β_1 receptors are localized primarily on Purkinje cells, it is not surprising these sites are spared. Indeed, if β_1 and β_2 receptors are located on different cell types within the cerebellum, this may also account for the changing ratio of β_1 to β_2 receptors during development since this would reflect the gradual change in the relative contribution of the Purkinje cells to the overall mass of this brain region.

In an attempt to examine the influence of innervation on adrenergic receptor development, studies have been conducted using animals deprived of their noradrenergic pathways (Harden, Wolfe, Sporn, Poulos, and Molinoff, 1977; Jonsson and Hallman, 1978). Following the administration to newborns of the neurotoxin 6-hydroxydopamine, there is a permanent and virtually complete destruction of noradrenergic nerve terminals projecting to the cerebral cortex and spinal cord, and an increase in outgrowth in the pons-medulla and cerebellum. These investigators found that 6-hydroxydopamine had no effect on the rate of development of β-adrenergic receptors in the cerebral cortex, but that there was an overall increase in the concentration of these receptors at all stages of development. In the hyperinnervated cerebellum, on the other hand, there was a decrease in the number of β-receptor binding sites (Wolfe et al., 1981). These data indicate that innervation has little effect on the rate of β-receptor development but appears to influence receptor number. Thus it seems that the target cell regulates qualitative aspects of receptor development, whereas innervation is important from a quantitative standpoint.

Brain β-adrenergic receptor development has also been found to be influenced by a variety of other treatments. For example, E. M. Smith and coworkers (1980) found that there are fewer β receptors in the cerebellum and forebrain of the hypothyroid animal than in a normal animal, suggesting that thyroid hormone influences the development of these sites. Moreover, monocular destruction leads to an increase in β-receptor number in visual areas, such as the lateral geniculate nucleus and superior colliculus, suggesting that environmental factors can also regulate receptor number (Schliebs, Burgoyne, and Bigl, 1982).

Forebrain α-adrenergic receptor development appears to be similar to that found for cerebral cortical β receptors. Although α-receptor binding is quite low at birth, it increases progressively with time, reaching a maximum concentration by three weeks of age. Also, the ratio of α_1 to α_2 receptors remains constant throughout development (Morris et al., 1980). Like that of cortical β receptors, the concentration of α receptors in the adult brain appears to be somewhat lower than it is in the brain at three weeks. In all cases the changes in receptor binding are due to modifications in the number of binding sites with no change in affinity. Since forebrain norepinephrine accumulation takes eight weeks to reach maximal levels (*vide supra*), it would appear that presynaptic development lags behind the appearance of α-adrenergic receptors. It is important to note that other studies have shown an excellent correlation between the rate of development of synaptosomal norepinephrine uptake and norepinephrine-stimulated ^{33}P incorporation into phospholipids, which is thought to be an α receptor–mediated phenomenon (Deskin et al., 1981). Moreover, these investigators found a decrease in ^{33}P incorporation after 6-hydroxydopamine treatment at a time when the number of α-receptor binding sites should be increased. Thus, while the number of α- or β-receptor recognition sites labeled with radioligands may be maximal prior to the complete development of the innervating neuron, the functional activity of these binding sites may mature along with synapse formation. Taken together these data suggest that while target cell recognition sites may develop somewhat independently of innervation, receptor function is directly influenced by the presynaptic neuron.

Dopamine. Although recent studies suggest that there may be a variety of pharmacologically and functionally distinct dopamine receptors, developmental studies have concentrated on examining only those designated as *D-1* and *D-2*. These receptors are distinguished from one another in that the D-1 is coupled to adenyl cyclase, whereas the D-2 is not (Snyder and Goodman, 1980; Beart, 1982). Whereas D-1 receptors are typically analyzed by studying dopamine-stimulated cyclic AMP (adenosine monophosphate) production in broken cells, the D-2 site is examined using a binding assay, usually with ^{3}H-spiroperidol as a ligand. Because of these technical differences, receptor function rather than recognition site number and affinity is analyzed in the case of D-1 studies, whereas for D-2 the situation is opposite.

Most studies aimed at examining the ontogenetic development of dopamine receptors have concentrated on the corpus striatum, a brain region rich in dopamine terminals and receptors. It is thought that a significant number, if not the majority, of dopamine terminals in the striatum are in synaptic contact with cholinergic and GABAergic (GABA, γ-aminobutyric acid) cell bodies or dendrites. As with the noradrenergic system, D-1 receptor activity and D-2 binding are quite low at birth, both increasing progressively until 30 days of age (Pardo et al., 1977). Presynaptic markers, such as dopamine concentration and accumulation, are only at 70 percent of maximum at 30 days, indicating that these parameters lag behind receptor production. Interestingly, however, the rate of appearance of endogenous acetylcholine is virtually identical to that found for dopamine receptors (Pardo et al., 1977). This indicates that the initial rate of development of the receptor systems is coupled more to the rate of maturation of the cells to which they are attached rather than to the influence of the innervating neuron.

Detailed pharmacological and kinetic analysis of the developing D-2 receptor has shown that the properties of these binding sites are modified during the maturation process (Nomura, Oki, and Segawa, 1982). Although detectable, ^{3}H-spiroperidol binding is not stereoselective in the fetal brain and acquires this characteristic only after the rat is born. Also, during the first two weeks of postnatal life, ^{3}H-spiroperidol binding to striatal D-2 receptors is to a single, low-affinity binding site, but after this time both high- and low-affinity binding sites can be detected. Both of these components are maximal by day 30. Moreover, although both sodium and GTP (guanosine triphosphate) can influence agonist interaction at the ^{3}H-spiroperidol binding site in adult rat striatum, this property is not apparent until the second week of life and reaches maximal levels by 21 days. Finally D-2 receptor binding in 70-day-old and one-year-old rats is significantly lower than in 30-day-old rats, suggesting that the number of binding sites may decrease with age. Thus, although quantitatively the D-2 recognition sites reach maturity by 30 days after birth, the pharmacological and kinetic characteristics of these sites appear to develop at different rates. It is conceivable, therefore, if not likely, that normal D-2 receptor function may also be accelerated or delayed relative to recognition site development. Such a situation would again explain the lack of apparent correspondence between innervation and recognition site development.

Serotonin. Ligand-binding assays have suggested the existence of two pharmacologically distinct types of brain serotonin (5-HT) receptors. These have been designated 5-HT_1 and 5-HT_2 (Peroutka, Lebovitz, and Snyder, 1979). Whereas 5-HT_1 recognition sites are detected using the agonist ^{3}H-5-HT as a radioligand, the cerebral cortical 5-HT_2 receptors are most easily identified with ^{3}H-spiroperidol. In addition to being pharmacologically distinct, the 5-HT_1 and 5-HT_2 binding sites also differ with regard to regulatory factors. Thus, GTP and other guanine nucleotides decrease the affinity of agonists for the 5-HT_1 site but appear to have no effect on agonist interaction at the 5-HT_2 receptor. Also, it has been proposed that 5-HT_1 receptors mediate inhibitory responses to 5-HT, whereas 5-HT_2 sites are responsible for excitation (Peroutka, Lebovitz, and Snyder, 1980).

There have been few studies aimed at examining the ontogenetic development of 5-HT receptors. Of those reported, all have concentrated on the 5-HT_1 system. These have revealed that 5-HT_1 binding sites develop at different rates in different regions of the central nervous system. In hippocampus and corpus striatum binding increases from birth, reaching adult levels by 21 days. In contrast, the full complement of 5-HT_1 sites is not detected in the cerebral cortex until 90 days of age. Regions associated with the visual system, such as the superior colliculus and visual cortex, show yet a different pattern in that binding increases sharply during the second postnatal week and remains constant thereafter (Uzbekov, Murphy, and Rose, 1979). Since, as discussed previously, receptor recognition site number is determined in part by the degree of innervation and subsequent synaptic activity, it seems likely that these differential rates are a consequence of different levels of activity within the nervous system.

Except for the corpus striatum there appears to be no temporal correlation between the rate of development of 5-HT–stimulated cyclic AMP accumulation and 5-HT_1 binding (Nelson et al., 1980). In addition, like in the other monoaminergic systems, there is no correlation between the rate of appearance of presynaptic markers and recognition site binding. Thus, for 5-HT, recognition site number and function appear to develop somewhat independently, and innervation seems to have little influence on the rate of appearance of these sites.

Acetylcholine. Two types of cholinergic receptors are known to be present in the mammalian brain, nicotinic and muscarinic. Although peripheral nicotinic receptors have been successfully labeled with the snake venom α-bungarotoxin, it is unclear whether this substance labels physiologically relevant nicotinic receptors in the central nervous system. In contrast, brain muscarinic cholinergic receptors can be studied using the antagonist ligand ^{3}H-quinuclidinyl benzilate (Yamamura, Kuhar, and Snyder, 1974). Using this ligand, researchers found that the number of muscarinic receptors in the whole chick brain increases between the embryonic ages of two days and 18 days and then remains constant until day 22 (Enna, Yamamura, and Snyder, 1976). The rate of recognition site development is relatively low between days 2 and 10 but increases dramatically between day 10 and day 18 of embryonic age (fig. 1). In rat brain, the medulla-pons and hypothalamus were found to have the highest concentrations of muscarinic cholinergic receptor binding at birth, and the number of these sites increased sharply between birth and 14 days (Coyle and Yamamura, 1976). Following two weeks of age the rate of increase slowed, and maximal values were attained by day 28. In contrast, the cortex and midbrain have much lower levels of binding at birth, and the rate of increase is quite slow during the first postnatal week. The period having the most rapid increase in binding is from day 7 to day 29, after which time no further increase in binding is noted. When these binding data are analyzed on the basis of binding sites per region

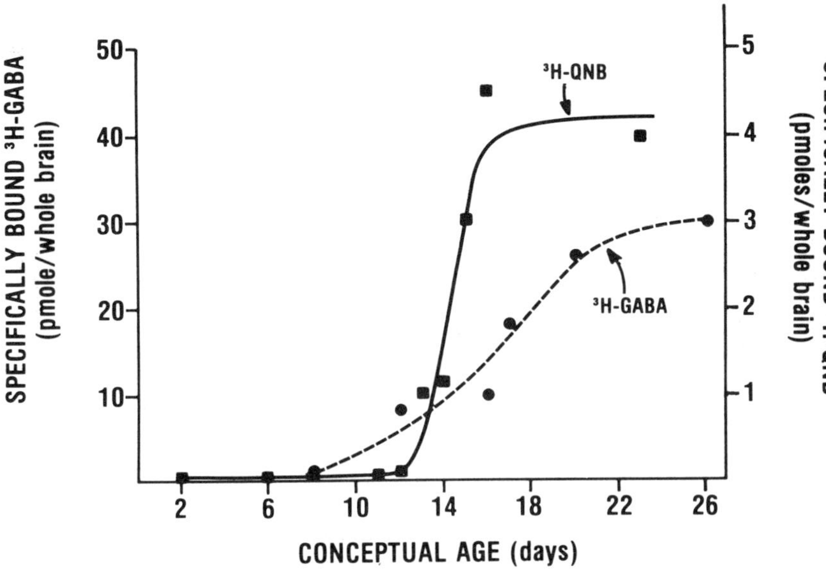

Figure 1. Development of GABA and cholinergic muscarinic receptor binding in chick whole brain using ^{3}H-GABA and ^{3}H-QNB as ligands, respectively. Each value is the mean of two to three experiments, each performed in triplicate (*GABA*, γ=aminobutyric acid; *QNB*, quinuclidinyl benzilate). (Adapted from Enna, Yamamura, and Snyder, 1976.)

rather than as a function of milligrams of protein, it is found that additional receptors appear after the 28-day period, even though neuronal cells have ceased developing by this time. Balazs and associates have reported a similar pattern for forebrain, but they indicated that the muscarinic receptor number in the cerebellum remains constant after 28 days regardless of how it is calculated (Smith, Kingsbury, Hunt, and Balazs, 1980). These workers also found that the cerebellum differs from the forebrain in that thyroid deficiency delays the rate of maturation of the former but not the latter.

Studies aimed at correlating presynaptic and postsynaptic parameters for the cholinergic system have revealed that choline acetyltransferase, an enzyme marker for cholinergic neurons, and muscarinic receptor binding develop simultaneously (Enna, Yamamura, and Snyder, 1976; Coyle and Yamamura, 1976) (fig. 2). In contrast, the endogenous levels of acetylcholine are quite high at birth relative to adult levels (Coyle and Yamamura, 1976). Because of the difficulties in analyzing the total (muscarinic and nicotinic) cholinergic receptor population in brain, correlations between presynaptic and postsynaptic phenomena are less mean-

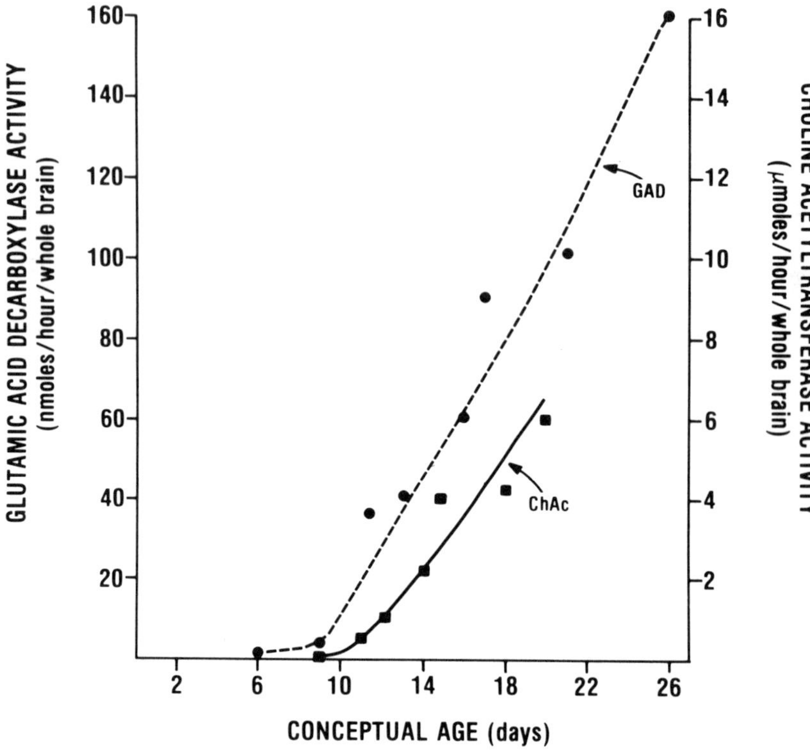

Figure 2. Development of glutamic acid decarboxylase (*GAD*) activity and choline acetyltransferase (*ChAc*) activity in chick whole brain. Each value is the mean of two to three experiments, each performed in triplicate. (Adapted from Enna, Yamamura, and Snyder, 1976.)

ingful. Also, unlike in the noradrenergic, dopaminergic, and serotonergic systems, few well-defined cholinergic pathways exist in brain, making it less easy to examine this issue using lesion studies. Therefore, little is known about the influence of cholinergic pathways on receptor development.

Histamine. Central nervous system histamine is most highly concentrated in the hypothalamus. Iontophoretic studies have revealed that this substance is excitatory on hypothalamic neurons but decreases firing rates in cerebral cortex. It is unknown whether these differential effects are related to the presence of two pharmacologically distinct histamine receptors, H_1 and H_2, or whether the receptors in these brain regions are coupled to different types of intracellular messengers or ion channels. It has been suggested that at least some histamine receptors are linked to adenyl cyclase (Brownstein, 1981).

Using the radioligand ^{3}H-mepyramine, an H_1-receptor antagonist, it has been reported that, on a milligram-wet-weight basis, the number of histamine receptors increases gradually in whole rat brain from birth to three weeks of age (Tran et al., 1980; Subramanian et al., 1981). This value remains more or less constant throughout adulthood. A similar developmental pattern has been reported for specific brain regions, such as cerebral cortex, thalamus-hypothalamus, pons-medulla, hippocampus, and cerebellum (Tran et al., 1980). When the data are expressed as total receptors per brain region, the value continues to rise beyond three weeks of age.

The presynaptic marker histidine decarboxylase develops at a rate quite similar to that found for ^{3}H-mepyramine binding (Tran et al., 1980). In contrast, the brain histamine content is higher at birth in all brain areas except the hypothalamus and decreases progressively during development. In the hypothalamus the histamine concentration rises sharply during the second week after birth and reaches adult levels by the beginning of the third week. Since it appears that a great deal of the brain histamine present at birth does not serve a neurotransmitter function, there is little point in attempting to correlate this parameter with receptor development. On the other hand, the finding that histidine decarboxylase activity correlates with receptor development may be taken as evidence that, as opposed to the innervation of catecholamines, histaminergic innervation reaches maturity at about the same time as the receptor for this substance. However, this proposal is speculative until more data can be accumulated to support the notion that the enzyme activity accurately reflects neuronal development. In any event, H_1-receptor binding appears to develop at a rate that is similar to that of other monoamines.

Amino Acid and Peptide Receptors

GABA. GABA is a major inhibitory neurotransmitter in the mammalian central nervous system (Enna, 1981a, 1981b). The pharmacological, biochemical, and physiological properties of the GABA receptor have been well documented (Enna and Gallagher, 1983). From these studies it would appear that there may be pharmacologically and physiologically distinct types of GABA receptors (Enna and Gallagher, 1983). One type appears to be coupled to a recognition site for benzodiazepines, a family of anxiolytic and hypnotic drugs (Olsen and Enna, 1983). This relationship suggests that these drugs may be acting, at least in part, through an interaction with the GABA receptor system. Indeed, physiological, behavioral, and biochemical data support this hypothesis (Enna, 1982). While the precise manner in which benzodiazepine and

GABA receptors are coupled is unknown, it would appear that these drugs are capable of unmasking a higher affinity GABA receptor recognition site, perhaps by removing some endogenous substance that normally decreases the availability of this component (Toffano, Guidotti, and Costa, 1978). Such a substance, termed *GABA-modulin,* has been identified in brain extracts (Costa and Guidotti, 1979). Thus, in addition to the recognition site, this receptor possesses an ion channel, GABA-modulin, and a benzodiazepine binding component. The appropriate interaction of these various constituents is necessary for maintenance of receptor function (Enna, 1981a; Costa and Guidotti, 1979; Costa, Rodbard, and Pert, 1979; Martin and Candy, 1978).

Several studies have been undertaken to define the developmental characteristics of GABA receptor binding sites (Enna, Yamamura, and Snyder, 1976; Coyle and Enna, 1976; Aldinio, Balzano, and Toffano, 1980; Smith, Kingsbury, Hunt, and Balazs, 1980; Palacios, Niehoff, and Kuhar, 1979). In chick brain binding is detectable in the nine-day-old embryo (fig. 1). The number of binding sites rises sharply between days 9 and 13, plateaus briefly, and then increases again up to 26 days of age (Enna, Yamamura, and Snyder, 1976). This pattern of development is the same whether the data are expressed on a per-milligram-protein or whole-brain basis. In rats, GABA receptor binding in whole brain is less than 5 percent of the adult level at gestational day 15 but reaches 24 percent of the adult level by birth (Coyle and Enna, 1976). The number of binding sites increases progressively from birth to four weeks of age, after which time the value remains constant. However, the rate of development of GABA receptors is somewhat region specific. Thus, the concentration of receptors remains quite low in the midbrain and cortex during the first week after birth, increasing thereafter until adult levels are reached by the third week of age (Coyle and Enna, 1976). On the other hand, a significant increase in the rate of receptor development is not observed in the cerebellum until two weeks of age, and adult levels are not reached until week 4. These different rates of development are thought to reflect differences in the rate of cellular maturation in the various brain regions.

Though early studies were concerned chiefly with the development of the lower affinity GABA binding component, subsequent work with a higher affinity site revealed a similar pattern (Aldinio, Balzano, and Toffano, 1980; Smith, Kingsbury, Hunt, and Balazs, 1980; Palacios, Niehoff, and Kuhar, 1979). In rat brain both the high- and low-affinity binding components are detectable at birth, and the concentration of each becomes maximal during the fourth week of development. No developmental change in receptor affinities has been noted in any of these studies (Aldinio, Balzano, and Toffano, 1980; Massotti et al., 1980).

In comparing the development of presynaptic and postsynaptic parameters a parallelism was found between the rate of appearance of glutamic acid decarboxylase activity (a marker for GABA neurons), and GABA receptor binding (Enna, Yamamura, and Snyder, 1976) (fig. 2). In contrast, GABA levels and high-affinity GABA uptake (other markers for GABA-ergic terminals) are much closer to adult values at birth than the number of receptor binding sites. However, studies have indicated that uptake may be a poor marker for studying the development of GABA synapses since there is an avid glial uptake system for this amino acid that is most apparent during the early days of development (Redburn et al., 1978).

As found for other neurotransmitter receptors, hyperthyroidism accelerates the rate of development of GABA receptors, and a thyroid deficiency retards growth (Smith, Kingsbury, Hunt, and Balazs, 1980). It is unknown whether this hormone effect is due to a direct influence on receptor synthesis or is secondary to an action of the hormone on overall cellular maturation.

The benzodiazepine binding component of the GABA receptor has been detected as early as day 14 of fetal life and, like GABA receptor binding, is approximately 25 percent of the adult level at this time (Braestrup and Nielsen, 1978; Regan, Roeske, and Yamamura, 1980). Similar to GABA binding, the binding of benzodiazepine reaches maximal levels by four weeks of age, the increase owed entirely to a change in the number of sites with no alteration in affinity. Moreover, the interaction between GABA and benzodiazepine receptors is present at birth and remains constant throughout development (Regan, Roeske, and Yamamura, 1980). This suggests that the two components are coupled and functionally related from the outset of development.

The regional developmental pattern of benzodiazepine binding is similar to that found in the whole organ. At birth the concentration of benzodiazepine receptors is approximately 30 percent of adult levels in the cerebellum, cerebral cortex, and corpus striatum, and adult levels are attained by four weeks of age (Candy and Martin, 1979; Palacios, Niehoff, and Kuhar, 1979; Massotti et al., 1980). Benzodiazepine binding appears to remain fairly constant beyond four weeks of age for most brain regions, but the number of receptors declines slightly after this time in the cerebellum (Palacios, Niehoff, and Kuhar, 1979). The functional significance of this finding is unclear.

Because withdrawal symptoms have been noted in newborns delivered of dams who received high doses of benzodiazepines during pregnancy, it has been speculated that the development of these receptors may be altered by drug administration (Rementeria and Bhatt, 1977). Although the rate of development of benzodiazepine binding appears to be unaltered under this condition, there has been a suggestion that the coupling between the benzodiazepine and GABA binding sites is modified (Massotti et al., 1980). This finding illustrates the importance of examining receptor function as well as binding before making definitive statements regarding the state of the receptor system.

In summary, GABA and benzodiazepine binding appear to develop in concert, which is consistent with the notion that these two sites are components of the same receptor molecule. Moreover, the functional relationship between these components appears to be established early. No studies have yet been performed to define the influence of the presynaptic GABA terminal on the development of GABA and benzodiazepine receptor binding and function. Since there have been reports suggesting that, in the adult animal, the number of brain GABA binding sites can be modified by altering receptor activity (Enna and Strada, 1983), it seems likely that the extent of innervation is important in determining the total number of GABA receptors, if not the rate at which they develop.

Glycine

Like GABA, glycine appears to be an inhibitory neurotransmitter in the mammalian central nervous system (Werman, Davidoff, and Aprison, 1968; Enna and DeFrance, 1980). Unlike those for GABA and other neurotransmitters, glycine receptors have yet to have pharmacologically distinct subsets identified. Thus, from an electrophysiological standpoint, a receptor-mediated action of glycine is defined as any response that can be blocked by the convulsant strychnine. Using ^{3}H-strychnine, workers have been able to identify a saturable binding site that possesses the characteristics of a glycine receptor. For example, there is a significant correlation between the regional distribution of ^{3}H-strychnine binding and high-affinity glycine uptake, and the relative potencies of a series of amino acid analogs to activate glycine receptors is similar to their relative potencies as inhibitors of ^{3}H-strychnine binding (Young and Snyder, 1974).

With respect to development, strychnine binding has been studied in chick and rat spinal cord (Zukin, Young, and Snyder, 1975; Benavides et al., 1981). In the rat spinal cord there is a gradual increase in the number of ^{3}H-strychnine binding sites from birth to 30 days of age, with the maximum rate of increase observed between birth and five days of age. In contrast to the increase of sites in spinal cord, the number of brain strychnine binding sites is maximal by day 15 (Benavides et al., 1981). The rate of development of high-affinity glycine uptake, a presynaptic marker, is virtually identical to the rate of development of strychnine binding (Zukin, Young, and Snyder, 1975). However, it is possible that, like GABA accumulation, a significant amount of glycine accumulation may not be into neurons, making it impossible to accurately assess the relationship between the presynaptic and postsynaptic development using this marker. Overall, progress in this area has been limited and will continue to be until a larger variety of glycine receptor agonists and antagonists are available for study.

Glutamic Acid. A great deal of evidence has been accumulated to suggest that glutamic acid itself, or a glutamate-containing substance, is a neurotransmitter in the central nervous system (Coyle, 1980). Electrophysiological studies have shown that glutamate will excite virtually all neurons when applied directly to the external surface of the cell. Two glutamate pathways have been extensively studied in brain (McGeer and McGeer, 1981). One of these has cell bodies in the cerebral cortex with projections terminating in the corpus striatum. The other is the granular cell population in the cerebellum. Using ^{3}H-glutamate, several workers have identified a saturable binding site for this amino acid in brain (Foster and Roberts, 1978; Coyle et al., 1980). While this binding component has characteristics expected of a glutamate receptor, the lack of a specific receptor antagonist has hindered the absolute identification of this binding site. Moreover, a specific binding site has been identified for ^{3}H-kainic acid, a glutamate analog (Simon, Conterera, and Kuhar, 1976; London and Coyle, 1979). However, the glutamate and kainate binding sites appear to be different entities, though they may in some way be related to one another (Slevin and Coyle, 1981).

Studies aimed at examining the corticostriatal glutaminergic system have revealed that glutamate accumulation and concentration, according to presynaptic markers, develop at a rate that is virtually identical to the rate of development of ^{3}H-kainate binding (Campochiaro and Coyle, 1978). Thus, the rate of development of the glutamate uptake system in the striatum is greatest between ten and 21 days after birth, reaching 80 percent of the adult level by 28 days. Similarly, the concentration of striatal glutamic acid itself is 90 percent of the adult value by three weeks, and ^{3}H-kainate binding is 80 percent of adult levels at this time as well. Thus, it has been proposed that the presynaptic and postsynaptic glutaminergic systems develop at a similar rate and may be causally related in this brain area.

In contrast, there appears to be a difference in the rate of development of presynaptic and postsynaptic glutaminergic markers in the cerebellum. With respect to ^{3}H-kainate and ^{3}H-glutamate binding, these values remain low, less than 10 percent of the adult values through the first week of life, but they increase dramatically thereafter until reaching the maximum by 21 days of age (Slevin and Coyle, 1981). On the other hand, high-affinity glutamate uptake is as much as 70 percent of the adult value at birth and reaches adult levels within 24 hours after birth. If glutamate accumulation in the cerebellum is restricted to glutaminergic neurons, this finding would suggest that receptor development lags considerably behind the development of glutamate nerve terminals.

Opiate. Although specific receptor sites for opiates had been suspected for a number of years, it has only been in the past decade that their existence has been firmly established. Subsequent work demonstrated the presence of endogenous peptides, the enkephalins and endorphins, that appear to be the normal substrates for these receptors (Childers, 1980). Pharmacologically distinct subsets of opiate receptors have been identified, although the significance of this classification is still debated (Snyder and Goodman, 1980). However, there appears to be general agreement on the existence of high- and low-affinity binding sites for opiates, and it has been suggested that opiate attachment to the high-affinity component mediates analgesia, whereas the low-affinity site may be responsible for other effects (Snyder and Goodman, 1980).

Numerous studies have been conducted to define the developmental characteristics of these peptide receptors (Coyle and Pert, 1976; Zhang and Pasternak, 1981). Using either ^{3}H-naloxone or ^{3}H-naltrexone, antagonists for these receptors, researchers have found that binding can be detected in brain as early as day 15 of gestation. The number of receptors increases dramatically from birth to 21 days of age, at which time it has reached adult levels. The increase in opiate binding is due entirely to an increase in the number of sites with no change in receptor affinity. There are, however, regional differences in the rate of development of opiate receptors, with brain regions that are more caudal (pons-medulla and hypothalamus) attaining adult levels earlier than the cerebral cortex does.

Studies aimed at examining the high- and low-affinity components have revealed that at birth only those of the higher affinity can be detected in brain tissue (Koch, Sakly, and Bucher, 1980; Zhang and Pasternak, 1981). It is not until several days later that the low-affinity component becomes apparent, after which both sites increase in number at a similar rate. Interestingly, there is a temporal correlation between the appearance and development of the high-affinity site and opiate analgesia and the appearance of the low-affinity component and the respiratory response to these agents (Zhang and Pasternak, 1981). These data provide further evidence for a functional differentiation between the two sites as well as suggesting that the appearance of the binding site corresponds with the appearance of receptor activity.

Like receptor binding, the endogenous level of enkephalins increases dramatically between birth and 21 days of age. However, when viewed in the context of the low-affinity site only, it would appear that the cells that produce and release these peptides are present long before this receptor. Therefore, it is impossible to determine whether the development of the low-affinity component is regulated by enkephalin levels or is more a function of the target cell.

Cautions, Correlations, and Conclusions

From the foregoing it would appear that the rate of development of all neurotransmitter receptors is quite similar in rat brain, with the most rapid increase occurring between birth and 21 days of age. This time frame is not surprising in light of the fact that it coincides with overall rate of neuronal cell proliferation and maturation. However, direct comparisons between studies are difficult to make given the fact that different investigators express data in different ways. Thus, in most studies, the amount of receptor binding is expressed as the amount of ligand bound per milligram brain protein. This is a valid way to calculate these data as long as the protein content of the tissue is not a variable. A complication arises, however, in that the absolute amount of protein per fraction does change with increasing age. For example, membrane protein in rat cerebellum has been estimated to increase from 38 mg/g wet weight at six days of age to 77 mg/g at 35 days of age (Smith, Kingsbury, Hunt, and Balazs, 1980; Smith, Patel, Kingsbury, Hunt, and Balazs, 1980). If this overall increase in protein content is not accompanied by a concomitant change in receptor protein, the developmental pattern, as judged on a per-milligram-protein basis, may be misleading.

A possible solution to this problem is to express the data as ligand bound per unit of wet weight. However, even this approach is limited if the relative proportion of gray and white matter varies significantly during development. Indeed, it has been reported that the proportion of white matter in brain is negligible in rat brain up to the second week of life. From that time it increases dramatically, such that by 21 days it represents 15–20 percent of the total, and by 60 days almost 30 percent (Reddy, Rajalakshami, and Ramakrishnan, 1982). In fact, the proportion of white matter continues to increase well beyond the second month of life until it reaches 50 percent of the total brain weight. This change in the proportion of gray and white matter would influence the data, especially in evaluating regions such as the pons, medulla, or midbrain, where white matter ultimately makes up a large proportion of the total tissue mass. Areas such as the striatum, hypothalamus, and cerebellum are less affected, given their higher concentration of gray matter in the adult.

A third way of expressing biochemical data is with respect to whole tissue. Such an analysis reveals the number of receptors per brain or brain region. Calculated in this way the data can be more easily compared though collected at different stages of development. Unfortunately, very few workers use this approach.

Another difficulty in comparing data between studies is the variation in the definition of an adult animal. Whereas some define adulthood on the basis of animal weight (Pardo et al., 1977), others use age. In the latter case some workers define adulthood as 160 days of age (Aldinio, Balzano, and Toffano, 1980), though others use 40, 50, or 60 days as the end point; in some cases, no definition is provided (Salvaterra and Moore, 1973; Uzbekov, Murphy, and Rose, 1979; Harden, Wolfe, Sporn, Perkins, and Molinoff, 1977; Patey et al., 1980; Nelson et al., 1980). Clearly, without a fixed definition of what constitutes an adult animal, it is difficult to determine when development has ceased.

Even with these limitations, however, some generalizations may be made with respect to receptor development. It would appear that in rat the most critical phase is during the first three weeks of postnatal life. Any insult to the nervous system during this time is likely to affect the orderly progression of receptor development for a variety of neurotransmitter systems. Moreover, while the precise relationship between synapse formation and receptor development has not yet been defined, the available evidence supports the notion that both the target cell and the incoming axon play a role in regulating receptor development. In general, it would appear that the neuron that synthesizes the receptor may be most important with regard to qualitative characteristics, whereas the synapsing terminal may be more important with regard to regulating total receptor number. Also, it is known that input is an important determinant for receptor regulation in the adult (Enna and Strada, 1983). On the other hand, what role, if any, is played by the receptor with regard to directing the developing axon to the appropriate target organ is unknown.

Although electrophysiological studies suggest that the receptor is capable of functioning prior to synapse formation, this conclusion does not rule out the possibility that innervation is crucial for the maintenance of this function. Indeed, it is possible that at early stages of development the neuronal membrane may be responsive to all neurotransmitter substances but retains sensitivity for only those agents utilized by terminals making synaptic contact with the cell.

Taken together these data indicate a reciprocal relationship between presynaptic and postsynaptic receptor elements during the course of development. Because of this interdependency, modifications in either parameter can be expected to ultimately affect the properties of both. Whereas in some cases this interplay is no doubt important with regard to adaptive mechanisms, in other situations the net result may be detrimental to the functioning of the organism. A better definition of the factors influencing receptor synthesis and development is necessary for understanding the fundamentals of neurotransmission in the central nervous system.

Acknowledgments

Preparation of this manuscript was made possible, in part, by the support of a U.S. Public Health Service Research Career Development Award (NS-00335) (to S. J. Enna) and by a grant from UNESCO through the Institute of International Education (to Shaila Telang). We thank Doris Rayford for her excellent secretarial assistance.

References

Aldinio, C., Balzano, M. A., and Toffano, G. (1980). Ontogenetic development of GABA recognition sites in different brain areas. *Pharmacol Res Commun* 12, 495–500.

Beart, P. M. (1982). Multiple dopamine receptors—New vistas. *Trends in Pharmacological Sciences* 3, 100–107.

Benavides, J., Lahoya, L., Validvieso, F., and Ugarte, M. (1981). Postnatal development of synaptic glycine receptors in normal and hyperglycemic rats. *J Neurochem* 37, 315–320.

Benjamins, J. A., and McKhann, G. M. (1981). Development, regeneration and aging of the brain. In *Basic neurochemistry*, ed. J. G. Siegel, W. R. Albers, W. B. Agranoff, and R. Katzman, pp. 445–470. Boston: Little, Brown.

Braestrup, C., and Nielsen, M. (1978). Ontogenetic development of benzodiazepine receptors in the rat brain. *Brain Res* 147, 170–173.

Brownstein, M. J. (1981). Serotonin, histamine and the purines. In *Basic neurochemistry*, ed. G. Siegel, R. Albers, B. Agranoff, and R. Katzman, pp. 219–231. Boston: Little, Brown.

Campochiaro, P., and Coyle, J. T. (1978). Ontogenetic development of kainate neurotoxicity: Correlates with glutamatergic innervation. *Proc Natl Acad Sci USA* 75, 2025–2029.

Candy, J. M., and Martin, I. L. (1979). The postnatal development of the benzodiazepine receptor in the cerebral cortex and cerebellum of the rat. *J Neurochem* 32, 655–658.

Childers, S. R. (1980). Enkephalin and endorphin receptors. In *Neurotransmitter receptors: Amino acids, peptides and benzodiazepines*, ed. S. J. Enna and H. I. Yamamura, pp. 105–147. London: Chapman and Hall.

Costa, E., and Guidotti, A. (1979). Molecular mechanisms in the receptor action of benzodiazepines. *Ann Rev Pharmacol* 19, 531–545.

Costa, T., Rodbard, D., and Pert, C. B. (1979). Is the benzodiazepine receptor coupled to a chloride anion channel? *Nature* 277, 315–317.

Cowan, W. M. (1978). Aspects of neural development. In *International review of physiology*. Vol. 17, ed. R. Porter, pp. 149–191. Baltimore, Md.: University Park Press.

Coyle, J. T. (1980). Excitatory amino acid receptors. In *Neurotransmitter receptors: Amino acids, peptides and benzodiazepines*, ed. S. J. Enna and H. I. Yamamura, pp. 3–40. London: Chapman and Hall.

Coyle, J. T., and Enna, S. J. (1976). Neurochemical aspects of the ontogenesis of GABAergic neurons in the rat brain. *Brain Res* 111, 119–133.

Coyle, J. T., and Pert, C. B. (1976). Ontogenetic development of ^{3}H-naloxone binding in rat brain. *Neuropharmacology* 15, 555–560.

Coyle, J. T., Slevin, J., London, E. D., Biziere, K., and Collins, J. (1980). Characterization of neuronal recognition sites for ^{3}H-kainic acid. In

Psychopharmacology and biochemistry of neurotransmitter receptors, ed. R. W. Olsen, H. I. Yamamura, and E. Usdin, pp. 501–513. New York: Elsevier-North Holland.

Coyle, J. T., and Yamamura, H. I. (1976). Neurochemical aspects of the ontogenesis of cholinergic neurons in the rat brain. *Brain Res* 118, 429–440.

Deskin, R., Seidler, F. J., Whitmore, W. B., and Slotkin, T. A. (1981). Development of α-noradrenergic and dopaminergic receptor systems depends on maturation of their presynaptic nerve terminals in the rat brain. *J Neurochem* 36, 1683–1692.

Enna, S. J. (1981a). GABA receptors. *Trends in Pharmacological Sciences* 2, 62–64.

Enna. S. J. (1981b) Neuropharmacological and clinical aspects of γ-aminobutyric acid (GABA). In *Neuropharmacology of central disorders*, ed. G. C. Palmer, pp. 507–537. New York: Academic Press.

Enna, S. J. (1982). The role of neurotransmitters in the pharmacologic actions of benzodiazepines. In *Biology of anxiety*, ed. R. Mathew, pp. 107–122. New York: Bruner/Mazel.

Enna, S. J., and DeFrance, J. F. (1980). Glycine, GABA and benzodiazepine receptors. In *Neurotransmitter receptors: Amino acids, peptides and benzodiazepines*, ed. S. J. Enna and H. I. Yamamura, pp. 41–70. London: Chapman and Hall.

Enna, S. J., and Gallagher, J. P. (1983). Biochemical and electrophysiological characteristics of mammalian GABA receptors. *Int Rev Neurobiol*, 24: 181–212.

Enna, S. J., and Strada, S. J. (1983). Postsynaptic receptors: Recognition sites, ion channels and second messengers. In *Clinical neurosciences*, ed. R. Rosenberg, R. Grossman, S. Schochet, E. R. Heinz, and W. Willis, pp. 145–170. New York: Churchill Livingstone.

Enna, S. J., Yamamura, H. I., and Snyder, S. H. (1976). Development of muscarinic cholinergic and GABA receptor binding in chick embryo brain. *Brain Res* 101, 177–183.

Foster, A. C., and Roberts, P. J. (1978). High affinity L-^{3}H-glutamate binding to postsynaptic receptor sites on rat cerebellar membranes. *J Neurochem* 31, 1467–1477.

Harden, K. T., Wolfe, B. B., Sporn, J. R., Perkins, J. P., and Molinoff, P. B. (1977). Ontogeny of β-adrenergic receptors in rat cerebral cortex. *Brain Res* 125, 99–108.

Harden, K. T., Wolfe, B. B., Sporn, R. J., Poulos, K. B., and Molinoff, P. B. (1977). Effects of 6-hydroxydopamine on the development of the beta adrenergic/adenylate cyclase system in rat cerebral cortex. *J Pharmacol Exp Ther* 203, 132–143.

Jacobson, M. (ed.). (1978). *Developmental neurobiology*, pp. 57–103. New York: Plenum Press.

Jones, E. G. (1981). Development of connectivity in the cerebral cortex.

In *Studies in developmental neurobiology*, ed. M. E. Cowan, pp. 354–394. Oxford: Oxford University Press.

Jonsson, G., and Hallman, H. (1978). Changes in β-receptor binding sites in rat brain after neonatal 6-hydroxydopamine treatment. *Neurosci Lett* 9, 27–32.

Koch, B., Sakly, M., and Bucher, L. (1980). Ontogeny of opiate receptor sites in brain: Apparent lack of low affinity sites during early neonatal life. *Horm Metab Res* 12, 342–343.

London, E. D., and Coyle, J. T. (1979). Specific binding of ^{3}H-kainic acid to receptor sites in rat brain. *Mol Pharmacol* 15, 492–505.

McGeer, P. L., and McGeer, E. G. (1981). Amino acid neurotransmitters. In *Basic neurochemistry*, ed. G. J. Siegel, W. R. Albers, W. B. Agranoff, and R. Katzman, pp. 233–254. Boston: Little, Brown.

Martin, I. L., and Candy, J. M. (1978). Facilitation of benzodiazepine binding by sodium chloride and GABA. *Neuropharmacology* 17, 993–998.

Massotti, M., Alleva, F. R., Balazs, T., and Guidotti, A. (1980). GABA and benzodiazepine receptors in the offspring of dams receiving diazepam: Ontogenetic studies. *Neuropharmacology* 19, 951–956.

Minneman, P. K., Pittman, N. R., Yeh, H. H., Woodward, J. D., Wolfe, B. B., and Molinoff, P. B. (1981). Selective survival of β_1 adrenergic receptors in rat cerebellum following neonatal X-irradiation. *Brain Res* 209, 25–34.

Morris, J. M., Dausse, J., Devynck, M., and Meyer, P. (1980). Ontogeny of α_1 and α_2 adrenoceptors in rat brain. *Brain Res* 190, 268–271.

Nelson, D. L., Herbert, A., Adrien, J., Bockaert, J. and Hamon, M. (1980). Serotonin-sensitive adenylate cyclase and ^{3}H-serotonin binding sites in the CNS of the rat—II. *Biochem Pharmacol* 29, 2455–2463.

Nomura, Y., Oki, K., and Segawa, T. (1982). Ontogenetic development of the striatal ^{3}H-spiperone binding: Regulation by sodium and guanine nucleotides in rats. *J Neurochem* 38, 902–908.

Olsen, R., and Enna, S. J. (1983). GABA and anxiolytics. In *Anxiolytics: Neurochemical, behavioral and clinical perspectives*, ed. J. B. Malick, S. J. Enna, and H. I. Yamamura, pp. 55–76. New York: Raven Press.

Palacios, J. M., Niehoff, D. L., and Kuhar, J. M. (1979). Ontogeny of GABA and benzodiazepine receptors: Effects of Triton X-100, bromide and muscimol. *Brain Res* 179, 390–395.

Pardo, J. V., Creese, I., Burt, D. R., and Snyder, S. H. (1977). Ontogenesis of dopamine receptor binding in the corpus striatum of the rat. *Brain Res* 125, 376–382.

Patey, G., Baume, S., Gros, C., and Schwartz, J. (1980). Ontogenesis of enkephalinergic systems in rat brain: Postnatal changes in enkephalin levels, receptors and degrading enzyme activities. *Life Sci* 27, 245–252.

Peroutka, S. J., Lebovitz, R. M., and Snyder, S. H. (1979). Serotonin receptor binding sites affected differentially by guanine nucleotides. *Mol Pharmacol* 16, 700–708.

Peroutka, S. J., Lebovitz, R. M., and Snyder, S. H. (1980). Two distinct central serotonin receptors with different physiological functions. *Science* 212, 827–829.

Pittman, R. N., Minneman, K. P., and Molinoff, P. B. (1980). Ontogeny of β_1- and β_2-adrenergic receptors in rat cerebellum and cerebral cortex. *Brain Res* 188, 357–368.

Redburn, D., Broome, D., Ferkany, J., and Enna, S. J. (1978). Development of rat brain uptake and calcium-dependent release of GABA. *Brain Res* 152, 511–519.

Reddy, T. S. R., Rajalakshami, R., and Ramakrishnan, C. V. (1982). Effects of nutritional rehabilitation on the content and lipid composition of brain gray and white matter of neonatally undernourished rats. *J Neurochem* 39, 1297–1301.

Regan, W. J., Roeske, R. W., and Yamamura, H. I. (1980). The benzodiazepine receptor: Its development and its modulation by γ-aminobutyric acid. *J Pharmacol Exp Ther* 212, 137–143.

Rementeria, J. L., and Bhatt, K. (1977). Withdrawal symptoms in neonates from intrauterine exposure to diazepam. *J Pediatr* 90, 123–126.

Salvaterra, P. M., and Moore, W. J. (1973). Binding of ^{125}I-α-bungarotoxin to particulate fractions of rat and guinea pig brain. *Biochem Biophys Res Commun* 55, 1311–1318.

Schliebs, R., Burgoyne, D. R., and Bigl, V. (1982). The effect of visual deprivation on β-adrenergic receptors in the visual centers of the rat brain. *J Neurochem* 38, 1038–1043.

Simon, J. R., Conterera, J. F., and Kuhar, M. J. (1976). Binding of ^{3}H-kainic acid, an analogue of L-glutamate, to brain membranes. *J Neurochem* 26, 141–147.

Slevin, J. T., and Coyle, J. T. (1981). Ontogeny of receptor binding sites for ^{3}H-glutamic acid and ^{3}H-kainic acid in the rat cerebellum. *J Neurochem* 37, 531–533.

Smith, E. M., Patel, A. J., Kingsbury, A. E., Hunt, A., and Balazs, R. (1980). Effects of thyroid state on brain development: β-Adrenergic receptors and 5′ nucleotidase activity. *Brain Res* 198, 375–387.

Smith, W. R. M., Kingsbury, A. E., Hunt, A., and Balazs, R. (1980). Effects of thyroid state on brain development: Muscarinic acetylcholine and GABA receptors. *Brain Res* 198, 389–402.

Snyder, S. H., and Goodman, R. R. (1980). Multiple neurotransmitter receptors. *J Neurochem* 35, 5–15.

Subramanian, N., Whitmore, W. L., Seidler, F. J., and Slotkin, T. A. (1981). Ontogeny of histaminergic neurotransmission in the rat brain: Concomitant development of neuronal histamine H_1 receptors and H_1 receptor-mediated stimulation of phospholipid turnover. *J Neurochem* 36, 1137–1141.

Toffano, G., Guidotti, A., and Costa, E. (1978). Purification of an endogenous protein inhibitor of the high affinity binding of γ-aminobutyric acid to synaptic membranes of rat brain. *Proc Natl Acad Sci USA* 75, 4024–4028.

Tran, V. T., Freeman, A. D., Chang, R. S. L., and Snyder, S. H. (1980). Ontogenetic development of histamine H_1 receptor binding in rat brain. *J Neurochem* 34, 1609–1613.

Uzbekov, M. G., Murphy, S., and Rose, S. P. R. (1979). Ontogenesis of serotonin "receptors" in different regions of rat brain. *Brain Res* 168, 195–199.

Werman, R., Davidoff, R. A., and Aprison, M. H. (1968). Inhibitory effect of glycine in spinal neurons in the cat. *J Neurophysiol* 31, 81–95.

Wolfe, B. B., Augustyn, B. H. P., Majocha, E. R., Dibner, D. M., Molinoff, P. B., Baldessarini, J. R., and Walton, G. K. (1981). Effects of isoproterenol on the development of β-adrenergic receptors in brain cell aggregates. *Brain Res* 207, 174–177.

Woodward, D. J., Hoffer, B. J., Siggins, G. R., and Bloom, F. E. (1971). The ontogenetic development of synaptic junctions, synaptic activation and responsiveness to neurotransmitter substances in rat cerebellar Purkinje cells. *Brain Res* 34, 73–97.

Yamamura, H. I., Enna, S. J., and Kuhar, M. J. (eds.). (1978). *Neurotransmitter receptor binding*. New York: Raven Press.

Yamamura, H. I., Kuhar, M. J., and Snyder, S. H. (1974). In vivo identification of muscarinic cholinergic receptor binding in brain. *Brain Res* 80, 170–176.

Young, A., and Snyder, S. H. (1974). Strychnine binding in the rat spinal cord membranes associated with the synaptic glycine receptor: Cooperativity of glycine interactions. *Mol Pharmacol* 10, 790–809.

Zhang, A., and Pasternak, G. W. (1981). Ontogeny of opiod pharmacology and receptors: High and low affinity site differences. *Eur J Pharmacol* 73, 29–40.

Zukin, S. R., Young, A. B., and Snyder, S. H. (1975). Development of the synaptic glycine receptor in chick embryo spinal cord. *Brain Res* 83, 525–530.

Cyclic Nucleotides

David J. Jones

This major section on developmental neurochemistry deals primarily with the ontogenic changes in neurotransmitter systems and receptor activities in the central nervous system (CNS). These components of neuron function influence secondary intracellular processes of information transfer that possibly regulate neural communication. The presence of intracellular messengers such as cyclic adenosine monophosphate (cyclic AMP) and possibly cyclic guanosine monophosphate (cyclic GMP) in neuronal tissue and linkage of their synthetic enzymes to receptor activation imply that these systems may also demonstrate an ontogenic pattern in response to various neurohormones. Moreover, although development of cyclic nucleotide systems may be governed by certain extracellular events, it is not unreasonable to propose that cyclic nucleotides may also be involved in the regulation of intracellular events that govern the further development of neuronal systems.

The purpose of this chapter is to provide a summary of the available data describing the ontogenesis of CNS cyclic nucleotide systems. In order to provide a framework on which to build and interpret the data, a description is provided of the components of the cyclic AMP system. Following this, a brief summary of evidence that implicates cyclic nucleotides in the physiology of the CNS is presented. Data on the ontogenesis of cyclic nucleotide systems are subsequently presented and divided into in vivo and in vitro studies of the various components of these systems.

While reading this chapter, it might appear that there is no consistent pattern to the ontogenesis of the various components of the cyclic nucleotide systems. However, a major point to consider is that each component is present to some degree at birth and thus capable of functioning in cellular regulation. Postnatal increases or decreases in either concentration or functional capacity occur that ultimately establish levels of activity in the adult animal. It is the pattern of these changes that will be summarized in this chapter.

Components of Cyclic Nucleotide Systems

Figure 1 is a diagram of the relationships of the various components of the cyclic AMP system. From this drawing, it is quite easy to follow the steps involved in the "second messenger" hypothesis whereby receptor activation and signal transduction result in a physiological or biochemical response (Sutherland and Rall, 1960). The following is a brief discussion of each component of this system. In order to not distract the reader with exhaustive notation, the references for this section frequently refer to reviews of each topic.

Transmission of information from the extracellular environment into the cell is accomplished through the receptor–adenyl cyclase complex. The adenyl cyclase system is associated with the plasma membrane and consists of a catalytic subunit facing the interior of the cell and a regulatory subunit exposed to the extracellular space (Limbird, 1981). This regulatory subunit composes the receptor that binds neurotransmitters, polypeptide hormones, and drugs. Transfer of the signal from the receptor subunit to the catalytic subunit is accomplished through a protein that binds guanine nucleotides. Through the binding of guanosine triphosphate (GTP) to this site and subsequent hydrolysis, both binding affinity of the hormone receptor (Maguire, Van Arsdale, and Gilman, 1976; Lefkowitz, Mullikin, and Caron, 1976) and cyclase activation (Rodbell et al., 1971a, 1971b) are accomplished. Cyclase activation results in the conversion of Mg^{++}–adenosine triphosphate (ATP) to cyclic AMP and inorganic phosphate (PPi). Once formed, cyclic AMP either binds to the regulatory subunit of a cyclic AMP–dependent protein kinase resulting in its activation or is hydrolyzed by phosphodiesterase.

The cyclic AMP–dependent protein kinase consists of a tetramer of two regulatory and two catalytic subunits, R_2-C_2 (Rubin and Rosen, 1975; Corbin and Lincoln, 1978). Binding of cyclic AMP to the regulatory subunits results in the dissociation of the complex releasing the free catalytic subunit. The catalytic subunit then transfers the terminal phosphate group of ATP to appropriate substrate proteins. Substrate proteins for this process have been identified in a large number of tissues.

Other components of the cyclic AMP system include cyclic nucleotide phosphodiesterase and a phosphoprotein phosphatase specific for dephosphorylating the substrate protein. Cyclic nucleotide phosphodiesterase converts cyclic AMP to adenosine 5′-monophosphate and thus can terminate reactions initiated by receptor activation (Appleman, Thompson, and Russell, 1973; Wells and Hardman, 1977). It is regulated by the heat-stable protein calmodulin (CaM), which is dependent on the presence of calcium (Cheung, Lynch, and Wallace, 1978). In addition to its regulatory role for phosphodiesterase, this protein also regulates the enzymatic activity of adenyl cyclase (Brostrom et al., 1975).

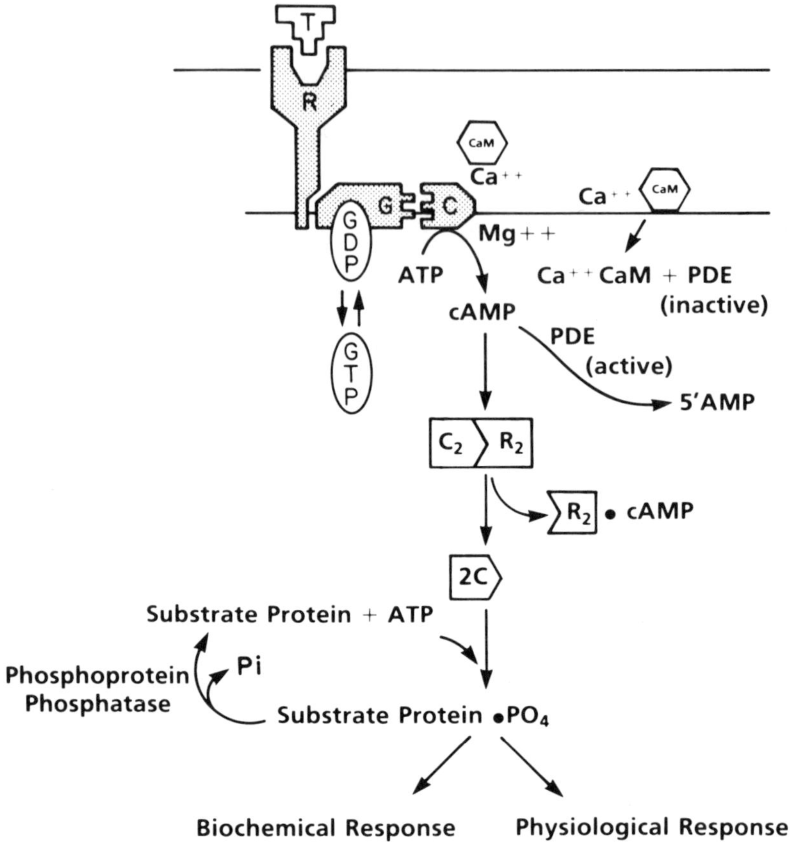

Figure 1. Diagram of the relationship between components of the cyclic AMP system (*T*, neurohormone agonist; *R*, regulatory subunit of cyclase; *CaM*, calmodulin; *G*, guanine nucleotide binding subunit of cyclase; *C*, catalytic subunit of cyclase; *GDP*, guanosine diphosphate; *GTP*, guanosine triphosphate; *PDE*, phosphodiesterase; C_2, catalytic subunit of protein kinase; R_2, regulatory subunit of protein kinase; *cAMP*, cyclic adenosine monophosphate; *ATP*, adenosine triphosphate).

The components of the cyclic GMP system are quite similar to those of the cyclic AMP system. In the presence of Mg^{++} or Mn^{++} guanyl cyclase synthesizes cyclic GMP from GTP (Murad et al., 1981). The exact mechanism whereby the synthesis of cyclic GMP is regulated by guanyl cyclase activation remains unknown. Guanyl cyclase is found to a substantial degree in the soluble fraction of the cell (Hardman and Sutherland, 1969; Goridis and Morgan, 1973) and unlike adenyl cyclase does not appear to contain a receptor complex for hormone activation. However, destruction of a receptor complex during membrane isolation may contribute to this finding (Goldberg, O'dea, and Haddox, 1973). Lack of a receptor complex mediating cyclic GMP formation has important implications relative to any proposed role for cyclic GMP as an intracellular second messenger.

Similar to kinases in the cyclic AMP system, cyclic GMP–dependent protein kinases mediate phosphorylation of substrate proteins (Kuo and Greengard, 1970; Nakazawa and Sano, 1975). Moreover, endogenous proteins that are specific substrates for cyclic GMP–dependent protein kinases have been identified (Casnellie and Greengard, 1974; Schlichter, Casnellie, and Greengard, 1978). Whether or not such phosphorylation is physiologically relevant remains to be determined.

Hydrolysis of cyclic GMP is accomplished by a cyclic nucleotide phosphodiesterase that also is regulated by the same calcium-dependent regulator noted earlier for adenyl cyclase and cyclic AMP phosphodiesterase (Brostrom et al., 1975). The sharing of this calcium-dependent regulator as well as the cyclic nucleotide phosphodiesterase suggests the possible dual regulation of the cyclic AMP and cyclic GMP systems. The fact that GTP, the substrate for guanyl cyclase, is part of the regulation of adenyl cyclase reinforces this possibility.

Cyclic Nucleotides in the Nervous System

Although a role for cyclic AMP in the physiology of the CNS has not been firmly established, a large number of studies provide us with evidence of intimate involvement of this system in regulating neuronal function. Much of this work relates to the role of cyclic AMP in synaptic transmission, whereby release of neurotransmitter and receptor activation are linked to cyclic AMP accumulation. However, a major difficulty in assigning a role for cyclic AMP as a mediator of receptor activation is that the brain is a functionally complex organ and the identity and precise role of neurotransmitters themselves in neurophysiological processes have not been fully elucidated. As will become clear, much of the evidence linking cyclic nucleotides to CNS physiology is indirect and concerns (1) demonstrating the presence of each of the components of the cyclic nucleotide systems in the CNS and (2) regulating levels through physiological or pharmacological intervention. This section will provide a more detailed description of the data that imply a role for cyclic nucleotides in CNS physiology. Complete reviews of this topic have appeared (Bloom, 1975; Daly, 1977; Nathanson, 1977; Greengard, 1978a; Bartfai, 1980; Bartfai, Nordstrom, and Tjornhammar, 1980).

Foremost among the data implicating cyclic AMP in nervous system function is the fact that adenyl cyclase occurs in brain tissue in higher levels than in any other organ system (Sutherland, Rall, and Menon, 1962). Further suggestive are the data on a distinct regional brain distribution and the fact that adenyl cyclase activity is high in gray matter and low in white matter (Daly, 1977). Moreover, studies have demonstrated high levels of adenyl cyclase in synaptosomal and synaptic membrane fractions of brain homogenates (DeRobertis et al., 1967; Von Hungen and Roberts, 1973).

Immunocytochemical studies have demonstrated cyclic AMP staining in cerebellum to occur in neuronal cell bodies and dendrites rather than glia (Bloom, et al., 1972). Moreover, high concentrations of cyclic AMP have been detected in synaptosomal fractions and synaptic membranes of brain homogenates (Weller, Rodnight, and Carella, 1972) and also synaptic vesicles (Johnson et al., 1973). Variations in regional brain levels of cyclic AMP are also evident (Jones et al., 1974; Jones and Stavinoha, 1977; cf. Daly, 1977).

Brain contains quite high concentrations, similar to those of adenyl cyclase, of phosphodiesterase (Butcher and Sutherland, 1962), which appears localized primarily in the gray matter. Regional brain studies indicate a sixfold range of phosphodiesterase activities, the highest levels occurring in neocortex and the hippocampus (Weiss and Costa, 1968). In distinction to adenyl cyclase, both particulate and soluble forms of phosphodiesterase are present in brain tissue (DeRobertis et al., 1967; Weiss and Costa, 1968; Thompson and Appleman, 1971). Most of the particulate enzyme activity is located in synaptic membrane fractions (DeRobertis et al., 1967) as well as in synaptic vesicles (Cheung and Salganicoff, 1967; Johnson et al., 1973). Localization of the phosphodiesterase to postsynaptic membranes of dendritic processes has also been established by Florendo, Barrnett, and Greengard (1971) and Adenolfi and Schmidt (1974).

Phosphodiesterase in brain tissue is not a single enzyme but rather a group of isoenzymes with different affinities for cyclic AMP (high- and low-K_m enzyme forms) and cyclic GMP (Weiss and Strada, 1973). The cyclic AMP phosphodiesterases are distributed equally between soluble and particulate forms, whereas the cyclic GMP phosphodiesterase is primarily soluble (Kakiuchi et al., 1975). Similar to earlier studies that identified variable regional brain activities for a single phosphodiesterase, more recent studies show the activity of the phosphodiesterase isoenzymes vary in different rat brain regions (Weiss and Strada, 1973). Moreover, responses of the enzymes to activators or inhibitors vary as a function of the brain region (Weiss, Fertel, and Uzunov, 1974; Weiss, 1975).

Phosphodiesterase activity can be regulated by the calcium-dependent regulator protein CaM, which occurs in relatively high concentrations in brain tissue (Cheung, 1970; Kakiuchi et al., 1975). In addition, Gnegy, Nathanson, and Uzunov (1977) established this regulator protein to be concentrated in rat brain synaptic membranes. In the presence of calcium, this heat-stable protein forms a complex with phosphodiesterase leading to an increase in enzyme activity and hydrolysis of cyclic AMP and cyclic GMP. This protein activator of phosphodiesterase has also been shown to regulate adenyl cyclase (Cheung et al., 1975; Brostrom et al., 1975).

The physiological and biochemical actions of cyclic AMP have been proposed to be mediated by the phosphorylation of substrate proteins (Kuo and Greengard, 1970; Greengard, 1978a, 1978b; Walsh, 1978). As shown in figure 1, the phosphorylation step is dependent upon activation of a protein kinase by cyclic AMP and the presence of a protein that functions as an acceptor for the terminal phosphate of ATP. A specific phosphoprotein phosphatase dephosphorylates the substrate protein (Maeno and Greengard, 1972). Although both cyclic AMP–dependent protein kinase and phosphoprotein phosphatase exist in particulate and soluble fractions in brain, the particulate enzymes are concentrated in synaptic plasma membranes (Gaballah and Popoff, 1971b; Maeno, Ueda, and Greengard, 1975). In addition, synaptic membrane and synaptic vesicle fractions exhibit a high rate of incorporation of phosphate, which is stimulated by cyclic AMP (Johnson, Maeno, and Greengard, 1971). The phosphorylation of several endogenous proteins located in synaptic membranes is regulated by this cyclic AMP–dependent process. Of most importance, Protein I (Ueda and Greengard, 1977) appears unique to neuronal tissue, and cyclic AMP–dependent phosphorylation of this protein may well provide the molecular basis for cyclic AMP regulation of synaptic transmission (Greengard, 1976).

The evidence supporting a role for cyclic GMP in nervous tissue is not as complete as that for the cyclic AMP system. In contrast to the cyclic AMP system where levels are uniform across the various brain areas, levels of cyclic GMP vary eightfold to tenfold between brain regions, cerebellum containing the highest concentration (Steiner, Ferrendelli, and Kipnis, 1972; Jones and Stavinoha, 1977). This variation does not appear related to guanyl cyclase activity since the highest guanyl cyclase activity is found in striatum, an area with low cyclic GMP levels (Nakazawa and Sano, 1975; Nakazawa, Sano, and Saito, 1976). It may, however, reflect the activity of cyclic GMP phosphodiesterase since the values in cerebellum are fivefold to eightfold lower when compared with activity levels in cerebral cortex, striatum, or hippocampus (Nakazawa and Sano, 1975; cf. Daly, 1977).

Adenyl cyclase is a membrane-bound enzyme complex in almost all tissues studied, but guanyl cyclase exists in both soluble and particulate fractions (White and Aurbach, 1969; Hardman and Sutherland, 1969). In brain tissue, basal activity of the cytosolic enzyme is approximately fourfold to fivefold the activity of the particulate enzyme. Marked activation of the enzyme occurs upon solubilization (Nakazawa and Sano, 1975; Nakazawa, Sano, and Saito, 1976), with the highest specific activity of the enzyme located in the soluble fraction from lysed synaptosomes (Goridis and Morgan, 1973). The fact that the major portion of guanyl cyclase is soluble suggests that intracellular events may regulate this enzyme rather than receptor activation at extracellular sites.

Cyclic GMP phosphodiesterase activity is higher in brain than in other tissues and also appears distributed between soluble and particulate fractions (Nakazawa and Sano, 1975). As pointed out earlier, the quantitative differences in regional brain cyclic GMP phosphodiesterase seem to predict the basal value of cyclic GMP in the particular areas.

Kuo and Greengard (1970) have identified a specific protein kinase that is regulated by cyclic GMP. This enzyme in the presence of ATP phosphorylates a number of exogenous protein substrates. Although this enzyme has been isolated from rat cortex and cerebellum (Hofmann and Sold, 1972; Sold and Hofmann, 1974), a specific endogenous substrate protein has not been identified in this tissue. However, Schlichter, Casnellie, and Greengard (1978) have isolated a protein from the cytosol of rabbit cerebellum, the phosphorylation of which is stimulated by cyclic GMP. Recently, Boehme, Kosecki, and Marks (1978) reported that cyclic GMP enhanced the phosphorylation of several proteins in synaptosomal membranes prepared from human brain.

Cyclic nucleotides have been implicated in at least three different processes in neuronal tissue function: (1) mediation of the postsynaptic actions of certain neurotransmitters (cyclic AMP and cyclic GMP), (2) regulation of the synthesis of catecholamines (cyclic AMP), and (3) regulation of microtubule function (cyclic AMP). The strongest evidence centers around cyclic AMP mediating at least some of the effects of receptor activation in the process of synaptic transmission (Greengard, 1976). The brief summary above provides evidence to establish that the components of the cyclic AMP and cyclic GMP systems are located in neuronal tissue at sites relevant to synaptic transmission.

Ontogenesis of Cyclic Nucleotide Systems

This section is concerned with the temporal relationship between the ontogenesis of the various components of the cyclic nucleotide systems and that of other functional units involved in synaptic transmission. Examining such a relationship will provide insight about whether the functional development of the synapse is dependent on cyclic nucleotide systems or vice versa. The data, for purposes of organization and clarity, deal primarily with the development of rodent CNS from birth to early adulthood. The reader is directed to excellent reviews of this topic that have appeared previously (Weiss and Strada, 1973; Schmidt, Palmer, and Robison, 1980).

Cyclic Nucleotide Levels

Two types of studies contribute to data on the "endogenous" levels of cyclic nucleotides in intact cells during CNS development. The first type involves the measurement of cyclic nucleotides in brain that has been inactivated in vivo. The second method measures in vitro basal accumulation of the cyclic nucleotides in incubated tissue slices prepared from brain. While in vivo levels supposedly predict for steady-state concentrations of cyclic nucleotides immediately prior to sacrifice of the animal, there are a number of problems with this method. On the other hand, in vitro tissue slices provide a relatively stable environment for isolated tissue in which intact cellular membranes, cell-to-cell contacts, and cellular respiration are maintained (McIlwain and Bachelard, 1971).

In Vivo Studies. Following sacrifice, in vivo levels of cyclic AMP in brain increase rapidly, necessitating rapid enzyme inactivation. Methods used for enzyme inactivation include rapid freezing in liquid nitrogen or heating by microwave irradiation (Schmidt, Schmidt, and Robison, 1971; Schmidt et al., 1972; Jones et al., 1974; Jones and Stavinoha, 1977). Major difficulties in using quick-freezing techniques center around slow and nonuniform enzyme inactivation. Conduction of the freezing temperature from the surface of the brain requires a number of seconds (Jongkind and Bruntink, 1970; Swabb, 1971), and since the increase of cyclic AMP levels following sacrifice is dependent on time from sacrifice to enzyme inactivation (Jones and Stavinoha, 1976), regional brain concentrations will vary as a function of the depth of the structure from the surface. The brain of the young and therefore smaller animal will freeze much faster; consequently, potentially lower levels of cyclic AMP would be measured. In addition, the rate of increase in brain cyclic AMP following decapitation is significantly different in animals of different ages (Schmidt et al., 1970).

In contrast, heating by microwave irradiation inactivates tissue uniformly, avoiding the gradients and time factors of enzyme inactivation (Jones, 1974). Unfortunately, no studies on in vivo cyclic AMP levels in brain have been completed in developing animals using microwave irradiation. As a result, studies have utilized the rapid-freezing technique and therefore data may reflect the freezing gradient artifact.

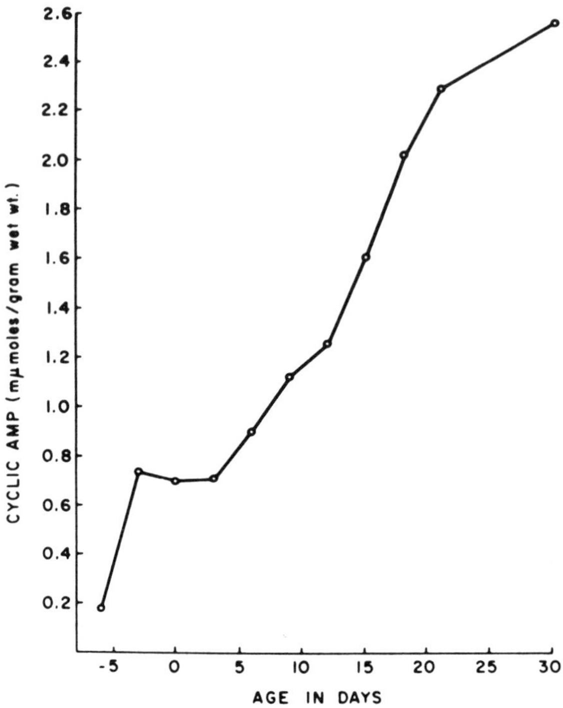

Figure 2. Concentration of cyclic AMP in rat whole brain as a function of age. (Reprinted with permission from Schmidt et al., 1970.)

Schmidt and two sets of coworkers (1970, 1972) utilizing the rapid-freezing technique established that levels of cyclic AMP in rat brain increase from fetal day 21 (six days before birth) through postnatal day (P) 30 (fig. 2). The greatest rate of increase occurred between P7 and P21 preceded and followed by rates that were 30 percent of the most rapid phase. Regional brain cyclic AMP levels also vary as a function of age (Weiss and Strada, 1973). In contrast to the data of Schmidt, Schmidt, and Robison (1971) from studies of whole brain, the data of Weiss and Strada (1973) showed brain stem and cerebral cortex cyclic AMP did not increase following birth (fig. 3). Cerebellar cyclic AMP levels did increase over fourfold from birth through P30. The rate of increase of cyclic AMP was fairly constant P1 through P30; thereafter, levels were constant through P65. In studies of cerebellar cyclic AMP levels in developing rats, Shambaugh et al. (1978) established similar increases; however, levels were stable P4 through P12, a sharp increase occurring P12 through P20.

Levels of cyclic AMP in mouse regional brain have also been shown to increase gradually from birth to adulthood (Steiner, Ferrendelli, and Kipnis, 1972). In forebrain, brain stem, and cerebellum levels were stable from birth through P8; thereafter, levels increased threefold in forebrain, fourfold in brain stem, and twofold in cerebellum. In agreement, Spinka, Farber, and Lolley (1980) demonstrated that mouse cerebellar cyclic AMP levels were stable from birth to P5 and subsequently increased threefold to adult age. In contrast to these studies, the studies of Orenberg et al. (1975) showed that levels of cyclic AMP in mouse brain did not change from birth to adulthood.

Following decapitation, cyclic GMP levels in brain do not increase rapidly (Steiner, Ferrendelli, and Kipnis, 1972; Jones and Stavinoha, 1976). In fact, a slight decrease at two minutes following decapitation has been reported (Jones and Stavinoha, 1976). Thus the concentration of cyclic GMP measured in regional brain is less a function of rapid en-

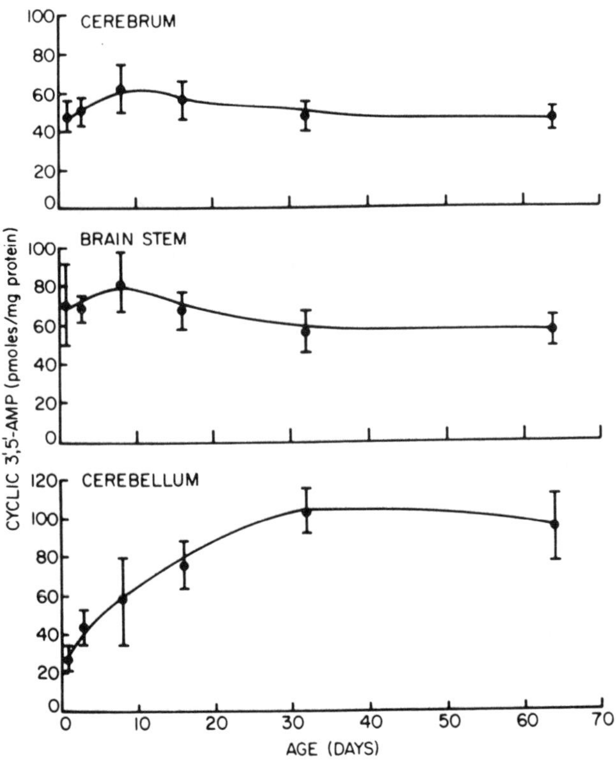

Figure 3. Concentration of cyclic AMP in rat regional brain as a function of age. (Reprinted with permission from Weiss and Strada, 1973).

zyme inactivation and therefore possibly more predictive of actual in vivo levels. Spano et al. (1975) demonstrated in rats that cerebellar levels of cyclic GMP at P3 and P6 were equivalent. However, on P12 cyclic GMP levels start to rise to values that are 25- to 30-fold greater at adult age than at P3 or P6. This study is the only one in which enzyme inactivation and animal sacrifice was accomplished by microwave irradiation.

Cyclic GMP also increases dramatically from birth to adulthood in cerebellum of mice (Steiner, Ferrendelli, and Kipnis, 1972; Spinka, Farber, and Lolley, 1980). A tenfold increase in the level of cyclic GMP in cerebellum at P25 was noted in the studies of Spinka and coworkers (1980) compared to a 14-fold increase in the studies of Steiner and coworkers (1972). Whereas levels of cyclic GMP in forebrain were relatively constant from birth to P21, levels of cyclic GMP in brain stem doubled from birth to P7, and thereafter levels did not change (Steiner, Ferrendelli, and Kipnis, 1972).

Conclusions from in vivo studies on regional brain cyclic nucleotides are somewhat difficult to draw. Although the time courses do differ, a consistent increase in cerebellar cyclic AMP does occur. The increase in this relatively large mass of brain tissue may account for the increase in whole brain levels noted by Schmidt et al. (1970). Consistent increases in cyclic GMP have also been measured in developing rat and mouse cerebellum. Thus the cerebellum appears unique in the responsiveness of intracellular systems to the development process.

In Vitro Studies. Basal levels of accumulation of cyclic AMP and cyclic GMP in slices of tissue from various regions of the CNS provide a measure of in vitro steady-state concentrations of cyclic nucleotides in intact cells. Following a period of equilibration in oxygenated buffer, respiratory status and levels of intermediary metabolites measured in tissue slices reflect those evident in brain tissue inactivated in vivo (McIlwain and Bachelard, 1971). Since membrane integrity is maintained and all the nutrients necessary for metabolism are provided, then basal accumulation may well predict a more normal state of cellular functioning than do other measures.

In tissue slices from cerebral cortex of rats, basal levels of cyclic AMP do not change significantly during the course of development (Perkins and Moore, 1973b; Harden, Wolfe, Sporn, Perkins, and Molinoff, 1977; Simmons and Jones, 1982). This has also been noted in slices from developing rabbit cerebral cortex (Schmidt and Robison, 1971; Palmer, Schmidt, and Robison, 1972). Whole rat brain levels of accumulation also do not change as a function of age (Schmidt et al., 1970; Schmidt et al., 1972). Similarly, basal levels of cyclic AMP accumulation in slices of rat spinal cord (Simmons and Jones, 1982), rabbit cerebellum, hippocampus, or hypothalamus (Schmidt and Robison, 1971) are not altered as a function of postnatal age. A slight (25 percent) increase in cyclic AMP accumulation from P2 to P70 has been reported in cerebellar slices by Garthwaite and Balazs (1978). In contrast, Daszuta, Pons, and Cadhilac (1979) demonstrated a fourfold increase in basal accumulation in slices from rat hypothalamus from birth to P90.

The lack of postnatal change in basal cyclic AMP accumulation in slices from cerebral cortex is in agreement with the in vivo data showing no change in cyclic AMP from this same area. However, the slight increase in basal accumulation of cyclic AMP in cerebellar slices contrasts sharply with the reported fourfold increase in in vivo levels during the same ontogenic period. Although no postnatal change in basal cyclic GMP accumulation occurs in cerebellum P1–P90 (Garthwaite and Balazs, 1978), in vivo levels of cyclic GMP in this tissue have been reported to increase sharply as a function of postnatal age (Spano et al., 1975).

Thus the question remains for both the cyclic AMP and cyclic GMP systems of whether brain levels increase or do not change as a function of postnatal age. Such variation in data compromises one's ability to assign a possible role to these intracellular messengers in the development process of the CNS. However, lack of consistency may also indicate that cyclic nucleotides are not participants in the development process and merely represent the ability of cyclic nucleotide systems to respond, to be activated, as a function of sacrifice and subsequent preparation of the tissue.

Adenyl and Guanyl Cyclase Levels

Most investigations concerning developmental alterations in the properties of adenyl cyclase systems have been carried out on broken cell preparations (homogenates) of whole brain or regional brain areas. In contrast to tissue slices in which cellular integrity is maintained, these preparations contain cellular and membrane fragments and cytosol in a buffered medium. Subsequent fractionation divides the preparation into cytosolic and particulate fractions.

Early studies by Schmidt et al. (1970) demonstrated the presence of adenyl cyclase activity at birth in particulate preparations from rat brain. Activity was not observed in the supernatant (soluble) fraction. Enzyme activity increased approximately 300 percent from birth to P9; thereafter, basal activity subsequently decreased to below that noted at birth (fig. 4, *solid bars*). Perkins and Moore (1973b) also reported substantial adenyl cyclase activity at birth in rat whole brain particulate fractions. This activity also increased approximately 200 percent by P16–P20 and declined thereafter to levels equivalent to those present at birth.

Regional brain studies have been completed by Weiss (1971). He noted an increase in adenyl cyclase activity in the pineal gland, cerebral cortex, brain stem, and cerebellum from P1 to P15 (fig. 5). While pineal adenyl cyclase activity continued to increase until P60, activity declined

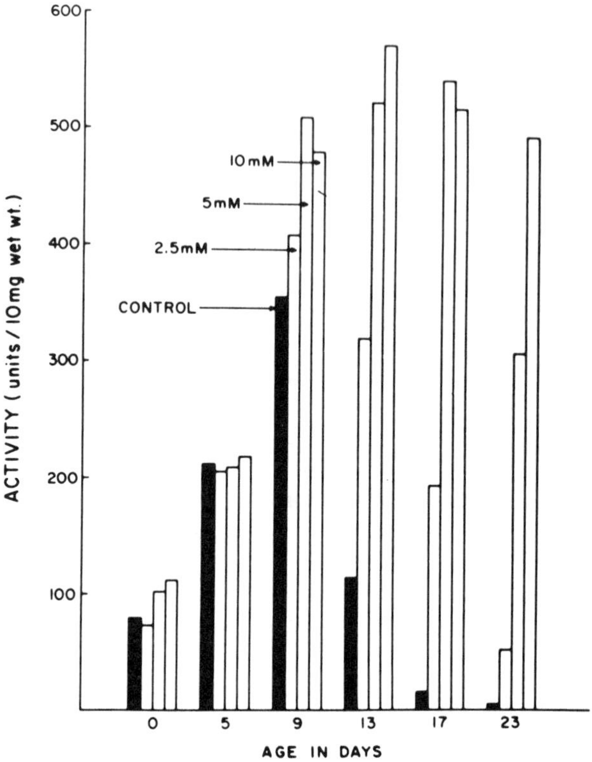

Figure 4. Adenyl cyclase activity in particulate fraction from whole brain in postnatal rats as a function of age. *Solid bars,* basal (control) activity; *unfilled bars,* activity in the presence of noted concentrations of NaF. (Reprinted with permission from Schmidt et al., 1970.)

somewhat (10–20 percent) on subsequent days in cerebral cortex and cerebellum. However, in contrast to the data of Schmidt et al. (1970), activity at P60 was still 200–300 percent above levels at birth. This developmental pattern in rat cerebral cortex has been confirmed by Kohrman (1973).

Von Hungen, Roberts, and Hill (1974) also demonstrated the presence of adenyl cyclase activity at birth in rat brain regional areas. As noted in figure 6, cerebral cortex and subcortex adenyl cyclase increased from birth to almost maximum activity at P14 (100 percent over birth activity); thereafter, levels were stable. Hindbrain-medullary adenyl cyclase, highest in activity at birth, did not demonstrate a dramatic postnatal increase in activity. In fact, activity was decreased at P42 in the hindbrain-medullary area. Although cerebellar levels also reached a

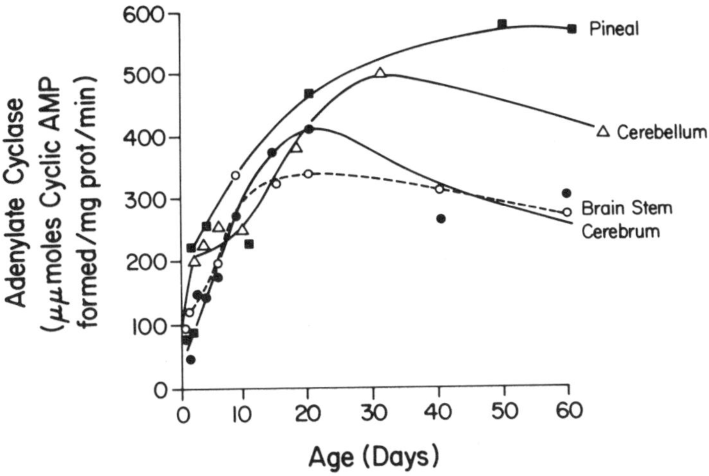

Figure 5. Adenyl cyclase activity in particulate fraction from regional rat brain areas as a function of age. (Adapted from Weiss, 1971.)

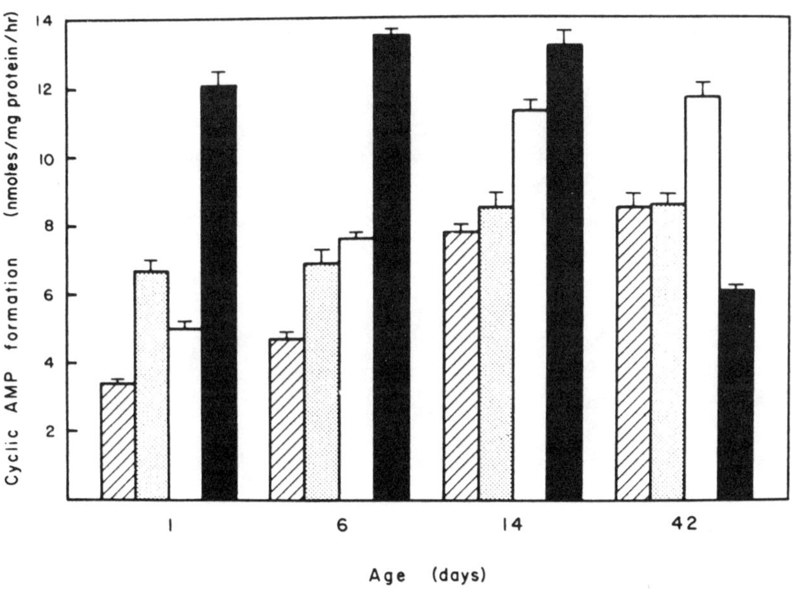

Figure 6. Regional distribution of basal adenyl cyclase activity in particulate fractions of rat brain during development. *Striped bars,* cerebral cortex; *stippled bars,* cerebellum; *unfilled bars,* subcortex; *filled bars,* hindbrain-medullary area. (Reprinted with permission from Von Hungen, Roberts, and Hill, 1974.)

Table 1. ***Summary Data for Ontogenesis of Rat Cerebral Cortical Adenyl Cyclase Activity***

Source	*Adenyl Cyclase Activity*[a]			
	P7	*P14*	*P21*	*Adult*
Weiss (1971)	400	650	700	500
Kohrman (1973)	150	300	420	280
Von Hungen, Roberts, and Hill (1974)	30	120	—	150

Note: P, postnatal day.
[a] Expressed as approximate percentage of activity present within 24 hours of birth.

maximum at P14, the increase was only 25 percent over values noted at birth. This contrasts with values from mouse cerebellum studies in which activity was shown to increase linearly over 100 percent from birth to adulthood (Spinka, Farber, and Lolley, 1980).

Enjalbert et al. (1978) reported an increase in adenyl cyclase activity in the rat corpus striatum following birth. Basal adenyl cyclase increased 400 percent from birth through P16 in particulate fractions from this tissue. Levels were stable thereafter. A similar 400 percent increase in basal activity has been measured from birth to P42 in colliculi from developing rat brain (Von Hungen, Roberts, and Hill, 1975). In cell-free homogenates prepared from rat striatum, basal dopamine-sensitive adenyl cyclase activity has also been reported to increase approximately tenfold from birth to adulthood (Coyle and Campochiaro, 1976).

A summary of the development of cerebral cortical adenyl cyclase activity from three different studies is shown in table 1. While there is a great deal of variability in the quantitative increases, the trend from birth to adulthood is consistent. As noted by Schmidt, Palmer, and Robison (1980) presumably uncontrolled ATPase activity might account for some of the quantitative variability. Differences in stability between neonatal and adult enzymes have also been reported (Von Hungen and Roberts, 1974).

Fluoride ions have been demonstrated to increase basal adenyl cyclase by a mechanism involving the catalytic subunit, thus bypassing the receptor complex. Interaction of NaF with this system demonstrates a developmental pattern that differs in both qualitative and quantitative aspects. As shown in figure 4, Schmidt et al. (1970) established that at birth through P5 stimulation of adenyl cyclase from rat whole brain was minimal or absent in the presence of 2.5–10 mM NaF. Activation was first present at P9 and, compared to control values, levels increased markedly thereafter. As is shown in figure 4, although the percentage increase in activity is significantly greater, the maximum increase does not really change P9–P23. The decrease in basal activity accounts for the greater percentage activation over control values. Subsequent to this study, Perkins and Moore (1973b), after studying rat cerebral cortex, reported a similar greater stimulatory effect of NaF relative to a decreasing basal activity with increasing age.

Further studies by Schmidt and Robison (1972), contrary to the above original report, established that NaF stimulated adenyl cyclase activity in whole rat brain at birth, and this stimulation increased approximately sixfold through P27. Von Hungen, Roberts, and Hill (1974) reported that NaF stimulated adenyl cyclase in cerebral cortex at birth, the increase becoming maximal P20–P25. Identical developmental patterns of NaF stimulation with peak activity at P20 (Weiss, 1971) and P45 (Harden, Wolfe, Sporn, Perkins, and Molinoff, 1977; Harden, Wolfe, Sporn, Poulos, and Molinoff, 1977) had previously been established in rat cerebral cortex. Weiss (1971) also reported that peak NaF stimulation in particulate fractions was reached at P10 in brain stem, at P30 in cerebellum, and at P20 in pineal gland. In this study, maximal stimulation in the presence of 10 mM NaF was fourfold in cerebral cortex, threefold in brain stem and cerebellum, and fivefold in pineal gland. Thus with the exception of activity in the brain stem, maximal activation of adenyl cyclase by NaF occurred within 20–30 days following birth.

Guanyl cyclase activity in rodent brain also reveals developmental patterns that, in comparison to those of adenyl cyclase, are remarkably distinct. Spano et al. (1975) reported that the cerebellum of immature rats (P8) has quite high basal activity of guanyl cyclase that subsequently decreased over 200 percent during the next three weeks. The decrease contrasts sharply with the 25- to 30-fold increase in in vivo cyclic GMP levels (Spano et al., 1975) and the absence of change in basal cyclic GMP accumulation (Garthwaite and Balazs, 1978) that occur during the same period.

In contrast to the above data, the findings of Spinka, Farber, and Lolley (1980) demonstrated that mouse cerebellar guanyl cyclase increased approximately fourfold from P5 to P25. Adult values were somewhat higher than P25 enzyme activity values. Levels of cyclic GMP increased eightfold during this same time period. Such differences between developing rat and mouse cerebellar guanyl cyclase activity may well reflect species-specific regulatory mechanisms for cyclic GMP levels in this tissue. However, the mechanism of regulation of this system is unclear.

Cyclic Nucleotide Phosphodiesterases

The catabolism of cyclic AMP and cyclic GMP to their respective 5′-mononucleotides is accomplished through cyclic nucleotide phosphodiesterases. These enzymes are present as isoenzymes with different affinities for cyclic AMP and cyclic GMP and are found in both soluble and particulate fractions from various brain regions. High- and low-affinity forms of cyclic AMP phosphodiesterases are present in both fractions, whereas the cyclic GMP phosphodiesterase is present primarily in the soluble fraction. Regulation of cellular cyclic nucleotide concentrations appears to occur through the low-K_m, high-affinity isozyme that has the capability of hydrolyzing both cyclic AMP and cyclic GMP (Weiss and Strada, 1973). Postnatal development patterns are unique for the particular form of the phosphodiesterase and also the brain area.

Initial studies by Schmidt et al. (1970) of rat whole brain demonstrated an increase in both total soluble and particulate forms of cyclic AMP phosphodiesterase from birth through P24. The P24 value was approximately the same as adult activity and was eightfold to ninefold greater than enzyme activity at birth. Smoake, Song, and Cheung (1974) reported a similar time course; however, maximal activity was approximately 20-fold greater than activity measured at birth. In addition, the studies of Gaballah and Popoff (1971a) demonstrated peak activity at P30 to be approximately twofold greater than activity at P10. In this study, similar changes were also found as a function of age in subcellular fractions of crude mitochondria, nerve endings, and nerve ending membranes.

Developmental patterns for cyclic AMP–phosphodiesterase in rat regional brain have been investigated by Weiss's laboratory (Weiss, 1971; Weiss and Strada, 1972, 1973; Strada, Uzunov, and Weiss, 1974). As shown in figure 7, while enzyme activity increased fourfold to fivefold in brain stem and cerebral cortex, cerebellar levels decreased 30 percent (Weiss, 1971). In measuring the contribution of high-K_m and low-K_m cyclic AMP phosphodiesterases to these ontogenic changes, it was shown that the ratio of high-K_m enzyme to low-K_m enzyme did not change in cerebellum with increasing age (Strada, Uzunov, and Weiss, 1974). However, the increase in cerebral cortex was due almost exclusively to a greater activity of the high-K_m phosphodiesterase. In the pineal gland, where total phosphodiesterase activity also increases with age, the low-K_m enzyme appears to account for the majority of the increase (Weiss and Strada, 1972). The low-K_m phosphodiesterases for cyclic AMP and cyclic GMP in the brain from developing guinea pigs is also greater in adults than in neonates (Davis and Kuo, 1976).

Studies by Kang (1977) in human cerebral cortex demonstrated that both cyclic AMP phosphodiesterase and cyclic GMP phosphodiesterase activities were greater in adult than in fetal tissue. The mature cortex had ten times more activity than the fetal brain for cyclic AMP hydrolysis and 15–20 times more activity for cyclic GMP hydrolysis. In addition, there occurred a shift in localization of low-K_m phosphodiesterase activities from supernatant to particulate fractions.

In consideration of the ontogenesis of brain phosphodiesterase and

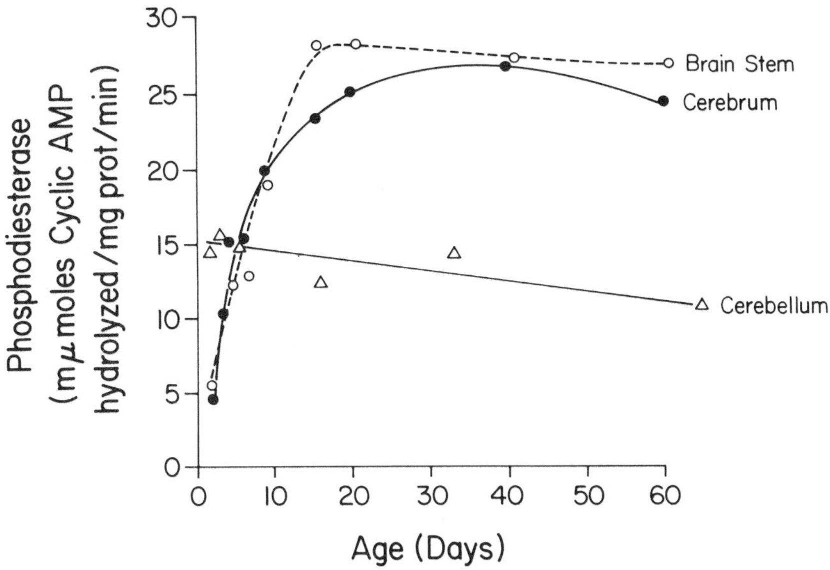

Figure 7. Ontogenic development of phosphodiesterase in rat brain regional areas. (Adapted from Weiss, 1971.)

the presence of regulatory proteins for this enzyme in brain (Kakiuchi et al., 1973), it might be expected that such regulatory proteins would also demonstrate postnatal development patterns. Smoake, Song, and Cheung (1974) have reported that from eight days before birth through adulthood the activity of CaM doubled in rat brain. This increase paralleled the increase in phosphodiesterase activity to P24. Studies of regional brain demonstrated that CaM activity varies from region to region and that cerebellum contained several-fold total less activity compared with activity in cerebral cortex or brain stem (Strada, Uzunov, and Weiss, 1974). However, the activity of the activator in cerebellum increased with age, but the levels in cerebral cortex and brain stem decreased.

In view of the developmental patterns of the cyclic nucleotide phosphodiesterases and CaM in regional brain, it is obvious that an extremely complex relationship exists involving the postnatal regulation of intracellular cyclic AMP and cyclic GMP levels. Independent genetic control of the various enzymes and activator may explain this complexity.

Cyclic Nucleotide-dependent Protein Phosphorylation

The phosphorylation of substrate proteins is generally accepted as the mechanism whereby cyclic AMP and cyclic GMP achieve their diversity of biologic effects in both neuronal and nonneuronal tissue (Greengard and Kuo, 1970; Greengard, 1978a, 1978b; Walsh, 1978). In the presence of ATP, phosphorylation is catalyzed by protein kinases activated by cyclic AMP or cyclic GMP. The presence of these enzymes is demonstrated by using various exogenous proteins as phosphorylating substrates and, in the presence or absence of cyclic nucleotide, measuring the incorporation of radiolabeled phosphate. In some instances, phosphorylation of endogenous proteins can be measured. The dephosphorylation step, catalyzed by specific phosphatases, is also vital to this final step in the cascade since this has been proposed as the method for termination of the biochemical or physiological event mediated by phosphorylation.

Using histone as a substrate protein for cyclic AMP–dependent and –independent phosphorylation, Gaballah, Popoff, and Sooknandan (1971) were the first to demonstrate that relatively high levels of cyclic AMP–dependent protein kinase were present in brains of rats at birth and at P3. At P6, activity increased and remained at this level to adult age. Since cyclic AMP–independent phosphorylation also increased during this time and the increase was proportionally greater compared with that of cyclic AMP–dependent phosphorylation, an actual decrease in the percentage activation of cyclic AMP–dependent phosphorylation was measured.

Subsequent studies by Schmidt and Robison (1972) demonstrated significant cyclic AMP activation of histone phosphorylation three to six days prior to birth. In contrast to the data of Gaballah and coworkers (1971), their data showed activity was maximal at birth and without any further postnatal increase in either cyclic AMP–dependent or –independent protein kinase activity. In both of the above studies, protein kinase activity more than doubled in the presence of cyclic AMP.

Takahashi, Mutsuzaki, and Nunez (1974), using both histone and casein as phosphorylating substrates in soluble fractions from brain, demonstrated that the rate of cyclic AMP–dependent histone phosphorylation was constant P4 to P30. Since the basal (cyclic AMP–independent) phosphorylation of histone increased with age, the net effect was a decrease in cyclic AMP–dependent phosphorylation until P30. These data confirm the findings of Gaballah and coworkers (1971). In contrast to findings from studies using histone as a substrate, the data of Takahashi, Mutsuzaki, and Nunez (1974) also showed that both the cyclic AMP–dependent and –independent phosphorylation of casein remained constant throughout development, resulting in no change in cyclic AMP–dependent phosphorylation. Thus, the availability of substrate protein appears to function as a major regulatory factor in determining the development of cellular functions involving cyclic AMP–dependent protein kinase.

An example of this fact is that nuclear material, which contains the major part of the cellular histones, contained little cyclic AMP–dependent protein kinase (Takahashi, Mutsuzaki, and Nunez, 1974; although see Shambaugh et al., 1978). Inasmuch as cyclic AMP–dependent protein kinase activity in the nuclear fraction does not change through development, it can be stated that exogenous substrate proteins used for measuring phosphorylation may well provide nonphysiological conditions upon which developmental data are based.

Schmidt and Sokoloff (1973) more completely evaluated the differences between endogenous and exogenous protein substrates for phosphorylation. In the absence of exogenous substrate protein, minimal cyclic AMP–dependent phosphorylation of microsomal protein was measured; however, in the presence of exogenous histone, maximal activity of microsomal protein kinase was expressed. Using this method, they found no changes in cyclic AMP–dependent phosphorylation from birth to early adulthood. Moreover, using brain ribosomal proteins as a phosphate acceptor, they found the same lack of developmental pattern. On the basis of their data they concluded that postnatal changes do not occur in the activities of soluble brain protein kinase, in microsomal protein kinase, or in the ability of brain ribosomes to serve as substrates for the kinase (Schmidt and Sokoloff, 1973).

In agreement with the above, Lohmann, Ueda, and Greengard (1978) established that relatively constant amounts of cyclic AMP–dependent protein kinases were present in both cytosol and membranes of rat cerebral cortex during development. However, incorporation of phosphate into certain endogenous substrate proteins by cyclic AMP–dependent protein kinases does increase in developing rat cerebral cortex (Lohmann, Walter, and Greengard, 1978). These endogenous substrates have been designated as proteins Ia and Ib and have been shown to be specific to nerve tissue and to be enriched in the synaptic membrane and synaptic vesicle fractions of rat brain (Greengard, 1978b). The concentration of these proteins increases threefold to fourfold in developing rat and guinea pig cerebral cortex (Lohmann, Walter, and Greengard, 1978). Since the amount of cyclic AMP–dependent protein kinase does not change during this period, then it follows that the changes in endogenous phosphorylation are due primarily to changes in the amount of substrate proteins.

The amount of substrate proteins Ia and Ib parallels the development of synaptic elements since both peak at approximately the same time. Since the guinea pig brain passes through this state *in utero* (Jones, Dittmer, and Reading, 1974), it would be expected that similar cyclic AMP–dependent processes involving phosphorylation would occur at this gestational stage in guinea pig brain. Lohmann, Walter, and Greengard (1978) did in fact confirm this relationship, showing not only that the amount of protein Ia and Ib increased in guinea pig brain prior to birth, but also that levels at birth were the same as those in the adult. Since the amount and activity of the protein kinase are constant in developing rat or guinea pig brain, the limiting factor in the expression of the developmental pattern of the cyclic AMP–dependent protein kinase appears to be the availability of endogenous substrate protein. An indirect link to synaptogenesis is also evident from this relationship.

Though similar substrate specificities have been demonstrated between cyclic AMP and cyclic GMP protein kinases (Krebs and Beavo, 1979), the versatility of action of the cyclic GMP–dependent protein kinase is quite limited. Part of this limitation may be due to the existence of the enzyme in only the soluble fraction of the cell (Kuo, 1975). The best evidence for a role of the cyclic GMP–dependent protein kinase in catalyzing physiologically important phosphorylation of proteins comes from studies of cerebellum. Here the levels of both cyclic GMP and guanyl cyclase are higher than in any other tissues (Daly, 1977). Bandle and Guidotti (1979) established that cyclic GMP–dependent protein kinase activity was quite low during the first ten days after birth. Over the next 15 days activity increased threefold and remained at this level thereafter. These data are in agreement with studies reported by Kuo, Kuo, and Shoji (1976) that show cyclic GMP–dependent protein kinase is lower in newborn than in adult guinea pig cerebellum.

The above studies demonstrate that the cyclic AMP–dependent protein kinase is fully operational in brain at birth. Availability of endogenous substrates appears to be the primary requisite for complete expression of cyclic AMP–dependent phosphate transfer. The fact that these substrate proteins develop at a time when synaptogenesis is occurring in brain implies functional relevance to this process. In contrast, the cyclic GMP system requires at least 25–30 days to mature. Possibly the cyclic GMP system may be linked to a unique cell population in the cerebellum that requires this period to become functionally mature.

Agonist Stimulation of Cyclic Nucleotide Formation

Stimulation of the formation of cyclic nucleotides has been demonstrated to occur with a large number of agonists, including compounds of diverse chemical structures. Most studies have used neurohumoral amines for activation inasmuch as these are the most relevant agonists for studying receptor responsiveness during neuronal development. Two types of tissue preparations are used for studies evaluating the responsiveness of cyclic nucleotide systems: (1) intact cell preparations (tissue slices) provide a reasonably accurate picture of regulation of intracellular cyclic AMP and cyclic GMP levels by maintaining the integrity of the cyclase-receptor complex; (2) broken cell homogenates and subcellular fractions isolate the cyclase from the effects of other cellular regulatory processes.

In Vitro Accumulation of Cyclic AMP and Cyclic GMP

As mentioned above, cyclic AMP formation in brain tissue slices occurs with a large number of agonists. However, in consideration of the predominant endogenous compounds available for activation of cyclic AMP accumulation, norepinephrine (NE) is probably the most important. Moreover, NE systems appear to be a dominant factor in the development of neurotransmission in regional brain (Coyle, 1977).

Schmidt et al. (1970), using tissue slices from whole brain, first demonstrated that NE-stimulated cyclic AMP accumulation was absent at birth and gradually increased to reach a maximum response at P9. NE stimulation on P22 was equivalent to that on P9. Subsequent studies from this laboratory demonstrated an approximate 50 percent decrease in NE responsiveness on P25 compared with that on P16 (Schmidt and Robison, 1972).

Perkins and Moore (1973b) used tissue slices from rat cerebral cortex to demonstrate that responsiveness to NE (3×10^{-5}M) did not occur until P11, reaching a maximum at P13. A slight decline occurred thereafter. In agreement, Simmons and Jones (1982), using tissue slices from rat cerebral cortex, demonstrated a lack of NE responsiveness prior to P10 followed by a maximum stimulatory response on P20 (fig. 8). Subsequently, on P30 and P45 a decrease in sensitivity resulted, confirming the findings of Schmidt and Robison (1972).

In addition to studies on rat cerebral cortical slices, studies were completed on the segmental development of NE responsiveness in regional spinal cord (Simmons and Jones, 1982). NE did not stimulate cyclic AMP accumulation prior to P5 in any spinal cord area. Thereafter, NE responsiveness reached a maximum in cervical and thoracic spinal cord at P10 and in lumbar cord at P15 (fig. 9). Whereas maximum NE-stimulated cyclic AMP accumulation in cervical and lumbar cord was significantly greater than that in cortex, the responses were equivalent in all regional CNS areas at P45 and P60 (D. J. Jones, Ph.D., unpublished observation, 1983). Consistent with the response noted in cerebral cortex, all areas decreased in responsiveness P20–P45.

In terms of CNS development, the guinea pig is much more mature at birth than is the rat (Davison and Dobbing, 1968; Jones, Dittmer, and Reading, 1974). If NE responsiveness of the cyclic AMP system is part of the development process of the CNS, then one would expect that in guinea pig brain NE responsiveness would be present at birth. Consonant with this relationship, Schmidt et al. (1970) demonstrated NE-stimulated cyclic AMP accumulation in slices from newborn guinea pig brain but not in slices from brains of newborn rats.

Schmidt and Robison (1971) also demonstrated marked regional differences in NE responsiveness in developing rabbit brain. In distinction to

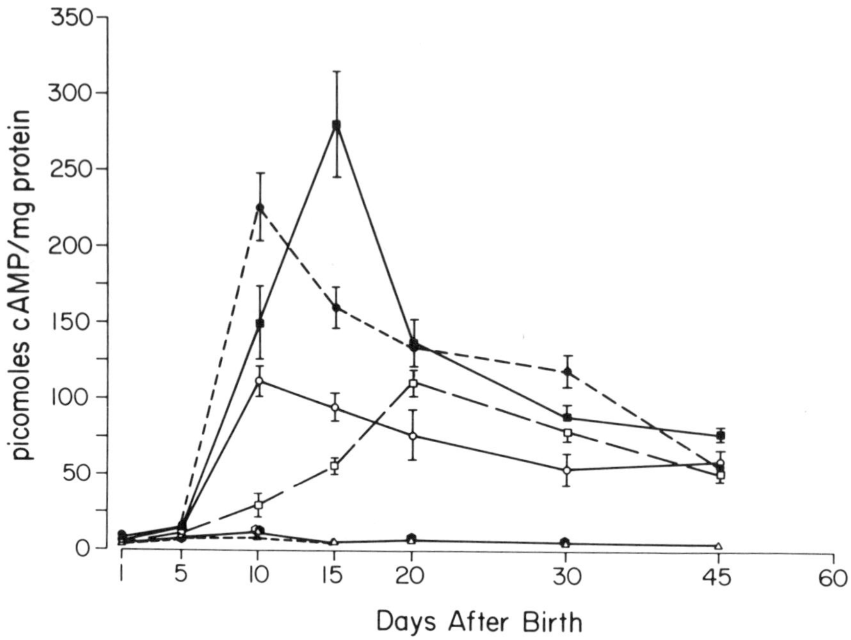

Figure 8. Norepinephrine-stimulated cyclic AMP accumulation as a function of age in tissue slices prepared from rat cerebral cortex (*open squares*), cervical spinal cord (*closed circles*), thoracic spinal cord (*open circles*), and lumbar spinal cord (*closed squares*). Incubation with norepinephrine (10^{-5} M) was for ten minutes at 37°C. (From Simmons and Jones, 1982; reprinted with permission of Pergamon Press, Ltd.)

findings in studies of the rat, NE-stimulated cyclic AMP accumulation was evident at birth in tissue slices from rabbit frontal cortex, hippocampus, cerebellum, and hypothalamus. NE responsiveness reached a maximum at P9 in frontal cortex, hippocampus, and hypothalamus and P14 in cerebellum. Although cyclic AMP accumulation increased 100–300 percent in frontal areas, only a 30 percent increase was noted in cerebellum. Decreased sensitivity followed the maximum response in each area. Except for in the cerebellum, NE-stimulated cyclic AMP accumulation was actually lower in the regions named above in adult animals than in newborn animals.

In most brain areas NE-stimulated cyclic AMP accumulation involves activation of both α-adrenergic and β-adrenergic receptors (Perkins and Moore, 1973a). Studies have investigated the development of the β-adrenergic component of this response by using the β-adrenergic agonist isoproterenol to stimulate cyclic AMP accumulation. Isoproterenol, similar to NE, did not stimulate cyclic AMP accumulation in rat cerebral cortex slices until P10 (Perkins and Moore, 1973b). Maximal responsiveness was evident at P18; thereafter, a slight decline resulted. Harden, Wolfe, Sporn, Poulos, and Molinoff (1977) also showed initial isoproterenol stimulation P10–P12 with maximal responsiveness at P16. No decrease in responsiveness occurred at later time periods. Comparative studies from this same laboratory demonstrated that isoproterenol first stimulated cyclic AMP accumulation on P7 with maximum stimulation at P14, followed by a slight decrease P17–P18 (Harden, Wolfe, Sporn, Perkins, and Molinoff 1977). Thus, the postnatal development of

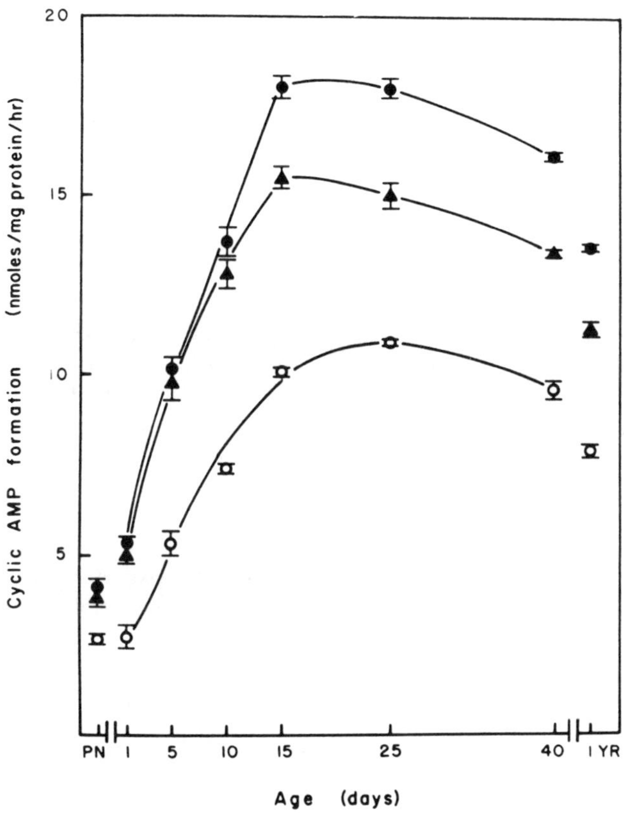

Figure 9. Catecholamine-sensitive adenyl cyclase activity in particulate fraction from rat cerebral cortex as a function of age. *Open circles,* basal activity; *closed circles,* activity in the presence of 5 × 10^{-5} M norepinephrine; *closed triangles,* activity in the presence of 5 × 10^{-5} M dopamine. (Reprinted with permission from Von Hungen, Roberts, and Hill, 1974.)

the β-adrenergic component of the NE response does not appear to be separate and contributes to the overall sensitivity to NE.

An interesting study by Walton, Miller, and Baldessarini (1979) points out the difficulties in studying an intact cell preparation from developing brain in which variable patterns of postnatal development of adenyl cyclase and phosphodiesterase are present. The above studies by Perkins and Moore (1973b) on developing rat cerebral cortex demonstrated that cyclic AMP accumulation was not stimulated by isoproterenol at birth. However, in the presence of the phosphodiesterase inhibitor isobutylmethylxanthine (IBMX), a significant increase in cyclic AMP accumulation is evident with isoproterenol in cortical slices taken at birth (Walton, Miller, and Baldessarini, 1979). The isoproterenol-stimulated increase in the presence of IBMX is constant until P8–P10 when a marked further increase occurs. At this time the ability of isoproterenol to stimulate cyclic AMP accumulation occurs in the absence of IBMX. The implication from these studies is that from birth until P8–P10 the balance of activity between adenyl cyclase and phosphodiesterase favors phosphodiesterase. Thereafter, adenyl cyclase is fully active. Thus the assay conditions can compromise the reliability of the data that predicts for the in vivo state.

The ontogenic development of the responsiveness of the brain cyclic AMP system to adenosine has also been studied (Perkins and Moore, 1973b). Adenosine responsiveness was absent at birth, but a significant increase in cyclic AMP accumulation occurred at P8 and a maximum response occurred at P16. Similar to that from NE responsiveness, adenosine-stimulated cyclic AMP accumulation was significantly less at P45 than it was at P16. Interestingly, the combined incubation of NE plus adenosine produced an increase in cyclic AMP accumulation at times when neither agent alone produced an increase. Moreover, a greater than additive response was evident with NE plus adenosine starting at P13. Thus once again, under varying conditions, quite different quantitative and qualitative patterns of development of the brain cyclic AMP system are produced depending upon pharmacological intervention.

Histamine-stimulated cyclic AMP accumulation is also dependent upon the developmental age of the animal. Palmer, Schmidt, and Robison (1972) demonstrated that during development of rabbit cerebral cortex, the first detectable increase of cyclic AMP in tissue slices in response to histamine occurred at fetal day 25. A twofold to threefold increase in cyclic AMP accumulation occurred in the presence of histamine at this time. A maximal (tenfold) response was observed P4–P8 and by P20 the response gradually decreased to adult levels.

Daszuta, Pons, and Cadhilac (1979) established that cyclic AMP accumulation in rat hypothalamic slices is stimulated by serotonin as early as fetal day 21. A twofold increase in accumulation was noted at this time followed by a fivefold increase on P7. The cyclic AMP response remained at this level until adult age. Since basal cyclic AMP accumulation also increased during the period P15 through P90, the percentage stimulation owed to serotonin actually decreased.

The effects of amino acids on cyclic AMP accumulation have also been studied in tissue slices from cerebral cortices from developing rats. Burgal, Jorda, and Grisiola (1982) demonstrated that N-acetyl aspartate stimulated cyclic AMP accumulation P5 through P180. Maximum effects were noted on P5 with a slight decrease thereafter. Additional studies demonstrated that aspartate stimulated cyclic AMP accumulation in slices from P10 to P180 and glutamate on P5 and P10. The glutamate response was absent in cortical tissue slices from animals 20 days or greater in age. It should be noted that the stimulatory response was extremely variable, depending on the duration of incubation with the particular amino acid.

The ontogenesis of responsiveness of the cyclic GMP system to various agonists has not been extensively studied. This is probably because of the scarcity of agents that have been shown to reliably stimulate cyclic GMP accumulation in a concentration-dependent manner. Garthwaite and Balazs (1978) have demonstrated a remarkable developmental pattern of glutamate responsiveness of the cyclic GMP system in developing rat cerebellum. Glutamate stimulated cyclic GMP accumulation as early as P2 (fivefold to tenfold increase), progressing to a maximum at P8 (130-fold). Following P8, the responsiveness declined until P70–P90 when glutamate no longer stimulated cyclic GMP accumulation. It was suggested from these studies that much of the glutamate-induced increase originates from cells formed prenatally, from cells such as Purkinje cells, Golgi cells, or other similar cells formed prior to P8, or from both.

Burgal, Jorda, and Grisiola (1982) recently demonstrated with rat cerebral cortex slices that glutamate did not stimulate cyclic GMP accumulation P5 to P20. However, in animals 40 days old and older glutamate increased cyclic GMP accumulation. In contrast, the closely related amino acids *N*-acetyl aspartate and aspartate did stimulate cyclic GMP accumulation in cerebral cortical slices from animals 40 days old and five to ten days old, respectively. Such variability severely limits one's ability to link the development of receptor systems to intracellular events that respond to receptor activation.

Adenyl Cyclase and Guanyl Cyclase

Studies with broken cell homogenates or subcellular fractions of these homogenates, in comparison to those of tissue slices, generally demonstrate much higher basal cyclic nucleotide formation. Neurohumoral stimulation of enzyme activity usually is less than a 100 percent increase with the requirement of various added cofactors or limited ion conditions in order to demonstrate the response. In spite of these requirements, dramatic differences are noted in the development of agonist responsiveness following birth.

The original work of Schmidt et al. (1970) demonstrated that although sensitivity of the cyclic AMP system to NE increased until P10 in brain cortex slices, NE had no significant stimulatory effect on adenyl cyclase measured in broken cell preparations. This is in contrast to the ability of NaF to stimulate adenyl cyclase activity P10–P20 (Schmidt et al., 1970). Kohrman (1973) also demonstrated that though NE did not stimulate adenyl cyclase in cerebral cortex homogenates from animals of any age, NaF stimulated adenyl cyclase activity as early as fetal days 20–21, maximal stimulation occurring on P21. Thus it appears that coupling of the receptor–adenyl cyclase complex is extremely sensitive to disruption in younger animals. The fact that NaF, which activates the enzyme independent of receptor occupation, stimulates enzymatic activity verifies this fact.

Von Hungen, Roberts, and Hill (1974) also demonstrated that NaF stimulated adenyl cyclase activity in particulate fractions from rat cerebral cortex. In contrast to the data of Schmidt et al. (1970), their findings showed that NaF activation was paralleled by both NE and dopamine stimulation of particulate adenyl cyclase (fig. 9). The percentage stimulation in the presence of either NE or dopamine increased from birth until P15, but thereafter the responsiveness remained stable to one year.

Adenyl cyclase in the particulate fraction from cortex and subcortex was stimulated at P1 by NE, dopamine, and serotonin (fig. 10) (Von Hungen, Roberts, and Hill, 1974). Quite different patterns of responsiveness are evident. The NE responsiveness was maximal at P6 in both areas with either a slight decline (cortex) or no change (subcortex) thereafter. Dopamine sensitivity in cortex was maximal at birth in cortex, declining subsequently on P14 and P42. The most dramatic change occurred in sensitivity of adenyl cyclase to serotonin, whose maximal stimulation is present at birth in both cortex and subcortex. At P14 the percentage stimulation is decreased, with a further decrease at P42.

Enjalbert et al. (1978) established similar decreases in percentage stimulation of both serotonin- and dopamine-sensitive adenyl cyclases in postnatal rat colliculus and striatum, respectively. In agreement, Von Hungen, Roberts, and Hill (1975) showed that in the presence of serotonin, the stimulation of rat colliculi adenyl cyclase decreased from 50 percent at birth to less than 10 percent at P42.

Significant regional differences are evident when comparing the responsiveness of adenyl cyclase to either NE, dopamine, or serotonin in the developing rat brain (Von Hungen, Roberts, and Hill, 1974). At P6, NE and dopamine stimulated adenyl cyclase approximately 130 percent in the corpus striatum, whereas serotonin stimulation was only 18 percent. In hypothalamus, serotonin stimulation was approximately double

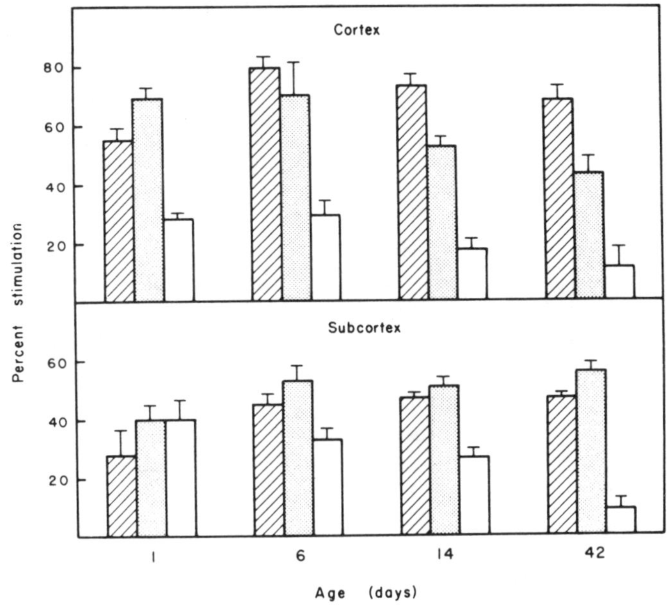

Figure 10. Developmental variations in the agonist responsiveness of particulate adenyl cyclase from rat cerebral cortex or subcortex as a function of age. Stimulation caused by 5×10^{-5} M norepinephrine (*striped bars*), 5×10^{-5} M dopamine (*stippled bars*), and 10^{-5} M serotonin (*unfilled bars*). (Reprinted with permission from Von Hungen, Roberts, and Hill, 1974.)

the stimulatory effect of NE and dopamine. Stimulation of adenyl cyclase in the hippocampus was equivalent for all three amine agonists. In cerebellum and hindbrain-medullary areas, the percentage stimulation in the presence of dopamine and serotonin decreased somewhat with age while responses to NE increased or decreased depending upon the area and postnatal age. Similar variations in regional brain serotonin- and dopamine-sensitive adenyl cyclase responsiveness have been reported in studies of rat cortex from newborns (Enjalbert et al., 1978).

Enjalbert et al. (1978) also compared agonist responsiveness of adenyl cyclase from regional brain areas of developing guinea pig. Amounts of cyclic AMP formed in response to serotonin in colliculi, hypothalamus, and lumbar spinal cord were the same in adults as they were in neonates. Since the basal adenyl cyclase activity remained constant during postnatal development in all areas, the stimulatory effect of serotonin expressed as a percentage of basal activity was identical for newborn and adult guinea pigs. This same relationship for dopamine stimulation of adenyl cyclase was present in all three areas. The above data are consistent with the *in utero* development of guinea pig CNS since the stimulation of adenyl cyclase did not decrease following birth (as with rat CNS). In agreement, serotonin stimulation of adenyl cyclase was threefold higher in fetal hypothalamus than it was in newborn or adult hypothalamus. Thus, downward regulation of this system is consistent for other differences in rat and guinea pig CNS, depending upon the developmental stage.

In consideration of receptor systems linked to cyclic AMP accumulation, it would be expected that functional maturity of these postsynaptic receptors precedes agonist-mediated stimulation of cyclic AMP formation. The data of Harden, Wolfe, Sporn, Perkins, and Molinoff (1977) confirm this proposition in that the development of cerebral cortical β-receptors correlates with data of isoproterenol-stimulated cyclic AMP accumulation. Both β-receptor binding and isoproterenol responsiveness of the cyclic AMP system are minimal at P8 and reach a maximum at P16. Whether or not this requirement must be satisfied for α, histamine, serotonin, and other receptor systems linked to cyclic AMP formation remains to be determined.

Implicit in the above discussion concerning the ontogenesis of cyclic AMP (and possibly cyclic GMP) systems in brain is their relationship to synaptic transmission. The question then is whether all the factors needed for functional activity of the synapse are required during development for neural communication. In consideration of studies that demonstrated spontaneous activity of NE neurons prior to the development of presynaptic input (Coyle, 1977), it follows that all components are not required. However, agonist-stimulated cyclic AMP accumulation does appear to require maturation of receptor systems. For instance, the development of maximal density of α- (Dausse, Bui, and Meyer, 1982) and β- (Harden, Wolfe, Sporn, Perkins, and Molinoff, 1977) receptor systems in rat brain is required for maximal responsiveness of the NE-stimulated cyclic AMP system. Thus, while all components of the cyclic AMP sytem may be present at earlier developmental stages, expression of its possible physiological role in neural communication must await receptor maturation.

Summary

The preceding review of the literature concerning the ontogenesis of cyclic nucleotide systems in the CNS reports data that are frequently inconsistent. However, there are a number of generalizations that can be made concerning this review. The first point is that although adenyl cyclase, phosphodiesterase, and cyclic AMP–dependent protein kinase display different quantitative patterns of postnatal development, all are present in brain at birth and thus capable of participation in the second messenger process. Second, as the requirement of neural communication becomes more prominent and the rate of synaptogenesis accelerates, postsynaptic receptor density increases to accommodate these demands. The cyclic AMP system subsequently becomes responsive to the extracellular environment, and enhancement of intracellular signaling is possible. Although we do not know the role or roles of cyclic AMP in the CNS, certainly the logic of this developmental scheme supports a role for cellular regulation dependent on neurohumoral receptor activation. The possible role of the cyclic GMP system in the process of neural communication is even less obvious, and therefore the relevance of ontogenesis to the overall pattern of CNS development remains in question.

Acknowledgments

I thank Kathy Angell and Mary Batterbarry for their assistance with the typing of this chapter. This work was supported in part by grants from the Morrison Trust Foundation and the National Institutes of Health (NINCDS 14546).

References

Adenolfi, A. M., and Schmidt, S. Y. (1974). Cytochemical localization of cyclic nucleotide phosphodiesterase activity at developing synapses. *Brain Res* 76, 21–31.

Appleman, M. M., Thompson, W. J., and Russell, T. R. (1973). Cyclic nucleotide phosphodiesterases. *Adv Cyclic Nucleotide Res* 3, 66–98.

Bandle, E., and Guidotti, A. (1979). Ontogenic studies of cGMP-dependent protein kinase in rat cerebellum. *J Neurochem* 32, 1343–1347.

Bartfai, T. (1980). Cyclic nucleotides in the central nervous system. *Curr Top Cell Regul* 16, 225–269.

Bartfai, T., Nordstrom, O., and Tjornhammar, M. L. (1980). Cyclic guanosine 3′,5′-monophosphate in the nervous system: Pre, post and transynaptic effects. *Prog Pharmacol* 4, 151–158.

Bloom, F. E. (1975). The role of cyclic nucleotides in central synaptic function. *Rev Physiol Biochem Pharmacol* 74, 1–103.

Bloom, F. E., Hoffer, B. J., Battenberg, E. F., Siggins, G. R., Steiner, A. L., Parker, C. W., and Wedner, H. J. (1972). Adenosine 3′5′-monophosphate is localized in cerebellar neurons: Immunofluorescence evidence. *Science* 177, 436–438.

Boehme, D. H., Kosecki, R., and Marks, N. (1978). Protein phosphorylation in human synaptosomal membranes: Evidence for the presence of substrates for cyclic nucleotide guanosine 3′:5′-monophosphate dependent protein kinase. *Brain Res Bull* 3, 697–700.

Brostrom, C. O., Huang, Y. C., Breckenridge, B. M., and Wolff, D. O.

(1975). Identification of a calcium-binding protein as a calcium-dependent regulator of brain adenylate cyclase. *Proc Natl Acad Sci USA* 72, 64–68.

Burgal, M., Jorda, A., and Grisiola, S. (1982). Effects of *N*-acetyl aspartate, aspartate and glutamate on cAMP and cGMP levels in developing rat cerebral cortex. *J Neurochem* 38, 1498–1500.

Butcher, R. W., and Sutherland, E. W. (1962). Adenosine 3′,5′-phosphate in biological materials: I. Purification and properties of cyclic 3′,5′-nucleotide phosphodiesterase and use of the enzyme to characterize adenosine 3′,5′-phosphate in human urine. *J Biol Chem* 237, 1244–1250.

Casnellie, J. E., and Greengard, P. (1974). Guanosine 3′,5′-cyclic monophosphate–dependent phosphorylation of endogenous substrate proteins in membranes from mammalian smooth muscle. *Proc Natl Acad Sci USA* 71, 1891–1895.

Cheung, W. (1970). Cyclic nucleotide phosphodiesterase—Demonstration of an activator. *Biochem Biophys Res Commun* 38, 533–538.

Cheung, W. Y., Bradham, L. S., Lynch, T. J., Lin, Y. M., and Tallant, E. A. (1975). Protein activator of cyclic 3′,5′-nucleotide phosphodiesterase of bovine or rat brain also activates its adenylate cyclase. *Biochem Biophys Res Commun* 66, 1055–1062.

Cheung, W. Y., Lynch, T. J., and Wallace, R. W. (1978). An endogenous Ca^{++}-dependent activator protein of brain adenylate cyclase and cyclic nucleotide phosphodiesterase. *Adv Cyclic Nucleotide Res* 9, 233–252.

Cheung, W. Y., and Salganicoff, L. (1967). Cyclic 3′,5′ nucleotide phosphodiesterase: Localization and latent activity in rat brain. *Nature* 214, 90–91.

Corbin, J. O., and Lincoln, T. M. (1978). Comparison of cAMP and cGMP dependent protein kinases. *Adv Cyclic Nucleotide Res* 9, 159–170.

Coyle, J. T. (1977). Biochemical aspects of neurotransmission in the developing brain. *Int Rev Neurobiol* 20, 65–103.

Coyle, J. T., and Campochiaro, P. (1976). Ontogenesis of dopaminergic-cholinergic interactions in the rat striatum: A neurochemical study. *J Neurochem* 27, 673–678.

Daly, J. (1977). *Cyclic nucleotides in the nervous system*. New York: Plenum Press.

Daszuta, A., Pons, F., and Cadhilac, J. (1979). Effect of serotonin on cyclic AMP level in rat hypothalamus slices during development. *Eur J Pharmacol* 56, 397–401.

Dausse, J. P., Bui, K. H., and Meyer, P. (1982). Effect of neonatal 6-hydroxydopamine treatment on $alpha_1$- and $alpha_2$-adrenoreceptors in rat cerebral cortex. *J Cardiovasc Pharmacol* 4, 586–590.

Davis, C. W., and Kuo, J. F. (1976). Ontogenic changes in levels of phosphodiesterase for adenosine 3′:5′monophosphate in the lung, liver, brain and heart from guinea pigs. *Biochim Biophys Acta* 444, 554–562.

Davison, A. N., and Dobbing, J. (1968). The developing brain. In *Applied neurochemistry*, ed. A. N. Davison and J. Dobbing, pp. 225–286. Philadelphia: F. A. Davis.

DeRobertis, E., DeLorez-Arnaiz, G. R., Butcher, R. W., and Sutherland, E. W. (1967). Subcellular distribution of adenyl cyclase and cyclic phosphodiesterase in rat brain cortex. *J Biol Chem* 242, 3487–3493.

Enjalbert, A., Bourgoin, S., Hamon, M., Adrien, J., and Bockaert, J. (1978). Postsynaptic serotonin-sensitive adenylate cyclase in the central nervous system: I. Development and distribution of serotonin and dopamine-sensitive adenylate cyclases in rat and guinea pig brain. *Mol Pharmacol* 14, 2–10.

Florendo, N. T., Barrnett, R. J., and Greengard, P. (1971). Cyclic 3′,5′-nucleotide phosphodiesterase: Cytochemical localization in cerebral cortex. *Science* 173, 745–747.

Gaballah, S., and Popoff, C. (1971a). Cyclic 3′,5′-nucleotide phosphodiesterase in nerve endings of developing rat brain. *Brain Res* 25, 220–222.

Gaballah, S., and Popoff, C. (1971b). Localization of adenosine 3′,5′-monophosphate–dependent protein kinase in brain. *J Neurochem* 18, 1795–1797.

Gaballah, S., Popoff, C., and Sooknandan, G. (1971). Changes in cyclic 3′,5′-adenosine monophosphate–dependent protein kinase levels in brain development. *Brain Res* 31, 229–232.

Garthwaite, J., and Balazs, R. (1978). Supersensitivity of the cyclic GMP response to glutamate during cerebellar maturation. *Nature* 275, 328–329.

Gnegy, M. E., Nathanson, J. A., and Uzunov, P. (1977). Release of the phosphodiesterase activator by cyclic AMP dependent ATP: Protein phosphotransferase from subcellular fractions of rat brain. *Biochim Biophys Acta* 497, 75–89.

Goldberg, N. D., O'dea, R. F., and Haddox, M. K. (1973). Cyclic GMP. *Adv Cyclic Nucleotide Res* 3, 155–223.

Goridis, C., and Morgan, I. G. (1973). Guanyl cyclase in rat brain subcellular fractions. *FEBS Lett* 34, 71–73.

Greengard, P. (1976). Possible role for cyclic nucleotides and phosphorylated membrane proteins in the postsynaptic actions of neurotransmitters. *Nature* 260, 101–108.

Greengard, P. (1978a). *Cyclic nucleotides, phosphorylated proteins and neuronal function*. Vol. 1 of *Society of General Physiologists Distinguished Lecture Series*. New York: Raven Press.

Greengard, P. (1978b). Phosphorylated proteins as physiologic effectors. *Science* 199, 146–152.

Greengard, P., and Kuo, J. F. (1970). On the mechanism of action of cyclic AMP. *Adv Biochem Psychopharmacol* 3, 287–306.

Harden, T. K., Wolfe, R. B., Sporn, J. R., Perkins, J. P., and Molinoff, P. B. (1977). Ontogeny of beta-adrenergic receptors in rat cerebral cortex. *Brain Res* 125, 99–108.

Harden, T. K., Wolfe, B. B., Sporn, J. R., Poulos, B. K., and Molinoff, P. B. (1977). The effects of 6-hydroxydopamine on the development of the beta adrenergic receptor/adenylate cyclase system in rat cerebral cortex. *J Pharmacol Exp Ther* 203, 132–143.

Hardman, J. G., and Sutherland, E. W. (1969). Guanyl cyclase, an enzyme catalyzing the formation of guanosine 3′,5′-monophosphate from guanosine triphosphate. *J Biol Chem* 244, 6363–6370.

Hofmann, F., and Sold, G. (1972). A protein kinase activity from rat cerebellum stimulated by guanosine 3′,5′-monophosphate. *Biochem Biophys Res Commun* 49, 1100–1107.

Johnson, E. M., Maeno, H., and Greengard, P. (1971). Phosphorylation of endogenous protein of rat brain by a cyclic adenosine 3′,5′-monophosphate dependent protein kinase. *J Biol Chem* 246, 7731–7739.

Johnson, G. A., Boukma, S. J., Lanti, R. A., and Mathews, J. (1973). Cyclic AMP and phosphodiesterase in synaptic vesicles from mouse brain. *J Neurochem* 20, 1387–1392.

Jones, D. G., Dittmer, M. M., and Reading, L. C. (1974). Synaptogenesis in guinea pig cerebral cortex: A glutaraldehyde-PTA study. *Brain Res* 70, 245–259.

Jones, D. J. (1974). Maximal field strength microwave irradiation as a tool for the study of brain cyclic adenosine 3′,5′-monophosphate. Ph.D. diss., The University of Texas Graduate School of Biomedical Sciences, San Antonio.

Jones, D. J., Medina, M., Ross, D. H., and Stavinoha, W. B. (1974). Rate of inactivation of adenyl cyclase and phosphodiesterase: Determinants of brain cyclic AMP. *Life Sci* 14, 1577–1585.

Jones, D. J., and Stavinoha, W. B. (1976). Regional brain cyclic nucleotide fluxes during tissue fixation in-situ. *Trans Am Soc Neurochem* 7, 116.

Jones, D. J., and Stavinoha, W. B. (1977). Levels of cyclic nucleotides in mouse regional brain following 300 msec microwave irradiation. *J Neurochem* 28, 799–763.

Jongkind, J. F., and Bruntink, R. (1970). Forebrain freezing rates and substrate levels in decapitated rat heads. *J Neurochem* 17, 1615–1617.

Kakiuchi, S., Yamazaki, R., Teshima, Y., and Uenishi, K. (1973). Regulation of nucleotide cyclic 3′,5′-monophosphate phosphodiesterase activity from rat brain by a modulator and Ca^{++}. *Proc Natl Acad Sci USA* 70, 3526–3530.

Kakiuchi, S. R., Yamazaki, R., Teshima, Y., Uenishi, K., and Miyamoto, E. (1975). Multiple cyclic nucleotide phosphodiesterase activities from rat tissues and occurrence of calcium plus magnesium ion-dependent phosphodiesterase and its protein activator. *Biochem J* 146, 109–120.

Kang, E. S. (1977). Cyclic nucleotide phosphodiesterase activities of the fetal and mature human cerebral cortex. *Pediatr Res* 11, 655–663.

Kohrman, A. F. (1973). Patterns of development of adenyl cyclase activity and norepinephrine responsiveness in the rat. *Pediatr Res* 7, 575–581.

Krebs, E. G., and Beavo, J. A. (1979). Phosphorylation-dephosphorylation of enzymes. *Annu Rev Biochem* 48, 923–959.

Kuo, J. F. (1975). Reciprocal changes in levels of guanosine 3′,5′-monophosphate–dependent and adenosine 3′,5′-monophosphate–dependent protein kinases in developing guinea pig lungs. *J Cyclic Nucleotide Res* 1, 151–157.

Kuo, J. F., and Greengard, P. (1970). Cyclic nucleotide dependent protein kinases: VI. Isolation and partial purification of a protein kinase activated by guanosine 3′,5′-monophosphate. *J Biol Chem* 245, 2493–2498.

Kuo, J. F., Kuo, W. N., and Shoji, M. (1976). Regulation by cyclic GMP and stimulatory protein kinase modulator of cyclic GMP–dependent protein kinase from brain and other tissues. *Adv Biochem Psychopharmacol* 15, 379–398.

Lefkowitz, R. J., Mullikin, O., and Caron, M. G. (1976). Regulation of beta adrenergic receptors by guanyl-5′-imidodiphosphate and other purine nucleotides. *J Biol Chem* 251, 4686–4692.

Limbird, L. (1981). Activation and attenuation of adenylate cyclase. *Biochem J* 195, 1–13.

Lohmann, S. M., Ueda, T., and Greengard, P. (1978). Ontogeny of synaptic phosphoproteins in brain. *Proc Natl Acad Sci USA* 75, 4037–4041.

Lohmann, S. M., Walter, U., and Greengard, P. (1978). Protein kinase in developing rat brain. *J Cyclic Nucleotide Res* 4, 445–452.

McIlwain, H., and Bachelard, H. S., eds. (1971). *Biochemistry and the central nervous system*, pp. 61–97. London: Churchill Livingstone.

Maeno, H., and Greengard, P. (1972). Phosphoprotein phosphatases from rat cerebral cortex. *J Biol Chem* 247, 3269–3277.

Maeno, H., Ueda, T., and Greengard, P. (1975). Adenosine 3′,5′-monophosphate dependent protein phosphatase activity in synaptic membrane fractions. *J Cyclic Nucleotide Res* 1, 37–48.

Maguire, M. E., Van Arsdale, P. M., and Gilman, A. G. (1976). An agonist specific effect of guanine nucleotides on binding to the betas adrenergic receptor. *Mol Pharmacol* 12, 335–341.

Murad, F., Lewicki, J. A., Brandwein, H. J., Mittal, C. K., and Waldman, S. (1981). Guanylate cyclase: Purification, properties, free radical activation, radiolabeling, and preparation of hybridoma antibodies. *Adv Cyclic Nucleotide Res* 14, 229–239.

Nakazawa, K., and Sano, M. (1975). Studies on guanylate cyclase: A new assay method for guanylate cyclase and properties of the cyclase from rat brain. *J Biol Chem* 249, 4207–4211.

Nakazawa, K., Sano, M., and Saito, T. (1976). Subcellular distribution and properties of guanylate cyclase in rat cerebellum. *Biochim Biophys Acta* 444, 563–570.

Nathanson, J. A. (1977). Cyclic nucleotides and nervous system function. *Physiol Rev* 57, 157–256.

Orenberg, E. K., Renson, J., Elliott, G. R., Barchas, J. D., and Kessler, R. (1975). Genetic determination of aggressive behavior and brain cyclic AMP. *Research Communications in Psychopharmacology* 1, 99–107.

Palmer, G. C., Schmidt, M. J., and Robison, G. A. (1972). Development and characteristics of the histamine-induced accumulation of cyclic AMP in the rabbit cerebral cortex. *J Neurochem* 19, 2251–2256.

Perkins, J. P., and Moore, M. M. (1973a). Characterization of the adrenergic receptors mediating a rise in cyclic 3′,5′-adenosine monophosphate in rat cerebral cortex. *J Pharmacol Exp Ther* 185, 371–378.

Perkins, J. P., and Moore, M. M. (1973b). Regulation of adenosine cyclic 3′,5′-monophosphate content of rat cerebral cortex: Ontogenic development of responsiveness to catecholamines and adenosine. *Mol Pharmacol* 9, 774–782.

Rodbell, M., Krans, M. J., Pohl, J. S., and Birnbaumer, L. (1971a). The glucagon-sensitive adenyl cyclase system in plasma membranes of rat liver. IV. Effects of guanylnucleotides on binding of ^{125}I-glucagon. *J Biol Chem* 246, 1872–1876.

Rodbell, M., Birnbaumer, L., Pohl, S. L., and Krans, M. J. (1971b). The glucagon-sensitive adenyl cyclase system in plasma membranes of rat liver: V. An obligatory role of guanylnucleotides in glucagon action. *J Biol Chem* 246, 1877–1882.

Rubin, C. S., and Rosen, O. M. (1975). Protein phosphorylation. *Annu Rev Biochem* 44, 831–887.

Schlichter, D. J., Casnellie, J. E., and Greengard, P. (1978). An endogenous substrate for cGMP-dependent protein kinase in mammalian cerebellum. *Nature* 273, 61–62.

Schmidt, M. J., and Ghetti, B. (1977). Studies on the cyclic AMP system in human brain during aging. *Society Neuroscience Abstracts* 3, 362A.

Schmidt, M. J., Hopkins, J. T., Schmidt, D. E., and Robison, G. A. (1972). Cyclic AMP in brain areas: Effects of amphetamine and norepinephrine assessed through the use of microwave radiation as a means of tissue fixation. *Brain Res* 42, 465–477.

Schmidt, M. J., Palmer, E. C., Dettbarn, W. D., and Robison, G. A. (1970). Cyclic AMP and adenyl cyclase in the developing rat brain. *Dev Psychobiol* 3, 53–67.

Schmidt, M. J., Palmer, G. C., and Robison, G. (1980). Cyclic nucleotide system in the brain during development and aging. In *Psychophar-*

macology of aging, ed. C. Eisdorfer and W. E. Fann, pp. 213–265. New York: Spectrum.

Schmidt, M. J., and Robison, G. A. (1971). The effect of norepinephrine on cyclic AMP levels in discrete regions of the developing rabbit brain. *Life Sci* 10, 459–464.

Schmidt, M. J., and Robison, G. A. (1972). The effect of neonatal thyroidectomy on the development of the adenosine 3′,5′-monophosphate system in the rat brain. *J Neurochem* 19, 937–947.

Schmidt, M. J., Schmidt, D. E., and Robison, G. A. (1971). Cyclic adenosine monophosphate in brain areas: Microwave irradiation as a means of tissue fixation. *Science* 173, 1142–1143.

Schmidt, M. J., and Sokoloff, L. (1973). Activity of cyclic AMP–dependent microsomal protein kinase and phosphorylation of ribosomal protein in rat brain during postnatal development. *J Neurochem* 21, 1193–1205.

Shambaugh, G. E., Hunzicker-Dunn, M., DeAngelo, A. G., and Jungman, R. A. (1978). Subcellular alterations in cyclic AMP-binding and protein kinase activities during ontogeny in the rat cerebellum. *Biochim Biophys Acta* 539, 294–304.

Simmons, K., and Jones, D. J. (1982). Postnatal development of the norepinephrine stimulated cyclic AMP response in rat spinal cord. *Proc West Pharmacol Soc* 25, 93–96.

Smoake, J. A., Song, S. Y., and Cheung, W. Y. (1974). Cyclic 3′,5′-nucleotide phosphodiesterase: Distribution and development changes of the enzyme and its protein activator in mammalian tissues and cells. *Biochim Biophys Acta* 341, 402–411.

Sold, G., and Hofmann, F. (1974). Evidence for a guanosine-3′,5′-monophosphate binding protein from rat cerebellum. *Eur J Biochem* 44, 143–149.

Spano, P. F., Kumakura, K., Govoni, S., and Trabucchi, M. (1975). Postnatal development and regulation of cerebellar cyclic guanosine monophosphate system. *Pharmacol Res Commun* 7, 223–237.

Spinka, K., Farber, D. B., and Lolley, R. N. (1980). Cyclic nucleotide content and granule cell degeneration in the developing cerebellum of staggerer mice. *J Neurochem* 34, 1531–1534.

Steiner, A. L., Ferrendelli, J. A., and Kipnis, D. M. (1972). Radioimmunoassay for cyclic nucleotides: III. Effect of ischemia, changes during development and regional distribution of adenosine 3′,5′-monophosphate and guanosine 3′,5′-monophosphate in mouse brain. *J Biol Chem* 247, 1121–1124.

Strada, S. J., Uzunov, P., and Weiss, B. (1974). Ontogenic development of a phosphodiesterase activator and the multiple forms of cyclic AMP phosphodiesterase of rat brain. *J Neurochem* 23, 1097–1103.

Sutherland, E. W., and Rall, T. (1960). The relation of adenosine 3′,5′-

phosphate and phosphorylation to the action of catecholamine and other hormones. *Pharmacol Rev* 12, 265–299.

Sutherland, E. W., Rall, T. W., and Menon, T. (1962). Adenyl cyclase: I. Distribution, preparation and properties. *J Biol Chem* 237, 1220–1227.

Swabb, D. F. (1971). Pitfalls in the use of rapid freezing for stopping brain and spinal cord metabolism in rat and mouse. *J Neurochem* 18, 2085–2092.

Takahashi, T., Mutsuzaki, S., and Nunez, J. (1974). Modifications in soluble protein kinase and cyclic AMP binding capacity of developing rat brain. *J Neurochem* 24, 303–309.

Thompson, W. J., and Appleman, M. M. (1971). Multiple cyclic nucleotide phosphodiesterase activities from rat brain. *Biochemistry* 10, 311–316.

Ueda, T., and Greengard, P. (1977). Adenosine 3′,5′-monophosphate regulated phosphoprotein system of neuronal membranes: I. Solubilization, purification and some properties of an endogenous phosphoprotein. *J Biol Chem* 252, 5155–5163.

Von Hungen, K., and Roberts, S. (1973). Catecholamine and Ca^{++} activation of adenylate cyclase systems in synaptosomal fractions from rat cerebral cortex. *Nature New Biol* 242, 58–60.

Von Hungen, K., and Roberts, S. (1974). Neurotransmitter-sensitive adenylate and cyclase systems in the brain. In *Reviews of neuroscience*. Vol. 1, ed. S. Ehrenpreis and I. J. Kopin, pp. 231–281. New York: Raven Press.

Von Hungen, K., Roberts, S., and Hill, D. F. (1974). Developmental and regional variations in neurotransmitter-sensitive adenylate cyclase systems in cell-free preparations from rat brain. *J Neurochem* 22, 811–819.

Von Hungen, K., Roberts, S., and Hill, D. F. (1975). Serotonin-sensitive adenylate cyclase activity in immature rat brain. *Brain Res* 84, 257–268.

Walsh, D. A. (1978). Role of the cAMP dependent protein kinase as the transducer of cAMP action. *Biochem Pharmacol* 27, 1801–1804.

Walton, K. G., Miller, E., and Baldessarini, R. J. (1979). Prenatal and early postnatal beta-adrenergic receptor-mediated increase of cyclic AMP in slices of rat brain. *Brain Res* 177, 515–522.

Weiss, B. (1971). Ontogenic development of adenyl cyclase and phosphodiesterase in rat brain. *J Neurochem* 18, 469–477.

Weiss, B. (1975). Differential activation and inhibition of multiple forms of cyclic nucleotide phosphodiesterase. *Adv Cyclic Nucleotide Res* 1, 435–453.

Weiss, B., and Costa, E. (1968). Regional and subcellular distribution of adenyl cyclase and phosphodiesterase in brain and pineal gland. *Biochem Pharmacol* 17, 2107–2116.

Weiss, B., Fertel, R., and Uzunov, P. (1974). Selective alteration of the activity of the multiple forms of adenosine 3′,5′-monophosphate phosphodiesterase of rat cerebrum. *Mol Pharmacol* 10, 615–625.

Weiss, B., and Strada, S. (1972). Neuroendocrine control of the cyclic AMP system of brain and pineal gland. *Adv Cyclic Nucleotide Res* 1, 357–374.

Weiss, B., and Strada, S. (1973). Adenosine 3′,5′-monophosphate during fetal and postnatal development. In *Fetal pharmacology*, ed. L. Boreus, pp. 205–232. New York: Raven Press.

Weller, M., Rodnight, R., and Carella, D. (1972). Determination of adenosine 3′,5′-cyclic monophosphate in cerebral tissue by saturation analysis. *Biochem J* 129, 113–121.

Wells, J. N., and Hardman, J. G. (1977). Cyclic nucleotide phosphodiesterases. *Adv Cyclic Nucleotide Res* 8, 119–143.

White, A. A., and Aurbach, G. D. (1969). Detection of guanyl cyclase in mammalian tissues. *Biochim Biophys Acta* 191, 686–697.

Brain Development, Genetics, and Neurochemistry as Bases for Neuropsychiatric Disorders

I. J. Butler

Implicit in an examination of an aspect of development such as the neurochemistry of the nervous system is the concept that a particular system or subsystem of cellular structure and function is evaluated for a prolonged period. It is important to differentiate such studies from those more usually performed at a specific time in which a number of processes are examined without emphasis being given to the preceding events leading up to that point. Developmental studies examine the growth and differentiation of the organism from conception to death, and processes occurring early in life may be important in understanding those degenerative events leading to death of the organism. This concept will be developed in this chapter by consideration of several disease states of the nervous system, many of which have a genetic component and for which current information indicates a disturbance of neurochemical functions.

Biochemical Genetics

In order to understand these mechanisms further it is necessary to briefly examine the relationship of genetic influences affecting neuronal development and possible environmental effects during early or late phases in the life of the organism. Disorders resulting from defects in expression of genetic material are considered as (Stanbury et al., 1983):

1. Obvious abnormalities in the number and structure of chromosomes (karyotype). For example in trisomy 21, or Down's syndrome, there is excessive genetic material that is obvious on chromosomal analysis. Recent neurochemical aspects of this syndrome will be discussed under Alzheimer's disease.
2. Disorders of single mutant genes, for which there is a great deal of information. Disorders occurring as a result of single mutant genes (Mendelian or monogenic disorders) are inherited as autosomal dominant, autosomal recessive, or X linked. The concept of a disorder being inherited in a "dominant" or "recessive" manner is sometimes

confusing and actually relates to the clinical expression of a genetic disorder.

3. Multifactorial genetic disorders in which there is a polygenic component.

Dominant Disorders

A disorder is said to be dominant when the mutation is clinically apparent in a patient with a single "dose" of the mutation and is thus in the heterozygous state. A recessively inherited disorder requires that a double dose of the mutation is necessary, and the patient is in the homozygous state. For example, in an autosomal dominant disorder of brain such as Huntington's chorea or Charcot-Marie-Tooth disease, the disease is manifest in the heterozygous state because there is one abnormal gene that is a mutant allele and the corresponding allele on the homologous chromosome is normal. Dominant disorders are inherited in a vertical manner and may also arise as a new mutation. Occasionally in isolated communities, asymptomatic carriers of the abnormal gene can marry and have offspring who are homozygous for the mutant gene and have a disorder much more severe than the heterozygous state of the parent, that will lead to early manifestation of the disorder and often a shortened life span. Examples of patients homozygous for Huntington's chorea and Charcot-Marie-Tooth disease (Killian and Kloepfer, 1979) have been described.

A prototypic disorder affecting the nervous system and occurring as a result of impaired genetic expression, abnormal neurochemistry, and environmental factors is the dominantly inherited hepatic porphyria of acute intermittent porphyria (AIP). Neuropsychiatric manifestations are initiated by exposure to various drugs, and the basis of the condition is approximately 50 percent of normal activity of porphobilinogen (PBG) deaminase in all tissues examined, including red blood cells, skin, cultured fibroblasts, mitogen-stimulated lymphocytes, amniotic cells, and liver (Stanbury et al., 1983). The gene locus encoding human PBG deaminase has been assigned to chromosome 11, and the decreased enzyme activity is similar in those patients latent for the disorder and others having overt symptoms.

Clinical expression is variable and is determined by other factors such as steroid hormones and certain of their metabolites, drugs, and nutrition, which appear to influence γ-aminolevulinic acid (ALA) synthase, the rate-limiting enzyme in heme synthesis from glycine and succinyl CoA. About 90 percent of subjects with PBG deaminase deficiency remain clinically latent, and the other 10 percent usually have onset after puberty. In this 10 percent, the disorder occurs more frequently in females, and neuropsychiatric manifestations may be precipitated by drug exposure. A motor-sensory neuropathy may occur; abdominal pain and autonomic nervous system manifestations (tachycardia, hypertension, urinary retention or incontinence) are common; and mental disturbances include hallucinations, disorientation, paranoia, and depression. Occasionally seizures occur, and inappropriate antidiuretic hormone secretion leads to hyponatremia.

The neurochemical basis for these manifestations is still not completely defined, but elevated levels of ALA and PBG have been demonstrated in cerebrospinal fluid (CSF) in patients during the acute attacks, and both precursors of heme can cause behavioral effects and convulsions when administered into the brain of an experimental animal. ALA is structurally similar to γ-aminobutyric acid (GABA) and could thus activate GABA receptors on brain and spinal cord neurons. ALA and PBG inhibit release of acetylcholine, and this action could contribute to the acute paresis observed during an attack. The increased circulation of catecholamines leading to tachycardia and hypertension could be owed to the inhibiting of reuptake of norepinephrine into neurons by ALA.

Environmental factors initiate the clinical manifestations by inducing the rate-limiting enzyme ALA synthase, thus increasing the precursor load for PBG deaminase, which has only 50 percent of normal activity. Barbiturates and most other anticonvulsants are among such agents, and thus clinical control of seizures in AIP patients may be difficult. Since heme production acts as a feedback control on ALA synthase activity, treatment with intravenously administered hematin has been used to decrease synthesis of ALA and PBG and hasten clinical improvement. Avoidance of other environmental precipitants can decrease future attacks and prevent the severe and permanent neuronal damage.

Recessive Disorders

Autosomal recessive disorders are much more common because about 7 percent of genes in individuals are mutant alleles. The disorder is usually only clinically apparent in the homozygous state for the mutant alleles. Many of these recessive disorders affect the brain and can be attributed to a specific enzyme defect. Phenylketonuria (PKU), in which phenylalanine levels are elevated in blood and the lysosomal storage disorders are examples of some recessive disorders that will be considered.

Of the many lysosomal storage disorders defined enzymatically in the last decade, many affect brain and peripheral nerve function, although the actual mechanism of neuronal dysfunction and destruction is usually unknown. Metachromatic leukodystrophy (MLD) is one such lysosomal storage disorder, is recessively inherited, and is due to a severe deficiency of the enzyme arysulfatase A (Stanbury et al., 1983). The gene locus responsible for expression of arysulfatase A has been mapped to chromosome 22. There is accumulation of galactosyl sulfatide in the white matter of central nervous tissue and peripheral nerves, and there are various clinical presentations, which usually relate to the age of onset of the manifestations. The disease may occur at any age but the late infantile form is usually recognized in the second year of life; the juvenile form appears during ages 5–7 years. In addition, there are rare congenital and adult forms. Ataxia, weakness (peripheral neuropathy), and spasticity with mental regression are seen in the childhood form, and progressive dementia with late spasticity are characteristic of the adult form. However, a colleague and I have recently observed two affected persons from the same family who presented with psychiatric disturbances and dementia at ages 5–7 years, and these signs preceded by several years overt neurological manifestations of ataxia and spasticity (I. J. Butler, M. B., and R. R. Howell, M. D., unpublished data, 1983). The degree of enzyme deficiency in patients is similar, and there is no obvious gradation of enzyme deficiency with age of clinical onset. There is no evidence to suggest regional neuronal enzyme deficits to explain the variable clinical manifestations observed in different families. The accumulation of sulfatide in Schwann cells of peripheral nerve and oligodendrocytes in the central nervous system leads to impaired maintenance of myelin and subsequent demyelination. Impaired catabolism of sulfatide from the innermost part of the myelin sheath may affect the normal acquisition of myelin and the increasing diameter of nerves during growth and development. The actual pathogenesis of MLD remains to be determined.

X-linked Disorders

X-linked disorders are due to abnormal genes or mutant alleles on the X chromosome, and the clinical expression will depend on the male having a single X chromosome and the female having an abnormal mutant gene on one X chromosome and a normal allele on the other. In this group of disorders the terms *X-linked dominant* and *X-linked recessive* refer to the clinical expression owed to a single mutant on the X chromosome. But in an X-linked recessive disorder the mutant allele is not clinically apparent in the female because of the effect of the normal allele on the other X chromosome. Characteristically, X-linked recessive disorders such as Duchenne type muscular dystrophy are clinically apparent only in the male, unless there is an inactivation of the X chromosome as a result of the Lyon effect. Should there be an imbalance in this inactivation process such that there is excessive inactivation of the normal X chromosome, it is possible for there to be expression in the female of the abnormal mutant. For example, in the X-linked recessive form of muscular dystrophy (Duchenne) the mother of an affected male may show clinical manifestations of weakness and muscular atrophy.

Another recently defined syndrome even more germane to this discussion is that owed to the fragile X chromosome and associated with men-

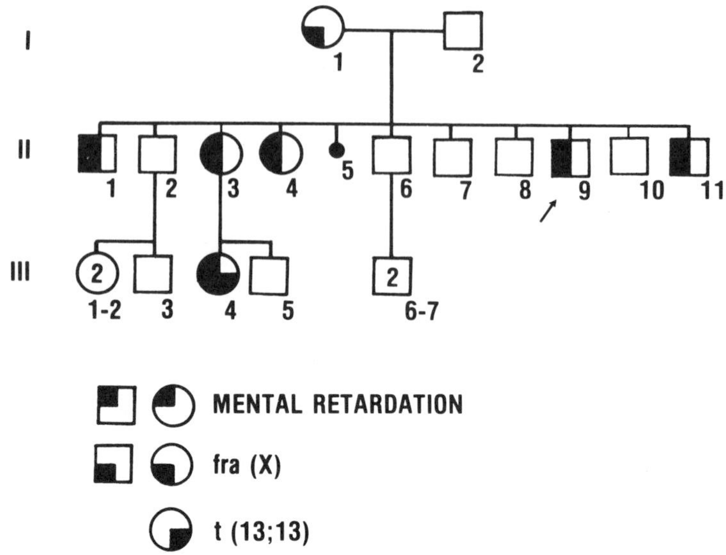

Figure 1. Pedigree of family with X-linked mental retardation in which males were retarded and females were retarded to a lesser degree. Affected males had large hands, ears, or both and macroorchidism. Fragile (*fra*) X chromosomes were identified by Q banding, and a female had an additional de novo translocation (*t*) between two number 13 chromosomes.

tal retardation and macroorchidism in the male. Although initially considered to be only manifest in the male, detailed family studies have revealed affected females less severely retarded than their brothers. A family demonstrating this phenomenon and defined by chromosomal analysis is outlined in an extended family tree (fig. 1). This family tree demonstrates the difficulty in defining the precise genetic nature of this brain disorder on clinical grounds alone without the use of a marker chromosome (Hecht et al., 1983). However, once a disorder is defined as X-linked recessive, there is considerable encouragement to look for a neurochemical basis of the retardation. Other X-linked disorders affecting the nervous system have been defined enzymatically and include a defect in α-galactosidase A in Fabry's disease (neuropathy, nephropathy, and accumulation of glycosphingolipids) and a deficiency of hypoxanthine-guanine phosphoribosyl transferase in Lesch-Nyhan syndrome, (choreoathetosis, spasticity, mental retardation, and hyperuricemia). X-linked dominant disorders are rare and often lethal in the hemizygous male, whereas the heterozygous female's condition is partially compensated by the normal allele on the other X chromosome. A disorder with X-linked dominant expression is that form of hyperammonemia owed to a deficiency of ornithine transcarbamylase. The hemizygous male infant usually dies early in coma and from convulsions and apnea owed to hyperammonemia, whereas the female heterozygote has a partial enzyme deficiency and can survive with appropriate treatment.

In order to develop more specifically this theme indicating a relationship between genetics, neurochemistry, and the development of the nervous system, I have selected several neurotransmitter systems that are defective in disorders of humans. In general, the genetic basis for these conditions has been definitely demonstrated or is strongly indicated; furthermore, there is now a growing tendency to consider that some of these disorders may originate much earlier in development than indicated by onset of the clinical manifestations.

Disorders Affecting Biogenic Amines

The biogenic amines, such as dopamine, norepinephrine, and serotonin, have been studied for many years, and as might be anticipated there are numerous disorders of the nervous system now recognized as results of defective metabolism of these amines. In looking critically at these disorders, it is possible to see the interaction of genetic principles and abnormal neurochemical metabolism.

PKU and Variants

Following the description of a defect in phenylalanine hydroxylase in liver in patients with classical PKU, investigators sought to understand the mechanism of the retardation, seizures, and abnormal behavior apparent in children who were not treated by diet and restriction of the amino acid phenylalanine. One explanation was that there was a defect in metabolism of the biogenic amines as demonstrated by low levels of the amines and their metabolites in body fluids, including blood, urine, and CSF (Butler et al., 1981). The deficiency in biogenic amines in brain during early phases of development could in part explain the clinical features, and the neurotransmitter deficiency can be corrected by removing phenylalaninc from diet. It seems that the excessive phenylalanine can decrease the synthesis of the biogenic amines by inhibiting rate-limiting enzymes tyrosine and tryptophan hydroxylase and also by inhibiting uptake of tyrosine and tryptophan from the gastrointestinal tract and into neurons, since these aromatic amino acids share membrane transport mechanisms with phenylalanine.

An illustration of the interaction of a genetic enzymatic defect and an environmental factor in the diet or external milieu is the disorder of maternal-fetal PKU. A mother with the classical, recessive form of PKU who is not or only partially treated by restriction of dietary phenylalanine can give birth to a retarded child because of an elevated maternal level of phenylalanine and passage of increased levels of this amino acid across the placenta into the fetal brain. The child generally does not have the genetic form of PKU, and the elevated levels of tissue phenylalanine subside following birth and removal from the maternal source of phenylalanine. The retardation in these children can be prevented by careful dietary management of the maternal PKU; furthermore, this situation is likely to become more frequent as an increasing number of successfully treated patients with classical PKU will enter the child-bearing age (Lenke and Levy, 1980).

Phenylalanine metabolism has become more complex with knowledge of the requirement of a cofactor for phenylalanine metabolism by phenylalanine hydroxylase. This cofactor is also required in the hydroxylation of tyrosine and tryptophan to dihydroxyphenylalanine (levodopa) and 5-hydroxytryptophan (5-HTP), respectively. These are rate-limiting and important steps in the synthesis of dopamine and norepinephrine from tyrosine and of serotonin from tryptophan. Furthermore, the active form of this cofactor, tetrahydrobiopterin, is synthesized from the precursor guanosine triphosphate (GTP) by various enzymes, including GTP-cyclohydrolase, dihydrobiopterin synthetase, and dihydrofolate reductase. Another enzyme, dihydropteridine reductase, is necessary for the reduction and activation of the intermediate form of the cofactor quinonoid dihydrobiopterin produced during the hydroxylation of phenylalanine. Recently, variant forms of PKU have been described as being owed to defects in the synthesis and reactivation of biopterin cofactor. In both types of disorders, there is abnormal metabolism of the biogenic amines because the active cofactor is also required in the rate-limiting steps of dopamine, norepinephrine, and serotonin synthesis (Butler et al., 1978; Kaufman et al., 1978). The defect in biogenic amine synthesis is not corrected by dietary restriction of phenylalanine but can be treated by administration of the hydroxylated precursor amino acids levodopa and 5-HTP together with a peripheral decarboxylase inhibitor such as carbidopa. These hydroxylated amino acids enter the respective metabolic pathways distal to the enzymatic defect. Because of the deleterious effect of low levels of neurotransmitters in the developing brain, specific diagnosis of amino acid precursor treatment should be instituted early in these children to prevent the seizures and mental retardation observed in children untreated or treated only by dietary phenylalanine restriction.

Hyperkinetic Disorders of Childhood: Hyperactivity and Gilles de la Tourette's Syndrome

Pediatricians and neurologists concerned with the developing child have been impressed by those patients, most frequently males, who are hyperactive, have poor impulse control and short attention spans, and are currently said to have an attention deficit disorder (ADD). These children have considerable difficulties socially and in coping at school, and frequently there is a family history of the father, or other close relative having a similar problem as a child. This disorder may vary in its severity (expression) and genetic penetrance within a family, and in many patients the symptoms become less prominent during adolescence. Although in many instances there is a considerable environmental (social) contribution that seems to exacerbate the condition, there is not infrequently a dramatic response and improvement in behavior and attention span in some children following administration of a sympathomimetic agent such as dextroamphetamine or methylphenidate, drugs known to affect biogenic amine metabolism. In order to explain the apparent developmental aspects of the clinical features, the male preponderance and genetic aspects, and the response to a narrow range of pharmacological agents, some investigators have suggested that ADD is a disorder of biogenic amines (Shaywitz, Cohen, and Bowers, 1980). The basis for the defect is not known but could be related to presynaptic or postsynaptic neuronal elements and owed to abnormal ontogenetic processes in biogenic amine neurons. It is of interest that several investigators have shown an inverse relationship between age and indicators of biogenic amine metabolism, including brain levels of amines or metabolites (Carlsson, 1976), biosynthetic enzymes in brain (McGeer, McGeer, and Suzuki, 1977), and CSF levels of biopterin cofactor (Williams et al., 1980), and metabolites (Seifert, Foxx, and Butler, 1980). It has been suggested that this inverse relationship with age may be related to specific partial defects in cellular metabolism becoming more apparent with advancing age and leading to programmed death of certain neurons, such as those synthesizing the biogenic amines. Any interference in these normal developmental processes could lead to an imbalance in biogenic amine neurotransmitter interactions and result in the clinical disorder of ADD. A study of this problem in the developing human may require the application of new, noninvasive investigative techniques of brain function and structure such as positron emission tomography (PET) or nuclear magnetic resonance (NMR) scanning.

Gilles de la Tourette's syndrome has some superficial similarities to ADD in that it is more common in males and there is often a familial component (possible dominant inheritance). In addition, the clinical manifestations suggest a hyperkinetic state of motor and verbal tics, and there is frequently a dramatic response to drugs, in particular, haloperidol, an agent that predominantly acts to block postsynaptic dopamine receptors. The developmental aspects of this syndrome are suggested by the onset in the first decade of life followed by the tendency for the symptoms to increase with advancing age during the second and third decade and to become less severe subsequently or even to regress almost completely. The hyperkinetic nature of the clinical manifestations does suggest overactivity of the motor system, which may possibly be owed to dopamine dysfunction. Direct studies on brain tissues have not been possible, but indirect studies analyzing CSF biogenic amine metabolites, particularly homovanillic acid (HVA) levels, have indicated a defect in dopamine metabolism (Butler, Koslow, and Seifert, 1979) that is corrected by administration of haloperidol (Singer et al., 1982). Explaining the use of the low levels of HVA in CSF as an indicator of dopamine metabolism in brain depends on our understanding the various feedback mechanisms utilized at the synapse to control release of neurotransmitter. Increased function of postsynaptic dopamine receptors could lead to feedback inhibition of dopamine synthesis or release and explain the decreased turnover of dopamine as observed by CSF analysis of the metabolite HVA. Restoration of the imbalance at the synaptic level by haloperidol's acting to decrease excessive dopamine receptor function was demonstrated by an increase in HVA levels in CSF after haloperidol treatment, an increase presumably owed to an increase in dopamine turnover in brain. The developmental aspects described in Tourette's syndrome might also indicate impaired ontogenetic mechanisms involving the dopamine synapse in specific areas of the brain, and such mechanisms might be investigated in vivo by the new techniques of PET and NMR scanning.

Parkinsonian Syndromes

Finally, at the other end of the spectrum of the developmental process of aging, there are those syndromes, and in particular Parkinson's disease, in which there is predominantly a deficiency of dopamine neurons. In Parkinson's disease the onset of clinical features is usually in the fifth or sixth decade of life with onset of rigidity, bradykinesia, impaired balance and stance, and limb tremors. Autopsy studies indicated a marked decrease in dopamine neurons in the substantia nigra and also changes in neurons of the locus coeruleus (norepinephrine-containing neurons) and brain cortex (Whitehouse et al., 1983). Although the death of dopamine-containing neurons may explain the motor manifestations, other clinical aspects such as depression and dementia may be due to the more widespread loss of biogenic amine-containing neurons. Furthermore, some of the peripheral clinical features—increased seborrhea, constipation, and postural hypotension—may be due to peripheral neuronal loss in the spinal cord. Despite the often dramatic response of these patients to administration of the dopamine precursor levodopa, in addition to a peripheral decarboxylase inhibitor, the effect of treatment is to postpone the inevitable, and these patients will continue to deteriorate and die with severe Parkinson's disease. The reason for the loss of neurons is not known, but there may be modest clues, such as a family history of a similar disorder in first-degree relatives and (rarely) early onset of Parkinson's disease in a family possibly because of a defect in the rate-limiting enzyme tyrosine hydroxylase (Martin, Resch, and Baker, 1971). A decrease in levels of the biopterin cofactor in CSF of patients with Parkinson's disease may reflect an abnormality in metabolism of this critical cofactor or the loss of biogenic amine-containing neurons (Williams et al., 1980). It seems likely that an increased understanding at a basic cellular level of the normal processes of cell development and death might eventually lead to an explanation for the early death of dopamine neurons in Parkinson's disease. A more intensive study of familial forms of Parkinson's disease or of patients at the beginning of the disorder, before the severe loss of neurons that occurs late in the disease, may indicate the basis in humans for the neuronal loss in specific brain regions.

Disorders Affecting the Cholinergic System

Investigations of cholinergic neuronal systems and acetylcholine metabolism have been in progress longer than studies of the biogenic amines, but only recently have these methods been applied to clinical problems. In part this delay relates to the difficulties in quantitating turnover of acetylcholine as was possible with the biogenic amines by measuring their metabolites in body fluids such as CSF. At this point, investigators must rely on analysis of tissues, such as brain and striatal muscle, to measure the levels of the synthetic enzyme choline acetyltransferase (CAT) and the degradative enzyme acetylcholinesterase (AChE) and to quantitate muscarinic (brain) and nicotinic (muscle) receptor activities and numbers. Quantitation in animals of tissue levels of acetylcholine or the precursor choline have not been as helpful since they do not necessarily reflect the actual flux of neurotransmitter in neurons. However, studies of the release of acetylcholine by tissue have been performed in brain (experimental) and in peripheral nerve-muscle preparations (experimental and human). Recent studies of human autopsy material that quantitate CAT levels by enzyme analysis or immunohistochemistry have been particularly revealing since CAT is a marker for cholinergic neurons.

Peripheral Neuromuscular Disorders

Neurobiological studies indicate that CAT is synthesized in the cell body of the motor or anterior horn cell neuron and transported along the axon to the terminal where it is involved in the synthesis of acetylcholine from acetyl-CoA and choline. The nerve terminal takes up choline after acetylcholine is released into the synaptic cleft and degraded by the postsynaptic enzyme AChE.

In young children and occasionally adults there are rare disorders in which there is degeneration of the motor neuron of the anterior horn of the spinal cord, and these disorders may be familial and recessively inherited. The terms *Werdnig-Hoffmann* (acute infantile) and *Kugelberg-Welander* (childhood onset) *disease* or *familial spinal muscular atrophy* (SMA) have been applied to such disorders and are defined by the age of onset and subsequent progression of the disorder. Death occurs frequently in the type having infantile onset, and severe skeletal deformities and inability to ambulate occur in the chronic childhood onset form. The cause of this disorder has not been determined, although the recessive inheritance might suggest an enzyme defect that is specific to the cholinergic motor neuron. Usually by the time of death there are few anterior cells remaining; thus, this is probably not the most opportune stage of the disorder at which to perform chemical analysis of these neurons. Establishment of an animal model of this recessive disorder may be necessary to determine whether there is a specific enzyme defect of cholinergic neurons.

Disorders of neuromuscular transmission have been extensively studied by neurophysiological techniques of repetitive nerve stimulation and motor action potential analysis, by histological techniques (including electron microscopy), and by quantitative analysis of enzymes and receptors following a muscle biopsy of the motor end plate. Such studies led to the demonstration of abnormal postsynaptic membrane structure of the nerve-muscle synapse and a decreased number of postsynaptic nicotinic receptors as the basis for the acquired form of myasthenia gravis (Drachman, 1978). This autoimmune disorder of the acetylcholine receptors produces weakness and fatigue in patients and can affect various muscle groups such as the extraocular or limb-girdle muscles. A defect in immune surveillance has been indicated by the clinical association with other autoimmune disorders, the elevated levels of acetylcholine receptor antibodies in blood, abnormal thymic tissue (including thyoma) obtained at thymectomy, and clinical responsiveness to immune-suppressive treatment such as thymectomy and adrenocorticoid steroids. Important in our understanding of the pathogenesis of this disorder was the demonstration of a transient defect in neuromuscular transmission in a proportion of children born to mothers with myasthenia gravis (Elias, Butler, and Appel, 1979). Elevated levels of acetylcholine receptor antibodies have been demonstrated in the affected children and occur as a result of transplacental passage of IgG protein from the mother.

As a result of intense interest and numerous studies in all clinical types of myasthenia gravis, there has been delineation of rare, familial, and sometimes congenital disease states having a defect in neuromuscular transmission. Persistent or episodic weakness in these young children may lead to impaired respiration, pneumonia, and even death. Neurophysiological studies show abnormal neuromuscular transmission, and this abnormality may or may not be corrected following parenteral

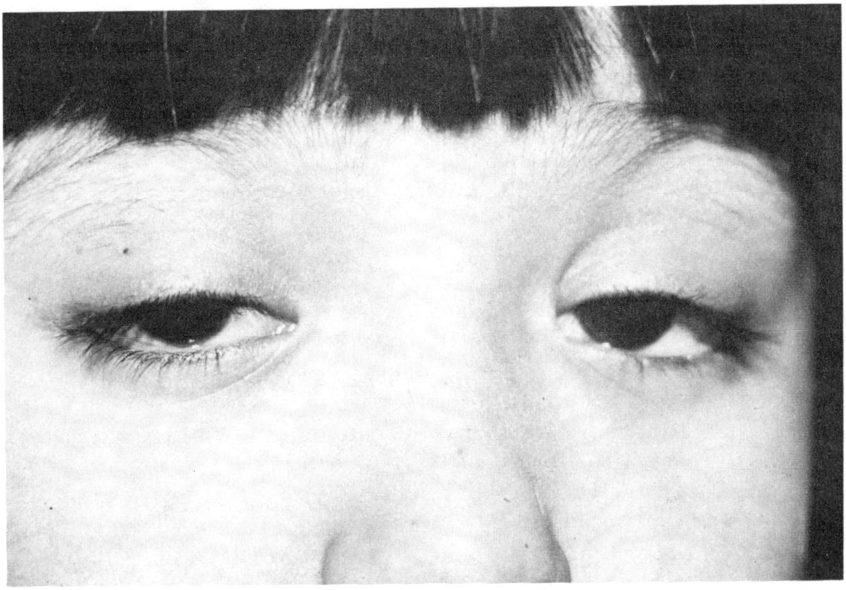

Figure 2. A four-year-old girl had congenital ptosis and ocular and facial weakness that improved dramatically following intravenous administration of edrophonium, an AChE inhibitor. Acetylcholine receptor antibody levels were not elevated in blood, and her subsequent clinical course was benign. The cause of the defect in neuromuscular transmission could be due to a defect in presynaptic synthesis or release of acetylcholine or a congenital defect in postsynaptic receptor function or numbers.

administration of an AChE inhibitor such as edrophonium. The familial nature may indicate recessive inheritance, and immune studies, including normal acetylcholine receptor antibody titers, do not indicate an acquired autoimmune disorder. Detailed examination of the neuromuscular junction following removal of an intercostal nerve-muscle preparation and analysis by all those techniques currently available indicate various defects in neural transmission. There may be a lack of AChE at the postsynaptic membrane and failure to degrade transmitter in the synapse. There may be impaired synthesis or release of neurotransmitter or a decreased amount of postsynaptic receptor substance, and in some patients there are mixed presynaptic and postsynaptic defects (Vincent et al., 1981). Although these familial examples are rare (fig. 2), an intensive study of their basic defect will certainly increase our understanding of the important events taking place at the neuromuscular junction and will assist in the management of the various types of myasthenia gravis.

Aging, Dementia, and Alzheimer's Disease

Because of the increase in the number of people of advanced age in the community in developed countries there has been considerable interest in understanding mechanisms of aging in the brain as well as other organ tissues. As a result of this activity, investigators have been able to quantitate specific regional neuronal losses and correlate these losses with impaired cognitive and memory functions. Recently during investigations of patients with Alzheimer's disease (Coyle, Price, and DeLong, 1983), a common cause of dementia in older persons, analysis of neuronal loss in an area of brain called *the nucleus basalis of Meynert* indicated that loss of this group of neurons may have caused the memory and cognitive defects. Furthermore, animal and human studies indicate that this small group of neurons sends axons to the brain cortex and hippocampal formation and that these neurons are cholinergic in type. There is a significant reduction (60–90 percent) of the cholinergic neuron marker enzyme (CAT) activity in brain cortex and hippocampal formation of patients who died with Alzheimer's disease. Histochemical and electro-microscopic studies indicate that the histological features of Alzheimer's disease—neurofibrillary tangles and senile plaques in the cortex and other brain regions—may be residue of the degenerating cholinergic neuron terminals (Price et al, 1982). Loss of neurons from the nucleus basalis of Meynert may not be specific for Alzheimer's disease but appears to be best correlated with dementia. Patients with Parkinson's disease and an associated dementia have loss of cells in the nucleus basalis of Meynert (Whitehouse et al., 1983). Patients with the chromosomal disorder of trisomy 21 syndrome (Down's syndrome) who survive into a third decade may show an additional deterioration of their previously impaired cognitive skills, and a dementing process becomes apparent. Histological changes of Alzheimer's disease are apparent in these patients (Ropper and Williams, 1980); CAT activity is reduced in the cortex and basal forebrain; and in one case of Down's syndrome, neurons were decreased in number in the nucleus basalis of Meynert. At this stage, there is no known causal relationship between this particular genetic disorder and the loss of cholinergic neurons in this specific area of brain. However, a similar defect in cholinergic neurons has been seen in the dominantly inherited form of Alzheimer's disease and does suggest that a genetic component may be important in the presumably programmed premature death of these specific cholinergic neurons.

Peptides

Since the 1970s there has been considerable interest in and growth of knowledge about the many peptides present in nervous tissue. Many of these substances were initially described in nonneuronal tissue and more recently in brain and spinal cord. Although some of these peptides, such as the enkephalins, are involved in control of central pain mechanisms, their role in specific brain functions and disease in humans has yet to be explored (Snyder, 1980). Other peptides, such as those in the hypothalamus, are concerned with pituitary control of release of hormones and include thyrotropin-releasing hormone (TRH), somatostatin, and luteinizing hormone–releasing hormone (LHRH). Somatostatin levels have been reported to be decreased in brain cortex from patients with Alzheimer's disease (Coyle, Price, and DeLong, 1983). Another peptide in the pituitary gland, antidiuretic hormone (ADH), in addition to its function in water homeostasis, has been implicated in affecting memory functions in animals (Reppert et al., 1981).

An example of a relevant peptide in human nervous tissue has been substance P, a undecapeptide found in the central and peripheral nervous system. Substance P is found in high concentrations in the substantia nigra, peripheral unmyelinated nerves, and the dorsal horn of the spinal cord and fulfills the neurochemical criteria for a neurotransmitter or neuromodulator substance. Substance P in the striatonigral neurons terminating in the substantia nigra has been determined to be decreased in Huntington's chorea, presumably as a result of the degeneration of neurons in the corpus striatum of these patients. Substance P has been implicated as a neurotransmitter of primary afferent neurons concerned with appreciation of pain sensation. In patients with clinical neuropathies having a sensory component, very low levels of substance P have been measured in CSF and probably relate to the loss of dorsal root ganglion neurons subserving pain sensation (Nutt et al., 1980). Examination of experiments of nature, such as the various inherited sensory neuropathies of both dominant (mild) and recessive (severe) type, by quantitation of substance P in CSF and neural tissue could be revealing and may lead to a better understanding of substance P synthesis and degeneration. Patients with the severe recessive form of sensory neuropathy have impaired pain sensation and appreciation of other sensory modalities and are therefore prone to fracture bones and to traumatize skin and joints (Charcot joint) with subsequent self-amputation and development of skin and bone infections. The hand of a young girl with a severe sensory neuropathy involving loss of pain sensation is pictured in fig. 3.

Conclusions

Although disorders studied in experimental animal models, either as they occur in life as genetic disorders or as produced by administration of a pharmacological agent or surgical procedure, have been of proven value in studying human neurological disease, so-called experiments of nature appear to have particular value in understanding the relationship of development and genetics to disease processes. In this chapter I have

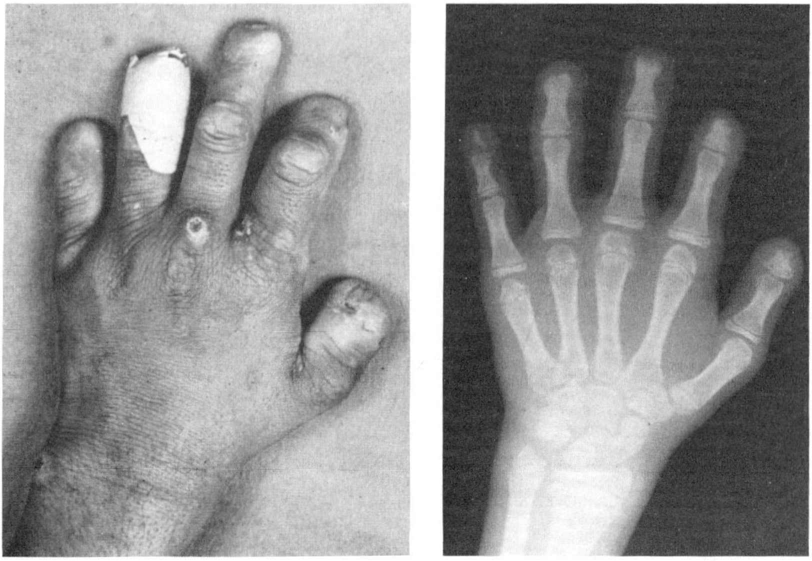

Figure 3. A ten-year-old girl had a fracture of her femur at seven months without appreciation of pain, and subsequently had repeated bone fractures. Her skin was readily traumatized and infected, and recurrent leg ulcers developed over pressure points. She had also lost the tips of her fingers and toes, and her lips and tongue were scarred from self-inflicted chewing injuries. Superficial pain appreciation and other sensory modalities were either decreased or lacking, and her deep tendon reflexes were absent. These and other findings indicated a congenital lack of sensory nerves, probably of recessive inheritance.

described a number of neurological disorders, many occurring in the developing human and frequently having a strong genetic component. The understanding of the neurochemical bases for many of these conditions is just beginning as techniques developed in basic neurochemistry laboratories are being applied to disorders of the human nervous system. Obviously, neurochemical disturbances may be particularly responsive to pharmacological manipulations using precursor agents, neurotransmitters, or receptor modulators, and such agents could thus be the basis of future therapy of these otherwise devastating neurological disorders.

References

Butler, I. J., Koslow, S. H., Krumholz, A., Holtzman, N. A., and Kaufman, S. (1978). A disorder of biogenic amines in dihydropteridine reductase deficiency. *Ann Neurol* 3, 224–230.

Butler, I. J., Koslow, S. H., and Seifert, W. E., Jr. (1979). Biogenic amine metabolism in Tourette syndrome. *Ann Neurol* 6, 37–39.

Butler, I. J., O'Flynn, M. E., Seifert, W. E., and Howell, R. R. (1981). Neurotransmitter defects and treatment of disorders of hyperphenylalaninemia. *J Pediatr* 98, 729–733.

Carlsson, A. (1976). Some aspects of dopamine in the basal ganglia. In *The basal ganglia*, ed. M. D. Yahr, pp. 181–189. New York: Raven Press.

Coyle, J. T., Price, D. L., and DeLong, M. R. (1983). Alzheimer's disease: A disorder of cortical cholinergic innervation. *Science* 219, 1184–1190.

Drachman, D. B. (1978). Myasthenia gravis. *N Engl J Med* 298, 136–142, 186–193.

Elias, S. B., Butler, I. J., and Appel, S. H. (1979). Neonatal myasthenia gravis in the infant of a myasthenic mother in remission. *Ann Neurol* 6, 72–75.

Hecht, J. T., Scott, C. I., Butler, I. J., and Moore, C. M. (1983). X-linked mental retardation with fragile site at band Xq2800. *Lancet* 1, 986.

Kaufman, S., Berlow, S., Summer, G. K., Milstien, S., Schulman, J. D., Orloff, S., Spielberg, S., and Pueschel, S. (1978). Hyperphenylalaninemia due to a deficiency of biopterin: A variant of phenylketonuria. *N Engl J Med* 299, 673–679.

Killian, J. M., and Kloepfer, H. W. (1979). Homozygous expression of a dominant gene for Charcot-Marie-Tooth neuropathy. *Ann Neurol* 5, 515–522.

Lenke, R. R., and Levy, H. L. (1980). Maternal phenylketonuria and hyperphenylalaninemia: An international survey of the outcome of untreated and treated pregnancies. *N Engl J Med* 303, 1202–1208.

McGeer, P. L., McGeer, E. G., and Suzuki, J. S. (1977). Aging and extrapyramidal function. *Arch Neurol* 34, 33–35.

Martin, W. E., Resch, J. A., and Baker, A. B. (1971). Juvenile parkinsonism. *Arch Neurol* 25, 494–500.

Nutt, J. G., Mroz, E. A., Leeman, S. E., Williams, A. C., Engel, W. K., and Chase, T. N. (1980). Substance P in human cerebrospinal fluid: Reductions in peripheral neuropathy and autonomic dysfunction. *Neurology* 30, 1280–1285.

Price, D. L., Whitehouse, P. J., Struble, R. G., Clark, A. W., Coyle, J. T., DeLong, M. R., and Hedreen, J. C. (1982). Basal forebrain cholinergic systems in Alzheimer's disease and related dementias. *Neurosci Comm* 2, 84–92.

Reppert, S. M., Artman, H. G., Swaminathan, S., and Fisher, D. A. (1981). Vasopressin exhibits a rhythmic daily pattern in cerebrospinal fluid but not in blood. *Science* 213, 1256–1257.

Ropper, A. H., and Williams, R. S. (1980). Relationship between plaques, tangles, and dementia in Down syndrome. *Neurology* 30, 639–644.

Seifert, W. E., Foxx, J. L., and Butler, I. J. (1980). Age effect on dopamine and serotonin metabolite levels in cerebrospinal fluid. *Ann Neurol* 8, 38–42.

Shaywitz, B. A., Cohen, D. J., and Bowers, M. B. (1980). Cerebrospinal fluid monoamine metabolites in neurological disorders of childhood. In *Neurobiology of cerebrospinal fluid*. Vol. 1, ed. J. H. Wood, pp. 219–236. New York: Plenum Press.

Singer, H. S., Butler, I. J., Tune, L. E., Seifert, W. E., and Coyle, J. T. (1982). Dopaminergic dysfunction in Tourette syndrome. *Ann Neurol* 12, 361–366.

Snyder, S. H. (1980). Brain peptides as neurotransmitters. *Science* 209, 976–983.

Stanbury, J. B., Wyngaarden, J. B., Frederickson, D. S., Goldstein, J. L., and Brown, M. S. (1983). *The metabolic basis of inherited disease*, pp. 3–38. New York: McGraw-Hill.

Vincent, A., Cull-Candy, S. G., Newsom-Davis, J., Trautmann, A., Molenaar, P. C., and Polak, R. L. (1981). Congenital myasthenia: End-plate acetylcholine receptors and electrophysiology in five cases. *Muscle Nerve* 4, 306–318.

Whitehouse, P. J., Hedreen, J. C., White, C. L., and Price, D. L. (1983). Basal forebrain neurons in the dementia of Parkinson disease. *Ann Neurol* 13, 243–248.

Williams, A., Ballenger, J., Levine, R., Lovenberg, W., and Calne, D. (1980). Aging and CSF hydroxylase cofactor. *Neurology* 30, 1244–1246.

Contributors

R. Wayne Albers, Ph.D., chief of section, Laboratory of Neurochemistry, National Institute of Neurological and Communicative Disorders and Stroke, National Institutes of Health, Bethesda, Maryland.

Shirley A. Bayer, Ph.D., associate professor, Department of Biology, Indiana University–Purdue University, Indianapolis, Indiana.

I. J. Butler, M.B., professor of neurology and pediatrics and director of the Division of Pediatric Neurology, Department of Neurology, The University of Texas Medical School at Houston, Houston, Texas.

C. J. Cummins, Ph.D., chief of neurochemistry, surgical neurology, National Institute of Neurological and Communicative Disorders and Stroke, National Institutes of Health, Bethesda, Maryland.

S. J. Enna, Ph.D., professor of pharmacology and neurobiology, Departments of Pharmacology and of Neurobiology and Anatomy, The University of Texas Medical School at Houston, Houston, Texas.

Robert M. Gould, Ph.D., research scientist, Institute for Basic Research in Developmental Disabilities, Staten Island, New York.

Richard E. Hauhart, M.S., research associate, Washington University in St. Louis, School of Medicine, St. Louis, Missouri.

Michael V. Johnston, M.D., associate professor of pediatrics and neurology, University of Michigan Medical School; associate research scientist, Center for Human Growth and Development, University of Michigan, Ann Arbor, Michigan.

David J. Jones, Ph.D., associate professor, Departments of Anesthesiology and Pharmacology, The University of Texas Health Science Center at San Antonio, San Antonio, Texas.

David J. Loreck, B.S., medical student, The University of Texas Medical School at Houston, Houston, Texas.

David W. McCandless, Ph.D., professor of neurobiology, The University of Texas Medical School at Houston, Houston, Texas.

Alexander L. Miller, M.D., associate professor of psychiatry, Department of Psychiatry and Pharmacology, The University of Texas Health Science Center at San Antonio, San Antonio, Texas.

Shaila D. Telang, Ph.D., professor of biochemistry, Department of Biochemistry, Faculty of Science, M. S. University of Baroda, Baroda, India.

Jean Holowach Thurston, M.D., professor of pediatrics and professor of neurology (Neurochemistry), Washington University in St. Louis, School of Medicine, St. Louis, Missouri.

Richard C. Wiggins, Ph.D., associate professor, Department of Neurobiology and Anatomy, The University of Texas Medical School at Houston, Houston, Texas.

Index